Applied
Physical Pharmacy
Second Edition

Notice

Medicine is an ever-changing science. As new research and clinical experience broaden our knowledge, changes in treatment and drug therapy are required. The authors and the publisher of this work have checked with sources believed to be reliable in their efforts to provide information that is complete and generally in accord with the standards accepted at the time of publication. However, in view of the possibility of human error or changes in medical sciences, neither the authors nor the publisher nor any other party who has been involved in the preparation or publication of this work warrants that the information contained herein is in every respect accurate or complete, and they disclaim all responsibility for any errors or omissions or for the results obtained from use of the information contained in this work. Readers are encouraged to confirm the information contained herein with other sources. For example and in particular, readers are advised to check the product information sheet included in the package of each drug they plan to administer to be certain that the information contained in this work is accurate and that changes have not been made in the recommended dose or in the contraindications for administration. This recommendation is of particular importance in connection with new or infrequently used drugs.

Applied Physical Pharmacy

Second Edition

Editors

Mansoor M. Amiji, PhD, RPh
*Distinguished Professor and Chairman
Department of Pharmaceutical Sciences
Director, Laboratory of Biomaterials
   and Advanced Nano-Delivery Systems (BANDS)
School of Pharmacy
Bouve College of Health Sciences
Northeastern University
Boston, Massachusetts*

Thomas J. Cook, PhD, RPh
*Director of Program Assessment
Associate Professor
Department of Pharmaceutical and Biomedical
   Sciences Touro College of Pharmacy
New York, New York*

W. Cary Mobley, PhD, RPh
*Clinical Associate Professor
Department of Pharmaceutics
University of Florida
College of Pharmacy
Gainesville, Florida*

New York   Chicago   San Francisco   Athens   London   Madrid   Mexico City
Milan   New Delhi   Singapore   Sydney   Toronto

Applied Physical Pharmacy, Second Edition

Copyright © 2014 by McGraw-Hill Education. All rights reserved. Printed in the United States of America. Except as permitted under the United States Copyright Act of 1976, no part of this publication may be reproduced or distributed in any form or by any means, or stored in a data base or retrieval system, without the prior written permission of the publisher.

1 2 3 4 5 6 7 8 9 0 DOC/DOC 18 17 16 15 14

ISBN-13: 978-0-07-174750-9
MHID: 0-07-174750-8

This textbook was set in Times by MPS Limited.
The editors were Michael Weitz and Karen G. Edmonson.
The production supervisor was Richard Ruzycka.
Project management was provided by Shruti Chopra at MPS Limited.
The cover designer was Elizabeth Pisacreta.
RR Donnelley was printer and binder.

This textbook is printed on acid-free paper.

Library of Congress Cataloging-in-Publication Data

Applied physical pharmacy / [edited by] Mansoor M. Amiji, Thomas J. Cook, W. Cary Mobley.—Second edition.
    p. ; cm.
Includes bibliographical references and index.
ISBN 978-0-07-174750-9 (hardcover : alk. paper)—ISBN 0-07-174750-8 (hardcover : alk paper)
I. Amiji, Mansoor M., editor. II. Cook, Thomas J., 1965- editor. III. Mobley, W. Cary, editor.
[DNLM: 1. Technology, Pharmaceutical. 2. Chemistry, Pharmaceutical. QV 778]
RS403
615'.19—dc23

McGraw-Hill books are available at special quantity discounts to use as premiums and sales promotions, or for use in corporate training programs. To contact a representative, please visit the Contact Us page at www.mhprofessional.com.

Brief Contents

Contributors ix
Preface xi

1. **Introduction to Biopharmaceutics** 1
   W. Cary Mobley

2. **States of Matter Related to Pharmaceutical Formulations** 15
   Beverly J. Sandmann
   Edited by Ann Newman and Gregory T. Knipp

3. **Physical Properties of Solutions** 41
   Beverly J. Sandmann, Antoine Al-Achi, and Robert Greenwood

4. **Ionic Equilibria and Buffers** 61
   Beverly J. Sandmann, Alekha K. Dash, Antoine Al-Achi, and Robert Greenwood

5. **Solubility, Dissolution, and Partitioning** 87
   Beverly J. Sandmann and Mansoor M. Amiji

6. **Mass Transport** 117
   Mansoor M. Amiji

7. **Complexation and Protein Binding** 135
   Mansoor M. Amiji

8. **Dispersed Systems** 157
   W. Cary Mobley

9. **Interfacial Phenomena** 173
   Maria Polikandritou Lambros and Shihong Li Nicolaou

10. **Rheology** 197
    Maria Polikandritou Lambros

11. **Chemical Kinetics of Pharmaceuticals** 221
    Thomas J. Cook

Appendix Basic Mathematical Concepts 245

Endnotes 251
Index 257

Contents

Contributors   ix
Preface   xi

1. **Introduction to Biopharmaceutics**   1
   W. Cary Mobley

   Introduction   1
   Drug Disposition—The Fate of the Drug After Administration   3
   LADME   8
   Conclusion   13
   Key Points   14
   Clinical Questions   14

2. **States of Matter Related to Pharmaceutical Formulations**   15
   Beverly J. Sandmann
   Edited by Ann Newman and Gregory T. Knipp

   Intermolecular Forces   15
   States of Matter   19
   Stability of Solids   34
   Acknowledgment   37
   Problems   37
   Answers   38
   Key Points   39
   Clinical Questions   39

3. **Physical Properties of Solutions**   41
   Beverly J. Sandmann, Antoine Al-Achi, and Robert Greenwood

   General Considerations   41
   Concentration Expressions   42
   Classification of Aqueous Solution Systems   45
   Physical Properties of Solutions   46
   Physiologic Applications of Colligative Properties   47
   Isotonic Solutions   50
   Sodium Chloride Equivalent Values (E) 52
   Problems   54
   Answers   57
   Key Points   59
   Clinical Questions   59

4. **Ionic Equilibria and Buffers**   61
   Beverly J. Sandmann, Alekha K. Dash, Antoine Al-Achi, and Robert Greenwood

   Electrolytes versus Nonelectrolytes   61
   Importance of Ionization in Pharmacy   61
   Acids and Bases   62
   Ionization of Water   63
   Ionization of Electrolytes   65
   Ionization of Salts   70
   Buffers   72
   Ionization of Amphoteric Electrolytes   74
   Ionization of Polyprotic Acids   75
   Activity and Activity Coefficient   75
   Acid-Base Titration and Titration Curve   76
   Case Studies   77
   Problems   78
   Answers   79
   Key Points   80
   Appendix   80
   Suggested Readings   85

5. **Solubility, Dissolution, and Partitioning**   87
   Beverly J. Sandmann and Mansoor M. Amiji

   Solubility   87
   Dissolution   97
   Dissolution of Particles   97
   USP Dissolution Testing   99
   Significance of Dissolution Studies   102
   Factors Affecting Drug Dissolution   103
   Partitioning   106
   Solubility Problems   110
   Answers   111
   Dissolution Problems   113
   Answers   113
   Partitioning Problems   114
   Answers   114
   Key Points   114
   Clinical Questions   115

6. Mass Transport 117
Mansoor M. Amiji

Introduction 117
Transport Systems 118
Diffusion Through a Membrane 121
Significance of Diffusion 125
Significance of Osmosis 131
Problems 132
Answers 133
Key Points 134
Clinical Questions 134

7. Complexation and Protein Binding 135
Mansoor M. Amiji

Introduction 135
Types of Complexes 136
Metal-Ion Coordinate Complexes 137
Cyclodextrin Complexes 140
Ion-Exchange Resins 141
Protein-Ligand Interaction 143
Plasma Protein Binding 146
Problems 148
Answers 149
Key Points 150
Clinical Questions 151
Appendix 151

8. Dispersed Systems 157
W. Cary Mobley

Dispersed Systems Classified by their Phases 157
Dispersed Systems Classified by their Particle Size 159
Physical Stability of Dispersed Systems 162
Summary 169
Problems 169
Answers 169
Key Points 170
Clinical Questions 171
Suggested Readings 171

9. Interfacial Phenomena 173
Maria Polikandritou Lambros and Shihong Li Nicolaou

Surface Tension 173
Electrical Double Layer 179
Adsorption 180
Problems 190
Answers 190
Key Points 191
Clinical Questions 192
Appendix 192

10. Rheology 197
Maria Polikandritou Lambros

Introduction 197
Newtonian Flow 198
Viscosity of Newtonian Fluids 200
Effect of Temperature on Viscosity 201
Non-Newtonian Flow 202
Thixotropy 205
Viscoelasticity 206
Significance of Rheology 206
Viscosity Modifiers 210
Problems 215
Answers 216
Key Points 216
Clinical Questions 217
Appendix 217

11. Chemical Kinetics of Pharmaceuticals 221
Thomas J. Cook

Common Drug Degradation Reactions 222
Order of Reaction 223
Determination of the Order of a Reaction 227
Stability and Shelf Life of Drugs 229
Enzyme Catalysis Reactions 234
Pharmacokinetics: An Extension of Reaction Kinetics 236
Problems 238
Answers 241
Clinical Questions 243
Suggested Readings 243

Appendix Basic Mathematical Concepts 245

Equations 245
Logarithmic Function 245
Exponential Function 246
Differentiation and Integration Functions 246
Bibliography 247
Mathematical Tables 248

Endnotes 251

Index 257

Contributors

Mansoor M. Amiji, PhD, RPh
(Chapters 5, 6 and 7)
Distinguished Professor and Chairman
Department of Pharmaceutical Sciences
Director, Laboratory of Biomaterials and Advanced
 Nano-Delivery Systems (BANDS)
School of Pharmacy
Bouve College of Health Sciences
Northeastern University
Boston, MA

Antoine Al-Achi, PhD, (Chapter 4)
Associate Professor of Pharmaceutical Sciences
Campbell University College of Pharmacy &
 Health Sciences
Buies Creek, NC

Thomas J. Cook, PhD, RPh (Chapter 11)
Director of Program Assessment
Associate Professor
Department of Pharmaceutical and Biomedical
 Science
Touro College of Pharmacy
New York, NY

Alekha K. Dash, RPh, PhD (Chapter 4)
Professor
Department of Pharmacy Sciences
School of Pharmacy & Health Professions
Creighton University Medical Center
Omaha, NE

Robert Greenwood, RPh, PhD (Chapters 3, 4)
Associate Dean of Academic Affairs
Professor of Pharmaceutical Sciences
Campbell University College of Pharmacy &
 Health Sciences
Buies Creek, NC

Gregory T. Knipp, PhD (Chapter 2)
Associate Professor of Industrial and Physical
 Pharmacy
Associate Director, Dane O. Kildsig Center for
 Pharmaceutical Processing Research
Purdue University, Department of Industrial and
 Physical Pharmacy
West Lafayette, IN

Maria Polikandritou Lambros, PhD
(Chapters 9 and 10)
Associate Professor of Pharmaceutical Sciences
College of Pharmacy
Western University of Health Sciences
Pomona, CA

W. Cary Mobley, PhD, RPh (Chapters 1 and 8)
Clinical Associate Professor
Department of Pharmaceutics
University of Florida
College of Pharmacy
Gainesville, Florida

Ann W. Newman, PhD (Chapter 2)
Adjunct Professor
Purdue University, Department of Industrial and
  Physical Pharmacy
West Lafayette, IN

Shihong Li Nicolaou, PhD (Chapter 9)
College of Pharmacy
Western University of Health Sciences
Pomona, CA

Beverly J. Sandmann, PhD (Chapter 5)
Professor Emeritus
College of Pharmacy
Butler University
Indianapolis, IN

Preface

In the spirit of understanding how drugs work, this textbook explores the fundamental physicochemical attributes and processes important for understanding how a drug, usually in the form of a crystal, is transformed into a usable product that is administered to a patient to reach its pharmacological target, and then leaves the body. This is the discipline of physical pharmacy—the study of the physical and chemical properties of drugs and their dosage forms. When integrated with other critical knowledge of how drugs work, such as their pharmacologic effects, physical pharmacy forms part of the scientific foundation for the clinical sciences and, therefore, for clinical practice. A distinguishing feature of physical pharmacy is that, unlike pharmacology, which is learned to different degrees by other healthcare practitioners, physical pharmacy is a body of knowledge unique to the education of student pharmacists, for whom this textbook is written. It provides the physicochemical basis for rational formulation, manufacturing, compounding, drug delivery, product selection, and product usage. Therefore, it is knowledge indispensable for the ability of the pharmacist to comprehensively understand and explain how drugs work, in a manner and to an extent that is unparalleled by any other healthcare practitioner. In other words, it's a part of the body of knowledge that equips pharmacists with unique perspectives and insights in the provision of pharmaceutical care.

Significant revisions have been made from the first edition of *Applied Physical Pharmacy* published in 2003, including addition of clinical examples and applications that are relevant in contemporary pharmacy practice. Each chapter includes a set of *Learning Objectives* to guide the student's focus in learning, *Key Points* to delineate the critical concepts discussed in the chapters, *Problems* to apply and assess the understanding of chapter concepts, and *Clinical Questions* to apply the chapter concepts in a clinical context.

An overarching goal in the writing and editing of this edition was to improve its focus on the critical knowledge needed for the education of the student pharmacist. The number of chapters was reduced from 13 to 11. Chapters with conceptually common material were blended into new chapters. Each chapter was examined for relevance to pharmacy education, and was accordingly edited to help achieve this goal. Some of the material from the first edition that was considered important but more relevant to graduate work in physical pharmacy was moved to chapter appendices. This was done to improve the flow of material within the chapter, again to make it more conducive for learning by student pharmacists.

The textbook begins with a review of the key biopharmaceutics concepts of drug liberation, absorption, distribution, metabolism, and excretion. These concepts and others set the framework for subsequent chapters that describe physicochemical properties and processes related to the fate of the drug. These include states of matter (Chapter 2), solutions, (Chapter 3), ionization (Chapter 4), dissolution and partitioning (Chapter 5), mass transport (Chapter 6), and complexation and protein binding (Chapter 7). Concepts in these chapters are important not only for understanding a drug's fate in the body but also for providing the scientific basis for rational drug formulation. Other physical pharmacy topics important to drug formulation and usage are discussed in the three chapters that follow, which describe dispersed systems (Chapter 8), interfacial phenomena (Chapter 9), and rheology (Chapter 10). The textbook concludes with an overview of the principles of kinetics (Chapter 11) that are important for understanding the rates at which many of the processes discussed in previous chapters occur. We are very grateful to all of the contributing authors who shared with us their expertise in important physical pharmacy knowledge and skills. A special thanks goes to Michael Weitz, the McGraw-Hill Executive Editor for Medical, Pharmacy & Allied Health Textbooks, for his enormous patience in accommodating our busy schedules, while guiding us to this textbook's conclusion. Thanks also to all of the McGraw-Hill editors, including Karen G. Edmonson. We also thank our pharmacy students who continuously inspire us to improve our craft of teaching. Most importantly, we are each grateful to our families, whose love, support, and patience helped make the writing and editing of this textbook possible and worthwhile.

The Editors

Applied
Physical Pharmacy
Second Edition

1 Introduction to Biopharmaceutics

W. Cary Mobley

Learning Objectives

After completing this chapter, the reader should be able to:

- ▶ List the types and locations of drug targets.
- ▶ List common dosage forms and the nature of the drug within the listed dosage forms.
- ▶ Describe common routes of drug administration, including the administration locations and the absorption sites.
- ▶ Define bioavailability and describe important physiological and physicochemical factors that influence a drug's bioavailability.
- ▶ Describe the following physicochemical factors that can influence the fate of the drug in the body: solid state properties, ionization, solubility and dissolution, partition coefficient, mass transport and membrane passage, complexation and protein binding.
- ▶ Briefly describe each parameter in the LADME acronym.
- ▶ Describe the general roles of drug transporters (with examples) in the ADME processes of a drug's disposition in the body.

INTRODUCTION

Biopharmaceutics can be defined as the study of the physical and chemical properties of drugs and their proper dosage as related to the onset, duration, and intensity of drug action, or it can be defined as the study of the effects of physicochemical properties of the drug and the drug product, *in vitro*, on the bioavailability of the drug, *in vivo*, to produce a desired therapeutic effect. Both definitions imply the relationship between the physicochemical properties of the drug, the drug's biological fate in the body after its administration, and the resulting pharmacological action of the drug. Most of this textbook is focused on the details of important physicochemical properties. This chapter introduces fundamental concepts and processes related to the fate of the drug in the body in order to help provide a framework for many of the concepts in subsequent chapters. As a background to discussing the fate (or disposition) of drugs, a brief review of their targets, their dosage forms, and their routes of administration is given next.

Drug Targets

Most drugs exert their effects by interacting with targets, which are molecules or structures that are linked to a particular disease.[1,2] Most often the target is a macromolecule, such as a protein or nucleic acid. Protein targets dominate, and they include enzymes, ion channels, nuclear hormone receptors, structural proteins, membrane transport proteins, and G protein-coupled receptors. These targets have important biochemical or physiological roles, and drugs interact with them, either blocking, inhibiting, or activating them, with biochemical and physiological consequences. These consequences are desired to help make the patient well, but they can also be manifested as adverse effects in the patient. The targets are found in various areas of the body, and they may be within cells, embedded in cell membranes, or outside of cells in various body fluids.

Some targets can be accessed by the nonsystemic administration of the drug (i.e., without reaching the systemic circulation),

- Describe, in general, how drugs are liberated from their dosage forms.
- Describe gastrointestinal drug absorption into the systemic circulation, including the influences of a drug's physicochemical properties, and the steps taken by a dissolved drug in the small intestines to its entry into enterocytes, hepatocytes, and finally the bloodstream.
- Describe enterohepatic cycling, possible rate-limiting steps in drug absorption, and first-pass metabolism.
- Describe the distribution of a drug from the plasma to the interstitia of tissues, including the influences of a drug's physicochemical properties, and of binding to plasma proteins and to tissue components.
- List other distribution sites, including the CNS, fat, and placenta.
- Define the apparent volume of drug distribution and relate it to the distribution of body water.
- Give an overview of the necessity, phases, and locations of drug metabolism.
- Describe renal drug excretion, including the influences of a drug's physicochemical properties.
- List nonrenal excretory pathways.
- Define and describe clearance and half-life, and briefly relate them to their importance for drug dosing.

such as targets in the intestinal lumen (reached from oral drug administration),[3] the dermis (reached from topical drug administration), and the bronchioles (reached from oral inhalation). However, in most cases, drug targets are reached by the drug after it enters the systemic circulation and distributes to the specific areas or cells that contain them. A drug's entry into the systemic circulation and distribution to its target(s) will be under the influence of many factors, including anatomical and physiological factors based on the route of the drug's administration (e.g., gastric emptying for the oral route) and blood flow to the various tissues. Drug entry and distribution are also under the influence of the physiochemical properties of the drug, such as size, charge, and lipophilicity. The roles of physicochemical factors are the main focus of this chapter.

Dosage Forms

A drug is administered to a patient as a specific *drug product*, which is a particular *dosage form* that is most often a formulation of the drug with various excipients. Excipients can be present for a variety of purposes, such as for improving stability, dissolution, manufacturing speed and quality, and flavoring. Dosage forms include the familiar tablets, capsules, and suspensions, as well as many others described in Table 1-1. When the

TABLE 1-1 Examples of Drug Dosage Forms and the Common State of the Drug Within Them

Dosage Form	Common State of the Drug in the Dosage Form
Tablet	Crystals in a compressed powder
Capsules, powder-filled	Crystals in a noncompressed powder
Capsules, liquid-filled	Molecules or crystals in vegetable oil
Suppository	Crystals in waxy, water-miscible or water-immiscible base
Solution	Molecules
Suspension	Crystals in an aqueous or nonaqueous liquid (see Chapter 8)
Ointment	Crystals or molecules in a semisolid oleaginous base
Cream	Crystals or molecules in water-miscible or immiscible semisolid cream base
Gel	Crystals or molecules in water-miscible semisolid gel base
Aerosol	Crystals or molecules in a gas, liquid, or semisolid

drug is formulated into its dosage form, it may be present in different physical forms, with the two most common being crystals and molecules. An important concept, described below, is that the drug must be released, or *liberated*, from its dosage form for it to begin the process of its journey to its receptor, with the first step of that process (for an undissolved drug) being its dissolution in the fluid of its route of administration.

Routes of Drug Administration

In addition to the variety of drug products in which a drug may be formulated, it's also important to understand that there are a variety of anatomical pathways, or routes of administration, through which a drug product may be administered. Table 1-2 describes many of the common routes of administration. These include routes for which it is intended that the drug be localized to a specific area, such as the eye for an ophthalmic application, the lungs for an oral inhalation, the central nervous system for an intrathecal injection, or a joint for an intrasynovial injection. For many administration routes, the drug is intended to pass into the systemic circulation for distribution throughout the body. In these cases, the drug may be administered directly into a vein, or indirectly by a number of other routes, with the oral route being the most common. For the indirect routes, the drug must pass through membranes of various types and it may be subject to metabolism before it reaches the systemic circulation. Therefore, drugs administered via indirect routes may not be completely bioavailable to the systemic circulation.

Bioavailability

Bioavailability can be defined as the proportion (percent or fraction) of an administered dose of unchanged drug that reaches the systemic circulation. An intravenous drug is administered directly into the systemic circulation, so is 100% bioavailable. For other routes, the bioavailability can be decreased by incomplete dissolution, incomplete absorption through epithelia, and presystemic metabolism. These will be described later. In some cases, the bioavailability can also be reduced if the drug chemically degrades in the body. For example, the proton pump inhibitor omeprazole can degrade in the acidic environment of the stomach, so it must be formulated with a buffer that raises gastric pH or in a way (e.g., enteric coating) that allows it to bypass the stomach before dissolving. Membrane and metabolic barriers can differ significantly from one route of administration to another (see Table 1-2). These barriers in combination with properties of the drug, drug product, and other physiological phenomena (e.g., gastric emptying) can reduce the drug's bioavailability. Bioavailability is a fundamental component of drug product performance that is assessed in the drug development process.

DRUG DISPOSITION—THE FATE OF THE DRUG AFTER ADMINISTRATION

As mentioned, once the intact drug reaches the systemic circulation it will distribute to various regions of the body, including those where its target is found. It will also be subject to metabolism and will be excreted from the body as intact drug or as drug metabolite. Figure 1-1 illustrates different processes that can affect the fate of an administered drug product, before and after the drug's entry into the systemic circulation. These processes are embodied in the acronym LADME, which stands for liberation (release of the drug from its dosage form), absorption (into the bloodstream), distribution (to various parts of the body), metabolism (by enzymes), and excretion (through the kidneys or other routes). The net effects of all of these processes yield a profile of plasma drug concentration versus time exemplified in Figure 1-2 for an extravascular (e.g., nonintravenous) drug administration. The curve depicts the minimum effective concentration needed for the desired therapeutic effect, the minimum toxic concentration for toxicity to manifest, and the duration of being at or above the minimum effective

TABLE 1-2 Common Routes of Drug Administration

Route	Administration Site	Primary Absorption Site(s)/Purpose	Common Dosage Forms
Oral	Mouth (swallowed)	Epithelia of the upper small intestine, stomach; for a systemic effect	Tablet, capsule, solution, suspension, emulsion
Sublingual	Under the tongue	Epithelium under the tongue; for a systemic effect	Tablet
Buccal	Between the cheek and gums	Epithelial of the lining of the cheek or local action; for a local or systemic effect	Tablet, lozenge
Rectal	Rectum	Rectal epithelium or local action; for a local or systemic effect	Suppository, gel, foam aerosol
Vaginal	Vagina	Vaginal epithelium; mainly for a local effect	Suppository, tablet, ointment, cream
Intranasal	Nasal cavity	Nasal epithelium; mainly for a local effect	Solution, suspension
Pulmonary	Mouth (inhaled)	Bronchiolar epithelium; mainly for a local effect	Liquid solution or suspension
Ophthalmic	Eye surface	Corneal epithelium; for a local effect	Liquid solution or suspension
Topical	Epidermal surface	Stratum corneum; for a local effect	Powder, liquid solution or suspension, ointment, cream, gel
Transdermal	Epidermal surface	Stratum corneum, capillaries; mainly for a systemic effect	Ointment, cream, gel
Intravenous injection or infusion	Veins	None; for a systemic effect	Solution, emulsion (if nano-sized)
Intramuscular injection	Striated muscles (e.g., deltoid, thigh, buttocks)	Blood capillaries; for a systemic effect	Aqueous or nonaqueous solution or suspension
Subcutaneous injection	Subcutaneous fat	Blood and lymphatic capillaries; for a systemic effect	Aqueous solution or suspension
Intraperitoneal infusion	Peritoneal cavity	Peritoneum; mainly for organs of the peritoneal cavity (e.g., for ovarian cancer)	Aqueous solution
Intrathecal injection or infusion	Spinal cord (into the cerebrospinal fluid)	None; for local action in the CNS	Aqueous solution
Epidural injection or infusion	Outside of the dura mater of the spinal cord	Dura mater; for local action in the CNS	Aqueous solution
Intrasynovial injection	Synovial space of joints	None; for local action in the joints	Aqueous solution or suspension
Intraosseous injection or infusion	Bone marrow (e.g., of the tibia)	Sinusoidal capillaries; for systemic action (e.g., in emergencies)	Aqueous solution
Intravitreal injection	Vitreous humor of the eyeball	None; for local action within the eye	Aqueous solution

Introduction to Biopharmaceutics 5

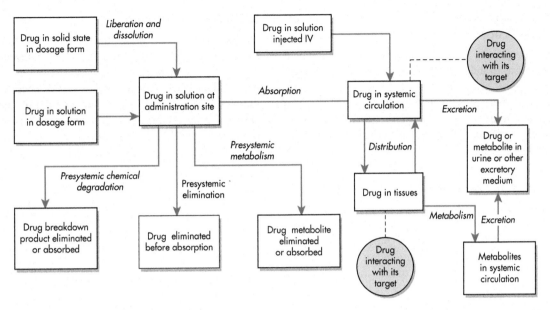

FIGURE 1-1 Common processes that affect the fate of a drug product for drugs intended to enter the systemic circulation.

concentration. The exact shape of a concentration versus time profile for a particular drug product, for example, how quickly and how high the plasma levels rise, depends on many factors including the dosage form (and the nature of the drug's release), the route of administration, the physicochemical properties of the drug, and the ADME processes, which are described next.

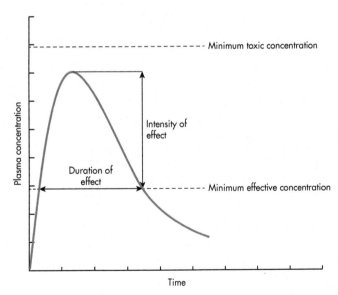

FIGURE 1-2 Plot of plasma concentration versus time for a single dose of a hypothetical drug administered by a nonintravenous route.

Physicochemical Factors Related to the Fate of the Drug in the Body

As mentioned, there are many physicochemical properties of a drug, which along with physiological factors will affect the fate of the drug in the body through their effects on the various LADME processes. Some of these properties are discussed in detail in different chapters of the textbook and are introduced here.[4]

Solid State Properties

Nearly all drugs reach the formulator in the solid state, most often as crystals. Many drugs can exist in different crystalline forms, or polymorphs, largely depending on crystallization conditions. The different polymorphs of a drug reflect the arrangement and interactions of the drug molecules within the crystal, and therefore different crystalline forms of the same drug will dissolve at different rates: Crystalline forms with weaker molecular interactions (within the crystal) will generally dissolve more quickly than forms with stronger interactions. In addition to crystalline forms, a solid drug may be amorphous (without crystalline form), with drug molecules generally exhibiting weaker interactions than crystalline forms. Therefore, amorphous forms of a particular drug generally dissolve faster than crystalline forms. Solid state properties are discussed in Chapter 2.

Ionization

Most drugs are weak acids or weak bases and therefore can exist in the ionized and/or the nonionized form, depending on the pH of the medium. The degree of a drug's ionization can be important for many biopharmaceutical properties, including water solubility, absorption, distribution, and excretion. This is largely because the two forms have different polarities, with the ionized form being the more polar form, which is the form that dissolves better in body fluids but will generally not pass through membranes as well as the less polar, nonionized form. Ionization and its implications are discussed in Chapter 4.

Solubility and Dissolution

Drugs must be in their molecular form to be absorbed through membranes, to be distributed throughout the circulation, to interact with their targets, and to be metabolized and excreted. Therefore, the *solubility* of the drug is a fundamental physicochemical property and its dissolution is a fundamental physicochemical process. Factors that tend to affect a drug's solubility in body fluids include its molecular nature (e.g., presence of polar functional groups), its state of ionization at the pH of the body fluid (with the ionized form being more water-soluble), and its crystallinity (with the weaker polymorph being more water-soluble). *Dissolution* rates in the body can be affected by the drug's water solubility and particle size (with smaller particles generally dissolving more quickly). Particle dissolution is discussed in Chapter 5.

Partition Coefficient

A drug's *partition coefficient* is a measure of its concentration in a nonpolar organic phase relative to that in a polar aqueous phase. The drug partitions between the two phases, largely based on its polarity. Therefore, the partition coefficient can be useful for predicting the passive diffusion of a drug across a lipid bilayer. Lipophilic drugs have higher partition coefficients and tend to have a better ability to pass through the cell membranes that they encounter during their passage through the body. Since most drugs are weak acids and weak bases, pH can play a significant role in passive diffusion. According to the pH-partition hypothesis, for ionizable drugs, the more lipid soluble nonionized form is the form that most readily crosses a lipid bilayer. Therefore, as the pH determines the extent of a drug's ionization, the pH at absorption or permeation sites plays a large role in determining the passive diffusion of a drug across a lipid bilayer and can be used, along with knowledge of a drug's pK_a, to predict drug absorption at different administration sites and the transport of drugs into different body fluids. Drug partitioning is discussed in Chapter 5.

Mass Transport and Membrane Passage

Mass transport refers to the movement of molecules of solutes (e.g., dissolved drug) or solvents from one region to another, and so it is a fundamental determinant of the drug's fate in the body. From their absorption to their elimination, drugs enter different regions of the body, largely by passing through membranes, a term that can be interpreted in different ways: It can refer to multiple layers of cells (e.g., the skin and eyeball), to single cell layers (e.g., intestinal and bronchial epithelia), and to the lipid bilayer of individual cells. Drugs cross membranes by passive mechanisms (where the membrane does not actively participate in drug passage) or by active mechanisms (where a membrane component actively participates in the transfer). *Passive diffusion* through cellular lipid bilayers is the most common way drugs cross biological membranes, and as mentioned, it favors drugs with sufficient lipophilicity. In addition, drugs may diffuse to various regions *between* cells, which is the typical way drugs pass through fenestrated capillary endothelia, such as with drug passage into many tissues and into the kidney glomerulus. In these passive processes, drugs will move down their concentration gradient (i.e., from a region of high concentration to a region of low concentration). Some drugs are actively transferred into and out of cells with the aid of drug transporters, which are membrane-bound proteins. If the active transfer is against the drug's concentration gradient the process is called *active transport*, and if it's with the concentration gradient, it is called *facilitated diffusion*. The processes of mass transport and membrane passage are discussed in Chapter 6.

Roles of drug transporters. Transporters are biologically critical to cellular homeostasis as they control the influx and efflux of many compounds, including nutrients, ions, and xenobiotics (i.e., foreign chemicals, including drugs). Their importance is supported by the fact that about 7% of the human genome codes for transporter-related proteins.[5] For many drugs, drug transporters are critical for determining the drug's fate, and can affect all ADME processes, as they are located in intestinal, renal, and hepatic epithelial cells, and in barrier endothelia of various organs, including the brain. Not only can they be important to a drug's fate in the body, transporters can also play roles in the resistance to drugs (e.g., some cancers and viral infections), and they can play a role in adverse drug effects. Transporters are classified into two superfamilies: ABC (ATP binding cassette) transporters, which are active transporters that utilize ATP for their energy; and SLC (solute carrier) transporters, which include facilitated transporters and ion-coupled active transporters, which require coupling with the transport of a second solute for their energy. P-glycoprotein, an ABC transporter, and organic anion transporting peptides (OATPs), SLC family members, are well-known examples of transporters that can be involved in the ADME processes. Their roles will be discussed further in subsequent sections of this chapter.

Complexation and Protein Binding

The binding of a drug to different molecules or macromolecules can be important for its fate in the body. For example, some drugs (such as tetracycline) can complex to calcium in the gastrointestinal tract, thereby limiting their absorption. In the bloodstream, drugs can bind to circulating plasma proteins, including albumin, which mainly binds acidic and neutral drugs, and alpha1-acid glycoprotein, which mainly binds basic drugs. Plasma protein binding can be an important factor for a drug's distribution, metabolism, excretion, and pharmacological activity. This is due to the fact that, generally, only free (unbound) drug can transfer from the bloodstream into the interstitial fluid that surrounds cells of a given tissue; only free drug can cross cell membranes and interact with other molecules, such as metabolic enzymes and drug targets; and only free drug can be renally excreted. In most cases, protein binding is a reversible equilibrium process, so as free drug is removed from an area, protein-bound drug is released to reestablish equilibrium. Details on drug binding are discussed in Chapter 7.

LADME

As mentioned, the discipline of biopharmaceutics encompasses an understanding of the factors that dictate a drug's fate in the body. *Pharmacokinetics* is a related discipline that can be described as the science of the kinetics of drug absorption, distribution, and elimination. It mathematically relates the ADME processes to parameters that are used to calculate and adjust dosing regimens for patients. Some of the mathematics of pharmacokinetics are described in Chapter 11 so will not be discussed here. However, some of the important pharmacokinetic concepts (such as bioavailability, volume of distribution, and clearance) will be introduced in the following review of LADME processes.[6]

Liberation

For most drug products, the drug is dispersed within its formulation, either in its molecular form or as solid particles (see Table 1-1). For it to be absorbed through biological membranes, it must be released (liberated) from its dosage form and, if it is in the solid state, it must dissolve in body fluids (i.e., it must be in its molecular form). (Drugs administered as aqueous solutions are ready to be absorbed.) Depending on the dosage form, there are different ways a drug can be released. If the drug is in an oily phase, such as an ointment that has been applied to the skin, an oil that has been injected into a muscle, or a fatty acid suppository base that has melted in the rectum, the drug, if undissolved, must partition to the aqueous biological fluid and dissolve. If the drug is an aqueous suspension, the suspended crystals must dissolve in the body fluid. If the drug is in a compressed tablet, the drug crystals are released and must dissolve, during and after a process of tablet disintegration, which functions to fully expose the drug crystals to the gastrointestinal fluids for dissolution. This process is depicted in Figure 1-3.

Absorption

Depending on the route of a drug product's administration, once the drug is dissolved in the fluid of the route, it faces different membrane barriers on its transfer into the fluid that will either contain the target or transport it to the target location (see Table 1-2). The transporting fluid is the bloodstream for most drug routes. A drug dissolved on the skin or eyeball must traverse several cell layers. A drug delivered epidurally (outside the spinal cord) must pass through the dura. Drugs injected intramuscularly and subcutaneously enter through blood capillaries and through lymphatic capillaries for subcutaneous injections. Drugs administered to the mucosal sites of the lungs, nasal cavity, mouth, and rectum must cross varying types of epithelia. This section will focus on drug absorption after oral administration, which is the most common route.

After oral administration and drug dissolution, absorption primarily occurs through the single layer of columnar cells that line the stomach and through the enterocytes of the upper small intestine (duodenum and jejunum). Of these absorptive pathways, absorption through the upper small intestines is usually far greater than through the stomach, primarily because of the larger small intestinal surface area (owing to the number of epithelial villi and enterocyte microvilli).

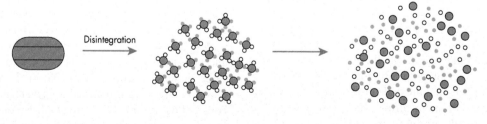

FIGURE 1-3 Depiction of the general process for tablet disintegration to release the drug for dissolution.

The most common pathway for drugs to cross the enterocyte lipid bilayers is by passive diffusion, although as discussed previously and below, transporters can play a significant role. Since passive diffusion dominates in most cases, and since most drugs are ionizable, gastrointestinal pH can play a critical role in drug absorption. The less polar, nonionized form of the drug is the form that can most readily traverse the enterocyte lipid bilayer. So, for example, a weak acid drug will be protonated and therefore predominantly nonionized at low gastric pH. The extent of ionization will depend on the actual pH and the drug's pK_a. Based on pH alone, a weak acid would be expected to be absorbed more in the stomach, but again, because of greater surface area, the small intestines will ordinarily be where most absorption occurs, whether the drug is a weak acid or a weak base. Another factor affecting the intestinal absorption of some drugs into the bloodstream will be the presence of drug transporters in the various cellular membranes encountered by the drug. Following is a general description of the intestinal absorptive pathway into the bloodstream, along with a brief introduction of the roles of drug transporters.

Until about the mid-1990s, a drug's oral bioavailability was considered to depend almost exclusively on its physicochemical properties of solubility and membrane permeability along with its susceptibility to hepatic first-pass metabolism. Since that time, it has become increasingly understood that for many drugs, the role of membrane-bound transporters can be significant.[6,7] These include *efflux* transporters, which as the name implies, move substances out of cells, and *uptake* (or *influx*) transporters, which move substances into cells. These transporters are also located in membranes in many other regions of the body, including the kidney and the blood–brain barrier, where they can serve roles on the disposition of drugs. In the typical sequence of cell membranes that a drug in intestinal fluids passes on its way into the bloodstream, the first encountered are the membranes of the enterocyte; first will be the apical membrane facing the intestinal lumen, and then the basolateral membrane facing the interstitium. On the apical membrane of the enterocyte is located P-glycoprotein, a major efflux transporter. It is believed that a role of this transporter is to move xenobiotics (foreign molecules) back into the intestinal lumen, followed by their reabsorption into the enterocyte. Repeated cycles of this process maximize the exposure of the xenobiotics to the metabolic enzymes that reside inside the cell. This interplay between efflux and metabolism can coordinate to reduce the amount of intact drug that enters the bloodstream, thereby decreasing its bioavailability. Drug that leaves the enterocyte through the basolateral membrane enters the interstitium, followed by its entry into the hepatic portal system, which transports the drug to the liver sinusoids where portal blood mixes with blood from the hepatic artery. The blood percolates through the large gaps in the sinusoidal endothelial cells to reach the hepatocytes. Drugs enter the hepatocyte on the basolateral side that faces the sinusoids. They will either enter by passive diffusion (e.g., if they are lipophilic enough) or with the aid of uptake transporters, such as organic anion transporting peptide (OATP), if, for example, it is charged or too hydrophilic to passively diffuse across the lipid bilayer. Metabolism occurs within the cell through the action of the metabolic enzymes that are located primarily on the endoplasmic reticulum. Drug and metabolites will leave the cell, again by passive diffusion or with the aid of transporters. Drug and metabolites that leave the cell through the basolateral side of the hepatocyte ultimately enter the hepatic veins, which drain into the inferior vena cava, followed by their distribution into the systemic circulation. Some drugs or metabolites may leave the hepatocyte through the apical side, which like the enterocyte, contains P-glycoprotein, which now functions to pump drug into the biliary canaliculi (bile capillaries), which eventually merge to form the hepatic bile duct that empties into the gall bladder, which ultimately deposits its contents into the intestinal lumen. Drugs that leave through the apical side of the hepatocyte are usually large (MW 300–500) and polar (such as glucouronide metabolites of some drugs).

Enterohepatic Circulation

If a drug (or metabolite) reenters the intestinal lumen by biliary excretion, some will be excreted in feces and some may be reabsorbed back into the enterocytes in a cycling process known as *enterohepatic circulation* or *enterohepatic cycling*. This process is favored for drugs with molecular weights greater than approximately 300 and can occur for some hydrophilic metabolites, such as glucuronide metabolites that can be restored to their parent drug by the metabolic activity of intestinal bacteria. The parent drug can then be reabsorbed and the cycle can be repeated. For a drug that experiences enterohepatic cycling, the process can prolong the time that it is in the body and therefore its duration of action. The process can also contribute to its elimination in the feces.

Rate-Limiting Step in Drug Absorption

The rate and extent of drug absorption into the systemic circulation are the net effects of processes that follow the administration of the drug product. Using an orally administered tablet as an example, the main processes are: (1) tablet disintegration, (2) dissolution of the solid drug particles in gastrointestinal fluids, (3) gastric emptying into the small intestine, and (4) absorption into the systemic circulation. The rate of each of these processes will differ, and the step with the slowest rate will be the *rate-limiting step* that will dictate how quickly a drug will enter the bloodstream. Note that for other routes and dosage forms, there may be different processes that will affect the drug's absorption. Usually, it will be the rate of dissolution or rate of absorption that will be rate-limiting. For example, a polar drug may dissolve quickly but have difficulty passing through enterocyte lipid bilayers. This would make absorption rate-limiting. A nonpolar drug may have the opposite problem, making dissolution rate-limiting. Gastric emptying (of the drug) can be rate-limiting for drugs that are very soluble (and dissolve quickly) and have high intestinal permeability. Having this information about a drug's solubility, dissolution rate, and intestinal permeability is critical basic information that is obtained during the drug development process. It has value for the development of formulations and for the design of clinical studies.

First-Pass Metabolism

For drugs given orally, following their absorption they are subject to metabolism, first in the enterocytes that make up the intestinal epithelium, and then in hepatocytes after being transported to the liver via the hepatic portal vein. This presystemic metabolism is called *first-pass metabolism*, because it occurs on the first pass of the drug from the intestinal lumen and through the liver, before it enters the systemic circulation. It can significantly limit the amount of active drug that enters the systemic circulation and therefore can significantly limit an oral drug's bioavailability. Although *hepatic* first-pass metabolism is the major type of first-pass metabolism, other drug absorption sites, such as the skin and lungs, can also exhibit first-pass metabolism, but usually to a much lesser degree.

Distribution

Once a drug is in the bloodstream, it will distribute fairly rapidly throughout the body. However, distribution is not equal to all parts of the body, both in terms of speed and extent.[8] Distribution is generally fastest to well-perfused organs, such as the liver, kidney, and brain, and slower to other tissues such as the muscle, fat, and skin. Upon reaching capillary beds of various tissues and organs, most drug molecules are able to traverse the capillaries into the interstitium, and if they have sufficient lipophilicity they can passively diffuse into the cells that are bathed by the interstitial fluid.

Plasma Protein Binding and Tissue Binding

An important point mentioned before is that only free drug is able to leave the vasculature. Because of this, the binding of drugs to plasma proteins, such as albumin and alpha1-acid glycoprotein, can play a critical role in drug distribution. In most cases, the binding of drugs to plasma proteins is reversible, so when free drug leaves the plasma during distribution to tissues or by excretion, protein-bound drug dissociates to maintain the equilibrium.

Upon reaching the tissues, drugs can also bind to tissue components, such as cell membrane phospholipids and proteins, which, along with active transport processes, can lead to tissue accumulation of the drug. This tissue-bound drug can act as a drug reservoir that can slowly release the drug and prolong its action, but it can also lead to adverse effects if the drug is toxic to the particular tissue.

Other Distribution Sites

Central nervous system. A notable exception to easy passage into interstitial fluid is found with the central nervous system (CNS), where the endothelial cells of the capillaries supplying the brain and spinal cord are held together by tight junctions, which cause the capillaries to restrict the easy passage of drugs into the CNS interstitium, unless they are small and lipophilic enough to pass through the membranes of the capillary endothelial cells. In some cases, even small lipophilic drugs may show poor penetration into the CNS, because like the enterocytes and hepatocytes discussed previously, the capillary endothelial cells in the CNS possess efflux transporters, such as P-glycoprotein, that function to limit the entry of xenobiotics (including drugs) that are their substrate. On the other hand, there are also uptake transporters that facilitate the entry of certain hydrophilic substances. Examples are glucose, which utilizes glucose transporter 1 (GLUT1) for its facilitated entry into the brain, and large neutral amino acids, which utilize the L-type amino acid transporter 1 (LAT1).

Fat. Some lipophilic drugs, such as certain general anesthetics, can accumulate in adipose tissue. In these cases, the distribution of the drug may be significantly affected in obese patients and must be accounted for in drug dosing. Note that it is important for the practitioner to be aware that many lipophilic drugs do not accumulate in fat, so any assumptions should be made with caution.[9]

Placenta. Drugs can cross the human placenta by passive diffusion, in which case, blood flow, protein binding, and the pH of maternal and fetal blood and their effects on the drug's ionization state can be important for placental drug transfer. Additionally, as with many other sites of distribution, there are numerous efflux transporters (including P-glycoprotein) and uptake transporters that are important for controlling transport of their substrates to the developing fetus.[10]

Apparent Volume of Distribution

Each drug has a unique, and typically uneven, pattern of distribution in the body based on its physicochemical properties, such as ionization and lipophilicity, its binding to plasma proteins and tissue sites, its affinity to transporters, and the differences in perfusion of different tissues. However, knowing that the distribution pattern of the drug in the body is typically uneven, it is often useful for the pharmacist to know what dose is required to produce a desired plasma or blood level to achieve the desired effect. The apparent volume of distribution, or Vd, is a key pharmacokinetic parameter that enables that calculation. In essence, the volume of distribution is a proportionality factor that relates the total amount of drug in the body to the plasma or blood level. Different equations can be used to calculate Vd, but one that reflects this proportional relationship is:

$$Vd = Dose/Cp_0$$

where Cp_0 is the initial plasma concentration achieved after an intravenous dose.

The volume of distribution is a hypothetical rather than an actual physiological volume, thus the term "apparent" is used. Although Vd doesn't indicate the volume of a specific body compartment, it can be used to rationalize the distribution behavior of the drug, based on its properties mentioned previously and on the approximate distribution of total body water in a patient. Figure 1-4 shows an approximate distribution of total body water, which is divided between intracellular and extracellular body water, the latter of which is further divided into plasma and interstitial body water. (Note that volumes in actual patients can vary.) A drug that is highly protein-bound may exhibit a low volume of distribution, approximating the plasma volume, because of the

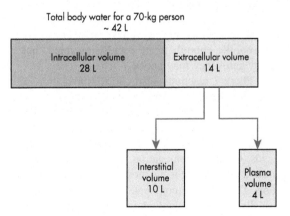

FIGURE 1-4 Approximate body water distribution for an individual with 42 L of total body water.

inability of the protein to pass into the interstitial fluid. On the other hand, a drug that has high interstitial tissue binding will have a relatively low plasma concentration, and therefore the calculated Vd can even exceed the total body water volume. Drugs with low protein and tissue binding and good ability to penetrate cell membranes will have an intermediate Vd closer to total body water, a reflection of their ability to distribute to total body water.

Metabolism

Most drugs are too lipophilic to be excreted unchanged in the urine. Though they will most often undergo glomerular filtration, their lipophilicity, which favored their passage through membranes elsewhere in the body, also favors their passive reabsorption from the filtrate (see "Excretion" section). The primary function of metabolism, of which the liver is the primary location, is to transform these compounds into more hydrophilic, and therefore more excretable metabolites.[11,12,13] The processes by which these metabolic biotransformations occur fall under two main types of enzymatic pathways, *phase 1* and *phase 2*.

Phase 1 reactions, also called *functionalization reactions*, either add or uncover a polar functional group—such as a hydroxyl, amine, or sulfhydryl group—on the parent compound. The most well known and most important enzyme family for phase 1 reactions is the cytochrome P450 (CYP) superfamily of enzymes, which catalyze the oxidation of organic compounds. Other possible reactive pathways in phase 1 reactions include hydrolysis and reduction. With the newly created or exposed polar functional group, the product of these phase 1 reactions may be hydrophilic enough to be rapidly excreted, but in many cases they may not be.

Phase 2 reactions, also called conjugation reactions, involve the covalent conjugation of a polar functional group on the parent compound or on a phase 1 metabolite, with endogenous polar compounds such as glucuronic acid, sulfuric acid, acetic acid, or amino acids. Of these, conjugation with glucuronic acid to form a glucuronide is the most common phase 2 reaction. The resulting conjugate in phase 2 reactions is usually hydrophilic enough to be rapidly excreted.

The products of drug metabolism usually have reduced or no pharmacological activity, but in some cases, the metabolites can have significant pharmacological effects or toxic effects. Pharmaceutical manufacturers may also create pharmacologically inert *prodrugs* that undergo metabolism to the pharmacologically active metabolite. Often, in these cases, the prodrug is more lipophilic and therefore better absorbed than its active metabolite.

Major Metabolic Sites

As mentioned, the liver is the primary metabolic organ. The most important extrahepatic metabolic site is the epithelium of the intestines, described in the section on absorption. Other important metabolic locations include the kidney, lungs, skin, and placenta.

Excretion

The kidneys are the major excretory organ for drugs or their metabolites. Renal excretion involves three major processes: (1) glomerular filtration, (2) secretion, and (3) reabsorption. Most drugs can be readily filtered through the fenestrated capillaries of glomeruli, regardless of their polarity or state of ionization. However, protein-bound drug cannot be filtered because the proteins (e.g., albumin) are too large to pass through the fenestrations. Those compounds that are polar or ionized tend to remain in the filtrate and are subsequently excreted in the urine. However, drugs that are nonpolar and relatively lipophilic can undergo extensive passive reabsorption in the distal tubules and will consequently be less readily excreted in the urine. Since reabsorption by passive diffusion of weak acid and weak base drugs favors the nonionized state, there can be significant variability in passive reabsorption for these drugs, depending on the urinary pH. Additionally, many compounds can be actively secreted into the proximal tubules by various efflux transporters, such as certain organic anion transporters (OAT) and organic cation transporters (OCT). Also, some compounds can be actively reabsorbed by uptake transporters in the proximal tubules.

Other Excretion Sites

Biliary/fecal. Though the kidney is the major excretory organ for drugs, other excretory pathways can be important, including biliary excretion. As discussed in the section on drug absorption, biliary secretion of drugs can be aided by efflux transporters and the process favors larger molecules. As also discussed, once drugs (or their metabolites) are secreted into the biliary canaliculi and ultimately reach the intestines, some will be eliminated in the feces and some may be reabsorbed back into the body through the intestines via the process of *enterohepatic recycling*.

Others: Saliva, sweat, tears, breath, breast milk. Drugs may also be excreted by other routes including saliva, tears, breath, sweat, and breast milk. Excretion by these routes is mainly by passive diffusion, thus it favors the nonionized forms of weak acid and weak base drugs (see Chapter 6). Though drug excretion by these routes is usually quantitatively insignificant for the patient, they can still be clinically important. For example, saliva can be used to assay some drugs when it is difficult to obtain plasma samples, and for many drugs, breastfeeding may be inadvisable because of the drug's effect on the child.

Clearance—The Primary Measure of Drug Elimination

The primary pharmacokinetic parameter that is used as a measure of drug elimination from the body is *clearance* (CLp), which can be defined as the volume of plasma that is completely cleared of drug per unit time (e.g., mL/min).[14] It describes the relationship between the rate of drug elimination from the plasma and its plasma concentration.

Clearance may also be defined with respect to other concentrations measured, for example, clearance from the blood (CLb) or clearance of unbound drug (CLu). It is also important to note the total body clearance or systemic clearance is a composite of the clearance from each of the drug elimination pathways, including the renal clearance of unchanged drug in the urine (CLren), the clearance by drug metabolism and/or biliary secretion in the liver (CLhep), and the clearance by the other elimination pathways (CLother), including those mentioned previously. The total systemic clearance (CLsystemic) for a drug is a composite of the contributing elimination processes. For drugs that are eliminated predominantly by one organ (e.g., renal excretion or liver metabolism), a reduced capacity of that organ can affect the systemic clearance and would require dosing adjustment. Whereas knowledge of the volume of distribution enables calculation of the drug dose required to reach a certain plasma concentration, knowledge of a drug's clearance allows calculation of the *dose rate* required to maintain a *target steady state plasma concentration* (Cp_{ss}). For example, dose rate (mg/hr) = Cp_{ss} (mg/L) × CL (L/hr). (Note: Steady state concentration is the concentration when the rate of drug administration equals the rate of drug elimination, which occurs when the drug stops accumulating with dosing at regular intervals or with continuous intravenous infusion.)

Elimination Half-life

The elimination half-life of a drug in the plasma is the time required for the plasma concentration to decrease by 50%. So, after one half-life 50% of the drug will be eliminated, 75% will be eliminated after two half-lives, and 97% will be eliminated after five half-lives. The half-life is a derived parameter that is dependent on the drug's clearance and volume of distribution at steady state and can be calculated from these parameters according to the following equation:

$$t_{1/2} = V_{ss} / CL$$

Knowing the drug's half-life can be useful for estimating appropriate dosing intervals and for estimating the time to reach steady state with multiple dosing, which requires about four to five half-lives. The concept of half-life is further discussed in Chapter 11.

CONCLUSION

This chapter discussed many of the important biopharmaceutical factors that determine a drug's fate after the administration of a drug product to a patient. Understanding these factors, integrated with a thorough knowledge of the principles and mathematics of pharmacokinetics, will enable pharmacist practitioners to competently initiate and modify dosing regimens for their patients. The discipline of pharmacokinetics will be touched on again in Chapter 11 but not fully covered. The reader is referred to the many excellent textbooks on pharmacokinetics. The remainder of this textbook is largely dedicated to elucidating the physicochemical drug properties that are not only important to a drug's fate in the body but are also important for a full understanding of other areas of a pharmacist's knowledge base, including dosage formulations and pharmacodynamics.

KEY POINTS

- Drug targets are generally macromolecules that are located in tissues and fluids throughout the body. Some of these targets can be accessed without drug absorption into the bloodstream. For most drug products, however, the targets are in tissues and cells that are accessed from the bloodstream.
- There are a variety of routes and dosage forms that can be used for drug administration to a patient, depending on the dosage form and route. The drug may have to first be released (liberated) from the dosage form, then cross membranes to enter the systemic circulation.
- Biopharmaceuticals relate the physicochemical properties of a drug to its disposition, or fate, in the body, to its pharmacological effect.
- The fate of the drug after it is administered to the patient is embodied in the acronym LADME, which stands for liberation (release of the drug from its dosage form), absorption (into the bloodstream), distribution (to various parts of the body), metabolism (by enzymes), and excretion (through the kidneys or other routes).
- Important physicochemical properties related to the drug's fate in the body, and discussed in this textbook, include solid state properties, ionization, solubility and dissolution, partition coefficient, mass transport and membrane passage, complexation, and protein binding.
- Drug transporters can play significant roles in drug disposition.
- Absorption after a drug is administered orally, mostly occurs through intestinal epithelial cells (enterocytes) by passive diffusion and can be under the influence of p-glycoprotein, which moves some drugs back into the intestinal lumen.
- Some drugs can be metabolized within enterocytes.
- After drugs leave the enterocytes, they enter the portal system, which transports them to the sinusoids that enable drug access to the hepatocytes where they can be absorbed, again mostly by passive diffusion, then metabolized.
- Drugs (and metabolites) leave the hepatocytes, and depending on where they leave the hepatocyte, they will enter the systemic circulation or the bile.
- Once in the bloodstream, the drug will distribute in the body, usually unevenly, based on its physicochemical properties, affinity to transporters, and on its binding to different components, such as plasma proteins. The volume of distribution is an important pharmacokinetic parameter that reflects a drug's distribution.
- Drug metabolism occurs at different sites, with the liver being the primary site. The general role of metabolism is to increase the hydrophilicity of a drug, making it more easily excreted in the urine.
- The kidney is the major excretory pathway, but other pathways, including the biliary pathway, can be important.
- Clearance is a primary pharmacokinetic parameter used as a measure of drug elimination.

CLINICAL QUESTIONS

1. The ingestion of grapefruit juice and grapefruits are known to increase the plasma levels of many drugs taken orally. The increase is due to the effects of certain furanocoumarins (present in grapefruit) on metabolizing enzymes in enterocytes and hepatocytes, and possibly on p-glycoprotein. Explain the likely effects of grapefruit on these proteins and how these effects can lead to increased plasma levels.
2. It has been demonstrated in many critically ill patients that plasma albumin concentrations can decrease. Explain why this may require dosage reductions for some highly albumin-bound drugs in critically ill patients.

2 States of Matter Related to Pharmaceutical Formulations

Beverly J. Sandmann
Edited by Ann Newman and Gregory T. Knipp

Learning Objectives

After completing this chapter, the reader should be able to:

- Explain the potential energy diagram.
- Understand the types of intermolecular forces and their relative magnitudes.
- Identify the physical characteristics of each state of matter.
- Distinguish between the gaseous, liquid, and solid states of matter.
- Relate the heat of vaporization and the boiling point of liquids to the magnitude of intermolecular forces.
- Relate the heat of fusion and the melting point of solids to the magnitude of intermolecular forces.
- Demonstrate and understand the physical properties of gases, liquids, and solids.
- Describe the different types of solid materials used in pharmaceuticals.
- Differentiate chemical and physical stability.

INTERMOLECULAR FORCES

Intermolecular Binding Forces

When molecules interact with each other, they do so by the forces of both attraction and repulsion. Forces of attraction are essential for molecules to come together. The two types of attractive forces are called *cohesive forces* and *adhesive forces*. For example, when like molecules are attracted to each other this represents cohesive forces of attraction. When different molecules are attracted to one another, these are adhesive forces of attraction. A good analogy would be the opposite poles of a magnet, which attract one another when in close proximity. In addition to attracting each other, molecules are acted on by *repulsive forces*, which act to separate molecules. Using the magnet analogy, bringing two positive ends of a magnet in close proximity leads to repulsion, as would two negative ends.

Consider two molecules coming together. When they are attracted to each other (i.e., when unlike charges are closer than like charges), the forces of attraction pull the molecules together. Attractive forces (F_A) are inversely proportional to the distance separating the molecules (r), as shown here:

$$F_A \propto \frac{1}{r^n}$$

where $n \sim 6$ or 7 for the interaction of two hydrogen atoms or nitrogen atoms and $n \sim 3$ or 4 for chloroform molecules. The inverse relationship between the forces of attraction and the distance has been derived from the phenomena first described by Sir John Edward Lennard-Jones and is known as the Lennard-Jones potential.[1] As described by the Lennard-Jones potential theory, attractive forces can be represented by means of a potential energy function. As the forces of attraction between the molecules increase, the potential energy becomes increasingly negative, as illustrated in Figure 2-1.

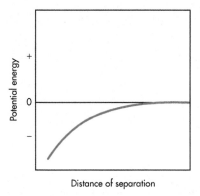

FIGURE 2-1 Potential energy diagram for attractive forces.

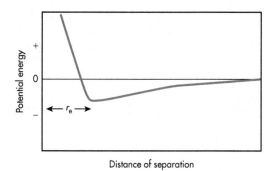

FIGURE 2-3 Potential energy diagram of two gaseous molecules.

Attractive forces operate over a greater distance than do repulsive forces. The long-range attractive component of intermolecular force is significant when the overlap of electron clouds is small. Short-range repulsive forces operate when the molecules come close together and electron clouds interact. Repulsive forces (F_R) are proportional to an exponential relationship with the reciprocal of the distance separating the molecules (r):

$$F_R \propto e^{1/r}$$

An exponential function changes more rapidly and curves much more at an angle on the potential energy diagram for repulsive forces (Figure 2-2).

By convention, the force F acting between two molecules that are close together is positive when it is repulsive and negative when it is attractive.

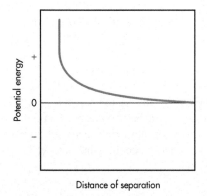

FIGURE 2-2 Potential energy diagram for repulsive forces.

Putting the interactive force components together gives the net potential energy diagram in Figure 2-3.

The quantitative details of the curve vary from one type of molecule to another, but the general form is the same. As two molecules approach each other, the energy changes are gradual and negative (attractive) to a point where a minimum in the potential energy occurs; then the energy starts rising rapidly as the intermolecular distances become smaller and repulsive forces begin to dominate. The distance between the molecules at which the attractive and repulsive forces just balance each other is the *collision diameter*. This is the most stable distance between the two molecules. The position of the molecules at equilibrium, r_e, is shown in Figure 2-3.

In summary, the Lennard-Jones potential allows for an equilibrium distance to be reached, where the balance between the attractive forces and repulsive forces is the same. Moving the molecules closer results in electron cloud repulsion, whereas separating the molecules apart increases the attraction. Cohesion results when the balance of the attractive and repulsive forces leads to bonding. The energy and type of bond is dictated by the nature of the molecules that are interacting, which is generally a function of each molecule's electronegativity. The concept of the equilibrium distance described in terms of attractive and repulsive forces is important for understanding molecular bonding and states of matter.

Intermolecular Attractive Forces

There are four main types of intermolecular attractive forces: the *van der Waals attractive forces* (*dipole-dipole, dipole-induced dipole, induced dipole-induced dipole*) and the *ion-dipole forces*. An additional and critical attractive force is the *hydrogen bond* that exists between an electronegative atom and a hydrogen atom. Hydrogen bonding is considered a uniquely strong type of dipole-dipole interaction. It is also responsible for the existence of another attractive force: *hydrophobic interactions*.

van der Waals Forces

Dipole-dipole forces (Keesom forces). Keesom forces occur when polar molecules possessing permanent dipoles, having both a partial positively charged end and a partial negatively charged end, interact. These molecules align themselves so that the negative pole of one molecule interacts with the positive pole of the other. The energy of this attraction ranges from 1 to 7 kcal/mole. Molecules that possess permanent dipoles include water, hydrochloric acid, alcohol, acetone, and phenol.

Permanent dipoles also form across the peptide (amide) bond in proteins and can act to stabilize protein secondary structure, particularly α-helices.

Dipole-induced dipole forces (Debye forces). A polar molecule can produce a temporary electric dipole in nonpolar molecules that are easily polarizable. The forces of attraction are weaker, being about half those of dipole-dipole forces. The energy of this type of attractive force is 1 to 3 kcal/mole. Easily polarized molecules include ethylacetate, methylene chloride, and ether:

Dipole induced-dipole forces are generally lower in energy, but multiple interactions can have a stabilizing effect on states of matter. Polar organic solvents typically possess these forces, as observed in the picture for methylene chloride.

Induced dipole-induced dipole or dispersion forces (London forces). These are forces that originate from molecular internal vibrations in nonpolar molecules to produce attraction that arises because of synchronized fluctuating dipoles in neighboring atoms. This attraction is produced by asymmetry in the distribution of the electrons around the nucleus. These forces are also temporary. This force is responsible for the liquefaction of gases. The energy of this attractive force is 0.5 to 1 kcal/mole. Nonpolar molecules exhibiting induced dipole-induced dipole forces of attraction include organic compounds such as carbon disulfide, carbon tetrachloride, and hexane. These forces are often found in the aliphatic regions of lipid bilayers as well, although other interactions also help to stabilize the bilayer structure, as illustrated in Figure 2-4.

In general, van der Waals forces of attraction account for the molecular interactions involved in solubility, complexation, and numerous other physical bonding phenomena. They are weak forces compared with the covalent bond, which is generally 50 to 100 kcal/mole, and the ionic bond, which ranges from 100 to 1000 kcal/mole.

Ion Dipole Forces

Ion-dipole forces. Molecules that are polar are attracted to either positive or negative charges. The energy of attraction is about 1 to 7 kcal/mole. An example is a quaternary ammonium ion with a tertiary amine:

$$R_4N^+ - :NR_3$$
$$\text{ion} \quad \text{dipole}$$

Pharmaceutical salts will have ion dipole forces holding the drug molecule and the counterion together. In the case of a carboxylic acid, the hydrogen atom will be removed resulting in a negative charge. The positive ion that bonds to form the salt

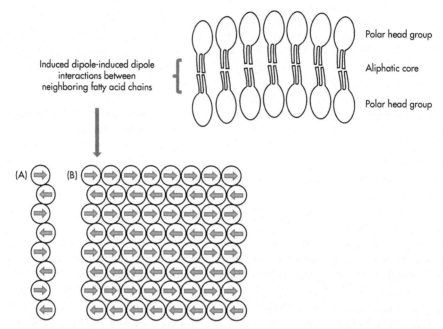

FIGURE 2-4 The nature of the induced dipole-induced dipole interactions in the phospholipid bilayer. In the bottom portion of the figure, the induced dipoles are illustrated by the arrows on the carbon attached to each fatty acid. In this model, (A) illustrates that the covalently bound hydrocarbons on the acyl chain induce opposite dipoles with one another. In (B), the alignment of multiple fatty acid chains with these alternating induced dipoles in a phospholipid bilayer all align in the same direction. While the forces are weak, cumulatively they add up and provide a large stabilizing force.

can be an inorganic species, such as sodium or potassium, or an organic species, such as arginine or choline.

Ion-induced dipole. The forces of attraction are induced by the close proximity of a charged ion to the nonpolar molecule, for example, iodine and potassium iodide:

$$K^+I^- - I-I \rightarrow K^+ + I_3^-$$

iodide iodine iodide
anion molecule complex

These attractive forces account for the solubility of ionic crystalline substances in water, where a cation attracts the large negative oxygen of water and an anion attracts the hydrogen atoms.

The Hydrogen Bond

A specific type of electrostatic attraction between molecules and parts of molecules is the *hydrogen bond*. Hydrogen bonding can be intermolecular or intramolecular. Hydrogen bonding is the attraction of a hydrogen atom for a strongly electronegative atom such as oxygen, nitrogen, fluorine, and, to some extent, sulfur. Hydrogen can form an electrostatic bond to these atoms because of its small size, generating an intense electrostatic field. As previously mentioned, the hydrogen bond is considered a uniquely strong type of dipole-dipole interaction. However, the hydrogen bond is also partly covalent in nature, exhibiting such "covalent" bonding properties as bonding strength, unidirectionality, and short interatomic distances. Hydrogen bonding accounts for the unusual properties of water relative to hydrogen compounds of other group 6A elements, such as sulfur (H_2S) and selenium (H_2Se). The boiling point of water is much higher than that of hydrogen sulfide—100°C (212°F), and −60.3°C (−76.5°F)—and so the hydrogen sulfide is a gas at room temperature. Hydrogen bonding is important for protein α-helix and β-sheet structures;

conformation of proteins; physical properties of alcohols compared to alkanes; carboxylic acids compared to esters, aldehydes, and ketones; and sugars. In a compound, the formation of a hydrogen bond in an aqueous solution with water or in the structure of the compound modifies both the physical properties and to some extent the chemical properties:

Salicylic acid intermolecular bonds (hydrogen bonds)

The hydrogen bond in proteins generally involves hydrogen with oxygen and nitrogen atoms. The hydrogen bonds of most importance in maintaining protein structure are those between the peptide backbone –NH groups and the backbone carbonyl groups, as shown here:

Peptide intermolecular hydrogen bonding

Hydrophobic Interactions

Hydrophobic interactions are forces of attraction between nonpolar atoms and molecules in water. They cause the nonpolar species to be driven together and are critical for the structure and stabilization of many molecules including proteins (with nonpolar amino acids) and aggregates of amphiphiles (containing polar and nonpolar moieties), such as micelles and bilayer membrane structures (e.g., cells and liposomes). Hydrophobic attractive forces are less due to interactions between nonpolar molecules than to the lack of ability of these molecules to bond with water molecules. When these non–hydrogen bonding nonpolar molecules are introduced into water, the hydrogen bonding network between water molecules is disrupted. To minimize this disruption and reestablish a hydrogen bonding network, the water molecules reorient to form a structured water "cage" around the nonpolar molecules, thereby stabilizing the interaction between them. Thus, nonpolar molecules are considered "water-fearing," leading to their "hydrophobic" interactions, which are the source of the *hydrophobic effect*.

STATES OF MATTER

In addition to their potential energy, molecules of gases, liquids, and solids have kinetic or thermal energy. Kinetic energy is proportional to temperature. For example, for 1 mole of an ideal gas, the kinetic energy (KE) is given by KE = 3/2 RT, where R is the ideal gas constant and T is the temperature in degrees Kelvin. It is worth noting that kinetic energy under nonideal conditions is governed by the forces mentioned previously. The transfer of potential energy into kinetic energy is an important concept throughout this section and across many different disciplines, including understanding transition state or reaction pathways in organic chemistry.

The Gaseous State

Gases are described as molecules that have higher kinetic energy that produces rapid motion, are held together by weak intermolecular forces, have no regular shape, are capable of filling all available space, are compressible, and, for many gases, are invisible. Gases such as oxygen, nitrogen, and hydrogen at normal temperatures and pressures can be defined by the ideal gas laws such as Boyle's and Charles' laws. Boyle's law states that for 1 mole of gas at a fixed temperature, the product of pressure (P)

INTERESTING RELATIONSHIPS BETWEEN FORCES AND PROTEIN STRUCTURE

An interesting side note is that the forces previously discussed are also relevant to protein structure. Permanent dipoles also can be formed form across the stacked peptide bonds along secondary structures in proteins and can act to stabilize protein secondary structure, particularly in α-helices. For example, carbonyl and amide groups in amino acids are permanent dipoles that join to form the peptide bond in proteins and can act to stabilize protein secondary structure, particularly α-helices. This phenomenon has been detailed in studies performed by Akiyoshi Wada[2] and is commonly found in most biology and biochemistry textbooks.[3] In fact, protein folding is largely held together by dipolar interactions.

Let us consider an α-helix:

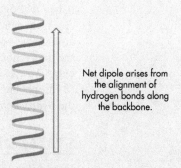

Net dipole arises from the alignment of hydrogen bonds along the backbone.

While there is a net dipole that occurs, the effects of the side chains are also important. When α-helices are grouped in a protein's secondary structure, the dipoles do align, however, each helix may be amphipathic. Amphipathic helices have polar groups aligned on one half and nonpolar side chains on the other:

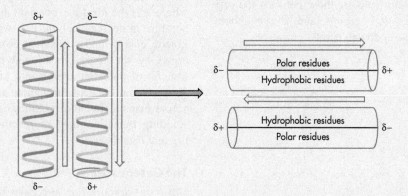

In cytoplasmic proteins, the hydrophobic residues will align in the core and the polar residues will face outward toward the aqueous cytoplasm. London forces play a significant role in the hydrophobic core's conformation in a cytoplasmic protein. That is because the aliphatic amino acid side chains populate the core and give rise to these interactions. They are weak and do not offer significant stability. The backbone will have normal hydrogen bonding and dipolar interactions that help stabilize the core by interacting in patterns commonly found in secondary structure. Dipolar forces and hydrogen bonding with water and its components are considerably more favored on the surface of the cytoplasmic protein, which is composed primarily of polar side chains. Again, the backbone will possess normal hydrogen bonding and dipolar interactions in patterns commonly found in secondary structure. The water accessible backbone regions will act like the polar side chains and interact with water. There are actually excipients that can be used to increase the tightness of the water-hydrogen bonding that occurs at the surface.

Protein storage and handling in the pharmacy is critical for maintaining the stability of the formulated product. Since many of the forces that stabilize a protein are weak, shear and stress encountered in shipping may impact the physical stability of the proteins. For example, excess shaking of a protein formulation during handling adds energy that could result in unfolding. Another potentially important factor that is often overlooked is the effect of temperature on the pKw for water. At room temperature, the pKw is 14, which corresponds to the pH range. However, the pKw increases to 14.94 if the temperature is lowered to 0°C, and 13.02 if it is raised to 60°C. While these changes may be slight, it could impact ionization of amino acid side chains and possibly impact preferential hydration. Therefore, parenteral and solution formulations must be stored at the temperature recommended by the manufacturer. One last consideration regarding proteins and other biological and many common pharmaceuticals is storage in the freezer. Frost-free freezers often use warmer temperatures than −20°C temperatures, which are run in cycles that prevent frost from accumulating into ice. During these cycles, the heating and cooling process can destabilize proteins that are susceptible to heat-thaw-induced physical and chemical instability. In summary, the critical point illustrated by these examples, and many others, is that as a pharmacist, you must pay careful attention to storage conditions for formulations, particularly proteins whose higher-order structure is maintained by a combination of weaker intra- and/or intermolecular forces and are therefore highly susceptible to denaturation, degradation, and potentially aggregation.

and volume (V) is a constant. Combined with Charles' law, which relates the volume of a gas to the temperature (T), the combined law may be written as:

$$\frac{PV}{T} = \text{constant}$$

Rewriting the combined law gives the general gas equation for an ideal gas:

$$PV = \text{constant} \times T$$

where the constant is the value R, the ideal gas constant. The value is determined when $P = 1$ atm, $V = 22.4$ L (volume of 1 mole of an ideal gas), and $T = 273°$K; in this case, it is 0.082 L · atm/deg · mole. R is in energy units/degree. The product of pressure and volume is given in energy units. Energy units for a gas are usually liters × atmospheres. The general gas equation for more than 1 mole of gas is:

$$PV = nRT$$

where n = moles of gas and T is the absolute temperature in degrees Kelvin.

The general gas equation is useful in calculating the properties of gases at atmospheric pressure and at temperatures above their boiling points. In a plot of the gas equation over a large range of pressures, it can be seen that at higher pressures the volume of a gas does not compact quite as much as it would if it were an ideal gas (Figure 2-5). This is in part due to the Lennard-Jones relationship with respect to repulsion at smaller distances. However, at near atmospheric pressure, which is usually less than 1 atm, the deviation of real gases from the ideal gas is very slight, and so it is reasonable to use for calculation of pressure, volume, and so forth. For example, what is the pressure of carbon dioxide gas when 1 mole of the gas occupies 0.5 L at 25°C (77°F)?

$$PV = nRT$$

where $T = (273 + 25)°$K;

$$R = 0.082 \frac{l \cdot \text{atm}}{\text{deg} \cdot \text{mole}}, n = 1, \text{ and } V = 0.5 l$$

$$P = 0.082 \frac{l \cdot \text{atm}}{\text{deg} \cdot \text{mole}} \times \frac{298°\text{K}}{0.5 \text{ L}}$$

$$P = 48.9 \text{ atm}$$

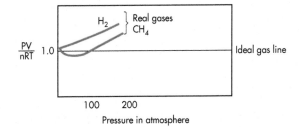

FIGURE 2-5 Plot of the gas equation.

The answer 48.9 atm is about 20 times the pressure in an automobile tire. Heavy containers called gas cylinders are used to contain gases under pressure, such as helium, air, and nitrogen, which are used in the research laboratory.

Blood Gases

One of the applications of gases in pharmaceutical sciences is in the interpretation of blood gases in patients. The important blood gases are oxygen (O_2) and carbon dioxide (CO_2). Gases that are in the blood are generally considered to be separated into hemoglobin-bound O_2 (found largely in the red blood cells), and CO_2, N_2, and other atmospheric gases that are primarily dissolved in the plasma based on the solubilities of each gas. Their plasma concentrations are related to atmospheric conditions and to biological and catalytic metabolic activity. The plasma gas concentrations are determined and expressed in terms of Henry's law of gas solubility, which states that the amount of gas dissolved in the plasma is proportional to the partial pressure of the gas in equilibrium with the plasma. The partial pressure is the pressure a gas would exert if it alone occupied the whole volume of the mixture, according to Dalton's law of partial pressures. For example, the partial pressure of a gas in the atmosphere is a fraction of the total atmospheric pressure, which can be given as $P_{TOTAL} = P_1 + P_2 + P_3 \ldots$. In reference to a patient, these pressures, as millimeters of mercury, are given as Po_2, the partial pressure exerted by oxygen in the blood, and Pco_2, the partial pressure exerted by carbon dioxide in the plasma (defined in terms of blood). Po_2 is the free oxygen in the blood. Since the majority of dissolved oxygen is carried by the blood as hemoglobin. The actual partial pressure of oxygen in the blood is on average about 80 millimeters of mercury (Hg), and the partial pressure of CO_2 is on average 35 to 45 mm Hg. It is important to remember that the oxygen content of blood is related to the fractional percentage of oxygen that is in the inspired air ($f = 0.21\%$). Table 2-1 shows the differences in arterial blood gases on average for an altitude difference of about 1 mile.

Oxygen enters the body through the lungs, which are able to transfer oxygen from air through the small alveoli branches of the lung across the capillaries into the blood. The dissolved oxygen exerts a pressure that can be readily measured as the Po_2. The amount of oxygen that is transported to the tissues is determined by hemoglobin and depends on the ability of the heart to pump blood containing oxygen to the tissues. Older patients have values for Po_2 and O_2 saturation near the lower part of the normal range, whereas younger patients tend to have higher normal range values.

Pco_2 refers to the pressure exerted by dissolved CO_2 gas in the blood. Pco_2 is influenced by respiratory function. In plasma, CO_2 may combine with water to form carbonic acid (H_2CO_3), which dissociates into bicarbonate (HCO_3^-) and hydrogen ion (H^+). There is actually about 800 times as much CO_2 in the form of dissolved gas in plasma as is converted to H_2CO_3. The CO_2 is formed by the oxidation of carbon ingested as a component of food. The CO_2 that comes from food is removed by the lungs, and expiratory efficiency is recorded by a measurement of Pco_2 in the blood. Pco_2 generally is considered to reflect the inverse relationship that exists between itself and lung ventilation. If Pco_2 is elevated, this reflects poor ventilation. If Pco_2 is low, this reflects excessive ventilation or hyperventilation. For example, if a patient is noted to have a low pH (indicating a high concentration of H^+ in the blood, also known as acidemia) and an elevated Pco_2 (indicating hypoventilation or poor ventilation of the lungs), this patient may be said to have respiratory acidosis, or a low pH in a particular tissue. Conversely, if pH is elevated and Pco_2 is decreased, the patient may have respiratory alkalosis, which could lead to alkalemia, which is a higher blood pH. Higher altitudes may actually influence the formation of these disorders in some patients based on changes in the Pco_2 concentrations.

TABLE 2-1 Arterial Blood Gases

	Denver	Sea Level
Po_2	70 mm Hg	> 80 mm Hg
O_2 saturation of hemoglobin	93%	> 95%
Pco_2	38 mm Hg	40 mm Hg

PULMONARY EXCRETION OF CO_2 CAN BE USED TO MEASURE METABOLISM

Interestingly, the drug metabolism activity of enzymes [such as Cytochrome P450 3A4 (CYP3A4)] can be measured by assessing changes in exhaled radiolabeled CO_2 from human lungs. Specifically, an *in vivo* erythromycin breath test uses carbon-14 [^{14}C]-radiolabeled erythromycin that will release [^{14}C]CO_2 upon metabolism.[4] Differing amounts of [^{14}C]CO_2 will be exhaled depending on the level of CYP3A4 expression. The *in vivo* erythromycin breath test can be also used to determine if other compounds can inhibit or enhance erythromycin metabolism *in vivo*. This subject will be covered more comprehensively in a pharmacokinetics class but does further illustrate how exhaled gases can be utilized in clinical settings.

The Liquid State

The liquid state is defined by comparison to the gaseous and solid states. A *liquid* occupies a definite volume and takes the shape of the container required to hold it. Liquids are denser than gases and possess less kinetic energy than do gases. They are considered less compressible than gases and more compressible than solids. Liquids flow very readily, and the flow is influenced by friction. They can also be frozen (become solids), have boiling points (become gases), and have vapor pressure and surface tension.

Vapor Pressure

Vapor pressure is a physical property of liquids. Equilibrium vapor pressure does not depend on the volume or weight of the liquid or on the atmospheric pressure or the presence of other vapors in the air. Vapor pressure does depend on the temperature. Molecules of a liquid possess a range of kinetic energies, often associated with where the molecules are located in the liquid. Molecules with the highest kinetic energy and proximity to a surface (e.g., air and water interface) break away from the surface of the liquid and pass into the gaseous state. Some molecules subsequently return to the liquid state. If the liquid is placed in an evacuated container, the rate of vaporization eventually equals the rate of condensation at equilibrium. The pressure of the saturated vapor above the liquid is known as the equilibrium vapor pressure (VP).

Vapor pressure is recorded in mm Hg. The relationship between temperature and vapor pressure is shown in Figure 2-6. Vapor pressure increases with temperature by a nearly exponential proportionality. According to the integrated form of the Clausius Clapeyron equation, given by log:

$$\frac{VP_2}{VP_1} = \frac{-\Delta H_v}{2.3R}\left(\frac{1}{T_2 T_1}\right)$$

where VP_1 and VP_2 are the equilibrium vapor pressures of a liquid at temperatures T_1 and T_2, respectively, and ΔH is the heat of vaporization, a positive value with a minus sign in front of it. The VP at any temperature can be calculated.

The boiling point (BP) is the temperature at which the vapor pressure of a liquid equals the atmospheric pressure. Vapor pressure and boiling point are inversely related: As VP rises, BP falls. At room temperature, the VP of water is about 20 mm Hg. Upon heating, VP increases. When the temperature reaches 100°C (212°F), the VP of water is 760 mmHg or less, depending on the elevation, and the water passes spontaneously into the vapor or gas phase. The heat of vaporization of

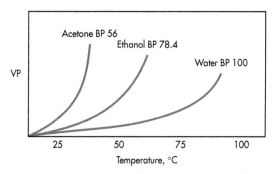

FIGURE 2-6 Vapor pressure (mm Hg) as a function of temperature.

water at boiling is given by ΔH_v (water)$_{BP}$ = 9720 cal/mole. This heat represents the thermal energy absorbed (therefore, a positive value) to cause the hydrogen bonding in liquid water to break and release the free molecules that transform into the gaseous state. A classic example of a therapeutic compound where vapor pressure at room temperature is an important consideration is nitroglycerin:

$$\begin{array}{c} CH_2\text{-}ONO_2 \\ | \\ CH\text{-}ONO_2 \\ | \\ CH_2\text{-}ONO_2 \end{array}$$
Nitroglycerin

Nitroglycerin is a liquid at room temperature and, as a solid, forms two different crystalline materials when it is cooled below 14°C. On heating, it begins to decompose at 50°C, is appreciably volatile at 100°C, and explodes at 218°C. It is very dense, having a density of 1.6 g/mL, concentrating a large amount of energy in a small volume. Its VP at 20°C is 0.00025 mm Hg. A problem associated with sublingual nitroglycerin tablets is that the nitroglycerin can diffuse and evaporate from the tablets, leading to changes in tablet strength and interactions with materials such as certain plastics. One approach to producing nitroglycerin tablets employed polyethylene glycol as an excipient (additive) in the formulation that reduced the VP and thereby reduced stability problems associated with nitroglycerin evaporation. Other vaporization-reducing excipients are used in current compressed nitroglycerin tablets.

Surface Tension

Surface tension is another physical property of liquids. The surface of a liquid is considered two-dimensional. Therefore, the dimensions of surface tension are force per unit length; generally, the units used are dynes $\cdot$ cm^{-1}. Molecules in the bulk of the liquid are attracted to each other by intermolecular forces. Molecules on the surface are attracted inward by the bulk and outward into the phase it is in contact with, thus creating an unequal distribution of attractive and potentially repulsive forces that is based on the chemical nature of the phases. This resulting force in the liquid surface tends to minimize the surface area. The surface tension of water is a particular concern in the pharmaceutical sciences. For example, the high surface tension of water can inhibit it from readily dissolving powders with hydrophobic crystal surfaces. However, there are means to reduce the surface tension that tend to lower the energy of attraction of the molecules in the liquid. This will be discussed in chapters 8 and 9, and is of considerable importance in pharmacy.

In general, the surface tension of liquids decreases with an increase in temperature. Consider the fact that temperature increases the kinetic energy, and consequentially the motion, of the molecules in the liquid. This offsets the energy of intermolecular bonding, for example, hydrogen bonding in water, and increases potential escapability of the molecule at an interface as discussed for vapor pressure. This relationship between surface tension of liquids and temperature is nearly linear. Therefore, surface tension values are reported at a given temperature. Some approximate values of surface tension at temperatures near room temperature are given in Table 2-2.

The Solid State

Solids are characterized as having a fixed shape and being nearly incompressible compared to gases and liquids. They have strong intermolecular forces and therefore very little kinetic energy. In solids, the

TABLE 2-2 Surface Tension of Pure Liquids Exposed to Air

Substance	Surface Tension, dynes $\cdot$ cm^{-1}	Temperature, °C
Mercury	487	15
Water	72	25
Glycerin	63.4	20
Triethanolamine	48	20
Mineral oil	33	25
Alcohol	23	20
Acetone	24	20
n-Hexane	18	20

Source: Data from the *Handbook of Pharmaceutical Excipients*, Pharmaceutical Press, 2012, and other sources.

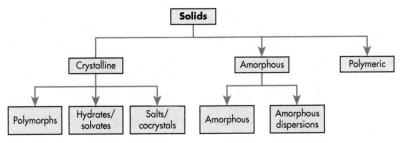

FIGURE 2-7 Overview of pharmaceutical solids.

atoms vibrate in fixed positions about an equilibrium position, and so there is little translational motion. Solids are characterized by shape, particle size, and melting point; some solids are volatile enough to have a sublimation point. Other characteristics of solids, such as those having importance in the manufacture of solid dosage forms such as tablets, are surface energy, hardness, elastic properties, compaction, and porosity.

There are three main types of solids: crystalline, amorphous, and polymeric. There are different types of crystalline and amorphous materials, as illustrated in Figure 2-7.

Crystalline Solids

In crystalline solids, the molecules or atoms are arranged in repetitious three-dimensional lattice units infinitely throughout the crystal. There are seven common lattice units, often referred to as the unit cell, as shown in Figure 2-8. They are defined by the length (a, b, c) and angles (α, β, γ) of the lattice. The cubic system exhibits the highest symmetry and the triclinic system exhibits the lowest symmetry. Crystals may be units of atoms, molecules, or ions (Table 2-3). Drug substance molecules are usually found in the lower-symmetry systems due to their

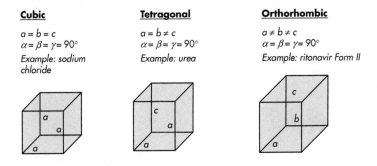

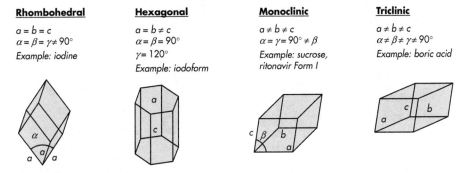

FIGURE 2-8 The seven crystal systems for crystalline materials with example compounds for each system.

TABLE 2-3 Types of Crystal Bonding

Unit	Example	Bonding	Physical Characteristics
Atom to atom	Carbon, diamond	Strong carbon covalent bonds	Hard large crystals
Metallic	Silver	Strong metal bond	Positive ions in a field of freely moving electrons
Molecular	Menthol, paraffin	van der Waals forces	Close packing, weakly held together, low melting point
Ionic	NaCl	Electrostatic ionic bond	Hard, close packing, strongly held together, high melting point

relatively large size compared to smaller inorganic systems, such as sodium chloride.

Crystalline materials usually have definite melting points. The melting point is reported as a single temperature when onset or complete melting is first observed or as a temperature range encompassing both the onset and melting of the material.

A variety of crystalline materials are found in pharmaceutics including *homomeric* crystals (composed of identical molecules or atoms) and *heteromeric* crystals (composed of more than one type of molecule or atom), such as *solvates*, salt crystals, and *cocrystals*. These crystalline materials are as illustrated in Figure 2-9. It should be noted that the salt cocrystal in Figure 2-8 (triclinic) is commonly referred to as an ionic cocrystal containing both charged and neutral species in the crystal lattice.

Polymorphs

Polymorphs are chemical entities, including pharmaceutical agents, that may exist in more than one crystalline structure, often termed *polymorphism*. The changes in the crystalline forms arise from changes in the intermolecular bonding patterns, conformational changes in the molecule, and/or molecular orientations between neighboring molecules in the solid. Figure 2-9 illustrates the intermolecular bonding pattern and molecular orientation differences observed for ritonavir polymorphs. Polymorphs have different physical properties, including different melting points, solubilities, and stability. For example, the two different forms of ritonavir have significantly different solubilities in ethanol/water mixtures, as shown in Figure 2-10[5] (see sidebar, Ritonavir Case Study). Other properties, such as density and melting point can also be different, as shown in Table 2-4 for aripiprazole. Other drug substances with multiple forms are summarized in Table 2-5. It is estimated that 89% of pharmaceuticals will exhibit different solid forms.[6]

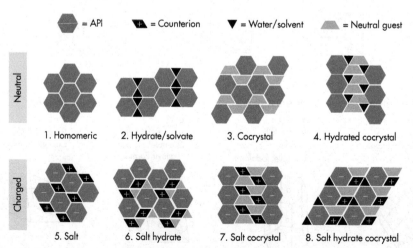

FIGURE 2-9 Different types of crystalline materials containing a variety of compounds in the crystalline lattice.

States of Matter Related to Pharmaceutical Formulations   27

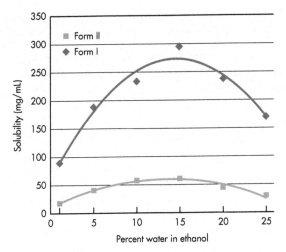

FIGURE 2-10 Solubility of ritonavir Form I (□) and Form II (◊) in different aqueous ethanol mixtures.

The crystal form of a compound can also affect the bioavailability of an oral formulation, such as a suspension, capsule, or tablet. An example is chloramphenicol palmitate, which exists in two crystalline forms known as Forms A and B. Form A exhibits a solubility that is four times lower than Form B. When oral suspensions of each form were dosed to humans, Form A showed a significantly lower plasma level than Form B, as shown in Figure 2-12. A mixture of both forms showed better plasma levels than Form A but was still significantly lower than that observed for Form B. This system illustrates the dependence of crystalline form on solubility, dissolution, and oral bioavailability.

Solvates and Hydrates

Solids can also incorporate other compounds into the lattice to produce new crystalline materials. When water is included in a lattice it is called a *hydrate*. When a solvent is incorporated into the lattice it is called a *solvate*. The water/solvent is usually present in stoichiometric ratios (a monohydrate contains one water molecule per drug molecule; a dihydrate contains two water molecules per drug molecule) but nonstoichiometric amounts are also possible in crystalline materials. Most solvates are not chosen as drug substances due to the possible toxicity of common solvents, however, hydrates are commonly used as drug substances. Hydrates and solvates may also exhibit polymorphism where, for example, two

TABLE 2-4 Some Characteristics of Aripiprazole Polymorphs and Solvates

Form	Melting Point, °C	Optics	Crystal System	Density, g cc^{-1}	Number of Molecules per Unit Cell	Hydration/ Solvation	Solubility, mmol/L[c]
I	149.1	Colorless plate	Monoclinic	1.308	2	Anhydrous	–
II	143.1	Colorless needle	Orthorhombic	1.290	4	Anhydrous	–
III	139.2	Colorless plate	Triclinic	1.297	4	Anhydrous	> 0.660
IV	134.9	Colorless rhomb-shaped plate	Triclinic	1.327	4	Anhydrous	–
X°	NA[a]	Colorless plate	Monoclinic	1.315	2	Anhydrous	0.495
H_1	NA[b]	Colorless plate	Monoclinic	1.333	4	1 mole of water	0.165
S_{Et}	NA[b]	Colorless plate	Triclinic	1.291	2	½ mole of ethanol	–
S_{Me}	NA[b]	Colorless plate	Triclinic	1.317	2	1 mole of methanol	–

[a]NA, not available
[b]Transforms to anhydrous Form III upon heating
[c]Solubility in 3:7 1-propanol:water

Source: Data from **Braun DE, Gelbrich T, Kahlenber V, Tessadri R, Wieser J, Griesser UJ:** Stability of solvates and packing systematics of nine crystal forms of the antipsychotic drug aripiprazole. *Crystal Growth Des* 9(2):1054-65, 2009; **Braun DE, Gelbrich T, Kahlenber V, Tessadri R, Wieser J, Griesser UJ:** Conformational polymorphism in aripiprazole: Preparation, stability, and structure of five modifications. *J Pharm Sci* 98:2010-26, 2009.

RITONAVIR CASE STUDY: POLYMORPHS AND AMORPHOUS DISPERSIONS

Ritonavir is a protease inhibitor used to treat AIDS. It was discovered in 1992 and was one of the first treatments on the market for AIDS patients. It was crystalline but was not bioavailable as the crystalline solid. It was marketed in 1996 as Norvir®, an oral liquid and semisolid capsules. Both formulations contained ritonavir dissolved in ethanol/water-based solutions. Only one crystalline form was identified during development. Two-hundred-forty lots of Norvir® capsules were produced without incident.

In 1998, Norvir® capsules failed dissolution. A new polymorph, Form II, was identified in the capsules and was found to be significantly less soluble than Form I (Figure 2-11). The ethanol/water solutions for the capsules were not saturated with respect to Form I but were 400% supersaturated with respect to Form II. Oral solutions could no longer be stored at the recommended 2° to 8°C without risk of crystallization. The presence of Form II made the original formulation unmanufacturable and limited supplies were available. The product had to be removed from the market. An oral solution was available in the interim, but due to the extremely poor taste of ritonavir, many patients were not willing to switch. A new formulation was developed using Form II. It is estimated that hundreds of millions of dollars were spent trying to recover Form I and $250 million lost in sales.

In 2000, Kaletra®, a combination product containing ritonavir and lopinavir, was marketed as a soft gelatin capsule and solution. Both needed refrigerated storage. In 2005, an improved tablet formulation was made using an amorphous dispersion. Patients could take fewer tablets (from six to four), the tablets need no refrigeration, and the product can be taken with or without food. The combination product not only overcame physical issues with the solid form but also resulted in a product that was easier and less expensive to use as well as providing better performance and better compliance.

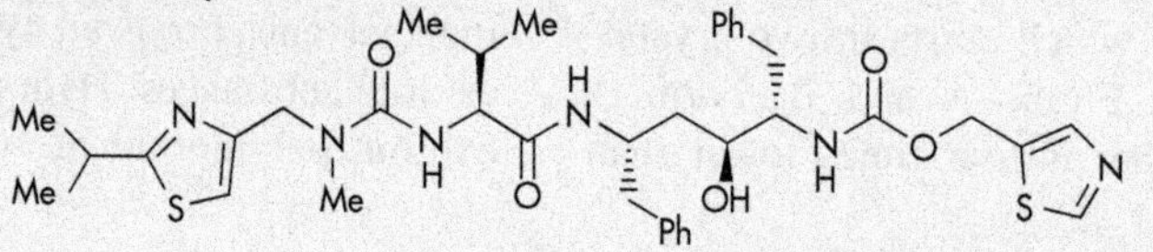

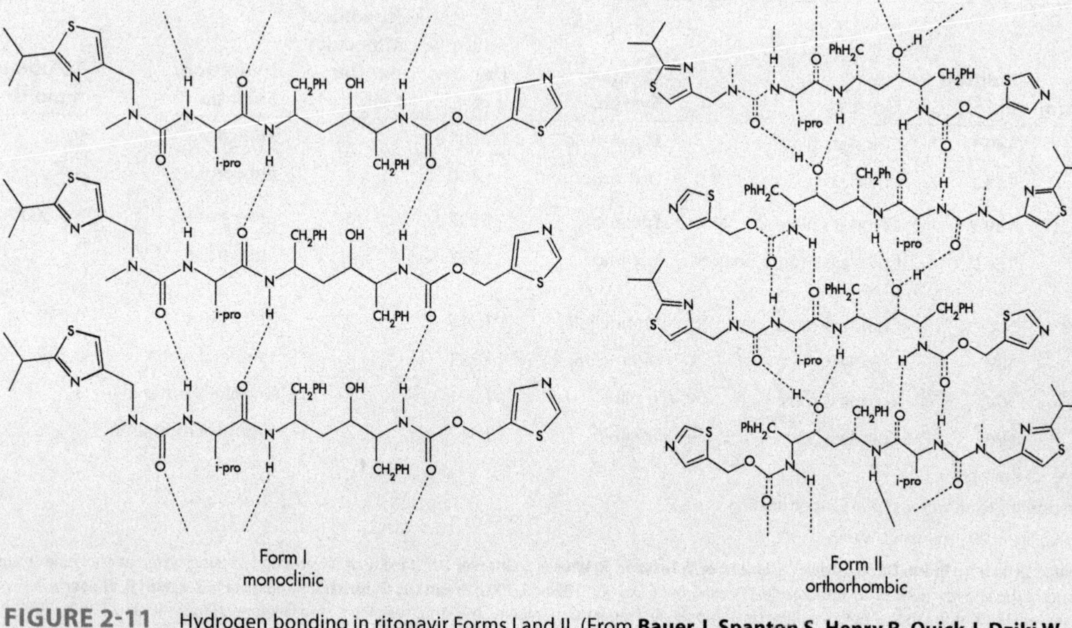

FIGURE 2-11 Hydrogen bonding in ritonavir Forms I and II. (From **Bauer J, Spanton S, Henry R, Quick J, Dziki W, Porter W, Morris, J:** Ritonavir: An extraordinary example of conformational polymorphism. *Pharm Res* 18:859–66, 2001.)

TABLE 2-5 Some Polymorphic Drugs

Drug	Number of Polymorphic Forms
Acetaminophen	3
Caffeine	2
Chloramphenicol palmitate	4
Cimetidine	3
Nifedipine	2
Phenobarbital sodium	2
Phenytoin	2
Progesterone	2
Theophylline	2

different crystalline forms can exist for a monohydrate material. Multiple hydrates can also exist for a drug substance such as a monohydrate and dihydrate. Hydrates will typically be less soluble in water or aqueous mixtures than anhydrous forms, as shown in Table 2-4 for aripiprazole. This difference in solubility, if significant, can lead to a lower bioavailability for the hydrate compared to the anhydrous form.

Salt Crystals

The lattice can also accommodate other molecules, such as acids and bases, to form salts. When there is a pKa difference of two between the molecules, a proton is transferred to form two ionized species, one with a positive charge and one with a negative charge. In the solid, the two ionized compounds will interact in the lattice to form a crystalline salt. The drug substance can be a weak acid or a weak base. The corresponding compound in a salt is called a counterion. The counterion has to be relatively nontoxic to be used in a drug substance, and the toxicity will be related to the final daily dose. A list of common counterions is given in Table 2-6.

Salts will have different properties from the free acid or free base that was used to make the salt and these properties can include melting point, solubility, dissolution, stability, and bioavailability. Crystalline salts can also exhibit different forms, such as polymorphs, hydrates, and solvates (see Figure 2-9) and these solid forms will have different properties, so it is important to characterize the material to understand what form is being used, the properties of that form, and how the properties fit the dosage form.

Cocrystals

A *cocrystal* is simply defined as a homogeneous, multicomponent phase of fixed stoichiometry where the chemical entities are held together in a crystal

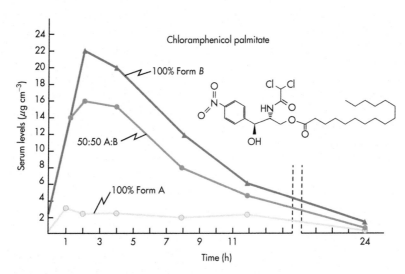

FIGURE 2-12 Comparison of serum levels (μg/cm^{-3}) obtained with suspensions of chloramphenicol palmitate after oral administration of a dose equivalent to 1.5 g of chloramphenicol. (Data from **Aquiar AJ et al:** *J Pharm Sci* 56:847–53, 1967. Figure: **Florence AT, Attwood D:** *Physicochemical Principles of Pharmacy*, 4th ed. 2005, p. 19.)

TABLE 2-6 List of Common Counterions Used for Pharmaceutical Salts and Frequency in Marketed Products

Anion Counterion for Weak Base Drugs	Frequency (%)	Cation Counterion for Weak Acid Drugs	Frequency (%)
Hydrochloride	53.4	Sodium	75.3
Sulfate	7.5	Calcium	6.9
Bromide	4.6	Potassium	6.3
Maleate	4.2	Meglumine	2.9
Mesylate	4.2	Tromethamine	1.7
Tartrate	3.8	Magnesium	1.2
Acetate	3.3	Zinc	1.2
Citrate	2.7	Lysine	0.6
Phosphate	2.7	Diethylamine	0.6

From **Paulekuhn GS, Dressman JB, Saal C:** Trends in active pharmaceutical ingredient salt selection based on analysis of the Orange Book database. *J. Med Chem* 50: 6665–72, 2007. Copyright 2007 American Chemical Society.

lattice by intermolecular forces. The formation of a cocrystal is an alternative approach to generate new crystalline phases between a drug and a *cocrystal former* molecule that may improve performance (see Figure 2-8). The drug and cocrystal former, which are usually both neutral, are held together by the intermolecular forces (e.g., hydrogen bonding) discussed previously. Cocrystals can also contain water and solvents to form cocrystalline hydrates. Another option is a salt cocrystal, which is a ternary system containing a salt and a neutral species (e.g., of a drug or other cocrystal former). Cocrystals are a good option to change properties when an ionizable group is not available. One example is itraconazole, a poorly soluble antifungal agent with poor bioavailability. A 2:1 itraconazole:L-malic acid cocrystal was produced that resulted in a solubility six times greater than the crystalline itraconazole material.[7]

Amorphous Solids

Solid material is referred to as *amorphous*, when there is no long-range order over many molecular units to produce a lattice or crystalline structure. These solids are referred to as glasses (nonequilibrium solid form) or possibly as supercooled liquids (a viscous equilibrium liquid form) because of the random order of arrangement and the distortion of the shape under pressure. It should be noted that a small amount of crystallinity may exist within amorphous materials, but it may be difficult to analyze. Amorphous materials do not possess a melting point but are defined by a glass transition (T_g) temperature, which is the temperature where an amorphous material converts from a glass to a supercooled liquid upon heating. Because of the relative weakness of the interactions between its molecules, amorphous materials are less physically stable than crystalline materials. The molecules in amorphous materials can have enough mobility to form bonds to create a more stable crystalline form under certain conditions such as high relative humidity or heat. For this reason, special handling and packaging conditions are often specified for amorphous materials.

The amorphous form of a drug will usually be more soluble than crystalline materials since there is no crystalline lattice that needs to be broken before the material can dissolve. Consider the energy required to break a number of bonds in a defined crystalline pattern in contrast to the reduced number of bonds that need to be broken in an amorphous form. The weaker energy of the amorphous form should require less energy for an individual molecule to be freed for solubilization. However, this is a general rule, as there are a number of other factors including lipophilicity, charge, polarizability, and dielectric constant of the media that may influence the degree of solubility. Higher solubility can translate to better bioavailability in some cases, but physiological factors will further influence the absorption of the drug and the degree of bioavailability.

Examples of amorphous materials used in marketed drugs include zafirlukast, cefuroxime axetil, quinapril hydrochloride, and nelfinavir mesylate.

A second form of amorphous material used in drug products is an amorphous dispersion. For these materials, an amorphous drug is stabilized by a polymer or a combination of polymer(s) and/or surfactants. This results in the increased solubility of the amorphous material with physical stability closer to that of a crystalline material. The amorphous material may still crystallize in the solid state or in solution, so extra studies are needed to determine the conditions where the material is stable. A number of amorphous dispersions have been used in marketed products including itraconazole, lopinavir/ritanovir, and etravirine. In the case of itraconazole, the low aqueous solubility of the drug (estimated as 1 ng/mL) resulted in no bioavailability. The marketed product is a 1:1 itraconazole:HPMC (hydroxypropyl methylcellulose) dispersion sprayed onto inert cores that are then filled into a capsule. The marketed product exhibits a bioavailability about 30% higher than the crystalline material, along with good physical stability.

Polymeric Solids

Polymers are large molecules formed by the covalent assembly of smaller molecules (*monomers*) into a chain or network of repeating structural units. Examples are listed in Table 2-7. The number of repeating units in the polymer (n subscript on the repeat unit) determines its size and molecular weight. There are an enormous variety of polymers and they are ubiquitous in pharmacy and in everyday life. They include natural, synthetic, and semisynthetic examples. Natural polymers include rubber (polyisoprene), polypeptides, and cellulose (and other polysaccharides or *gums*). Synthetic and semisynthetic polymers include the plastics used in packaging and devices, such as polyvinylchloride, polyethylene, and polystyrene and those used in controlled release devices, such a polyvinyl acetate, polylactides, and cellulose derivatives, such as methylcellulose and hydroxypropylmethylcellulose.

A variety of polymers are used to make the amorphous dispersions discussed in the previous section. The polymers help stabilize the amorphous drug in the solid state and may help prevent crystallization upon dissolution. The drug:polymer ratio is an important consideration when making dispersions and can be influenced by the polymer used. Dispersions made with different polymers and concentrations can exhibit different properties such as glass transition temperature, water uptake, and physical stability.

The properties of a polymer will change depending on the grade or molecular weight of the polymer, and this change in properties will determine their use. For example, low molecular weight (<400 g/mole) polyetheylene glycol (PEG) is a nonvolatile liquid at room temperature and is commonly used as a cosolvent for solution formulations. Higher molecular weight PEGs have a higher melting point and can have excellent solubility in water. These can be used as bases for suppositories and ointments, as well as binders during tablet processing.

Polymers are also used as excipients in solid, semisolid, and liquid formulations. For example, carbomer is used as a thickening agent or suspending agent or stabilizer in lotions, and at higher concentrations is commonly used to form topical gels. Eudragit [poly(meth)acrylate] and HPMC (hydroxypropylmethylcellulose) polymers are used as coatings on tablets.

Changes in State

Liquid to Gas

The boiling point is the temperature at which the VP = atmospheric pressure. For molecules to leave the surface of a liquid and pass into the air above it, the forces of attraction between nonsurface molecules in the liquid must be overcome. To overcome these attractive forces, energy must be supplied to the liquid in the form of heat. The heat absorbed when 1 g or 1 mole of liquid is vaporized is the heat of vaporization, ΔH_v. Intermolecular forces are related to heat of

TABLE 2-7 Common Pharmaceutical Polymers

Name	Formula	Common Name
Polyvinyl alcohol		PVA
Methylcellulose		Methocel A
Carbomer		Carbopol
Polyethylene (PE)		
Polyethylene glycol / Polyoxyethylene glycol		PEG
Polypropylene (PP)		
Polymethylmethacrylate (PMMA)		Plexiglass
Polymethacylate		Eudragit
Poly (d,l-lactide)		Polylactide

vaporization and—for a family of similar compounds—to molecular weight. For a family of structurally related compounds higher molecular weight represents more intermolecular points of contact, and therefore stronger intermolecular interactions, leading to higher heats of vaporization and higher boiling points while at the same time having lower VP.

Solid to Liquid

The melting point of a solid is the temperature at which the solid changes into a liquid. When a solid material is heated to its melting point, the temperature does not rise until the entire solid has passed into the liquid state. The heat required to increase the interatomic or intermolecular distance in the solid state to form the liquid state is called the heat of fusion, ΔH_f. Values for the heat of fusion are positive as heat is absorbed. For water at 0°C (32°F), the heat of fusion is given by ΔH_f (water)$_{MP}$ = 1436 cal/mole. Intermolecular forces are related to the heat of fusion and—for a family of structurally related compounds—to molecular weight. In general for structurally related compounds (such as the polyethylene glycols), stronger intermolecular forces are associated with higher molecular weight, leading to higher heats of fusion and higher melting points.

However, for compounds that are not structurally similar, melting points are dominated by intermolecular interactions rather than molecular weight. For instance, urea (MW = 60 g/mole) has a considerably higher melting point of approximately 134°C in contrast to stearic acid (MW = 284.5 g/mole), which melts at 69.6°C. Urea possesses strong hydrogen bonding interactions in the solid state, whereas stearic acid largely interacts through van der Waals forces in the aliphatic chain. Finally, compounds having higher melting points generally have decreased solubility because of the increase in the strength of the bonds in the crystal lattice of the solids.

Eutectic Mixture

A number of organic compounds, metals, and salts have a eutectic point. The eutectic point is the lowest temperature at which the existence of the liquid phase is possible. Several mixtures of pharmaceutical solids will liquefy at a given temperature and

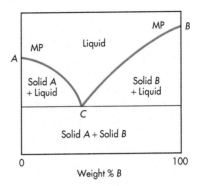

FIGURE 2-13 Phase diagram for a eutectic mixture.

composition. If one solid compound is A and the other one is B, a solid–liquid phase diagram shows the formation of the eutectic point at C (Figure 2-13).

The temperature (point C) is lower than the melting point of each component. Some examples of eutectics common in pharmacy are given in Table 2-8.

Eutectic mixtures can create problems when one is compounding powders. Mixing the eutectic components with the powders separately helps prevent the liquefaction. However, it may be preferable to form the liquid eutectic mixture prior to mixing with the powders, to allow the powder particles (e.g., starch) to adsorb the liquid eutectic. Another example of the utility of eutectics is found with the increased skin transport rates with the combination of lidocaine [MP 68° to 69°C (156.2°F)] and prilocaine [MP 37° to 38°C (100.4°F)], two local anesthetics. When they are mixed together, an oil is formed that is incorporated into a cream to provide dermal anesthesia/analgesia on intact skin within several minutes of application.

TABLE 2-8 Common Eutectics

Eutectic Mixture	Temperature of the Eutectic, °C
Testosterone-menthol	40
Cholesterol-menthol	40
Menthol, camphor, phenol thymol, and/or chloral hydrate	~25°C (room temperature)

STABILITY OF SOLIDS

Drug products need to be tested under a variety of conditions to determine parameters that will and will not change the molecule and solid form. There are two types of stability that need to be considered for stability in the solid state: chemical and physical. Chemical stability involves the molecule degrading into other products. Physical stability involves changes in the solid form such as a change from amorphous to crystalline or from one crystalline form to another.

The FDA has outlined stability conditions that are used to test drugs.[8] These include both temperature and relative humidity conditions (Table 2-9). Both drug molecules and drug products (in a dosage form) are tested for stability. In the case of drug products, potential interactions between the drug molecule and the excipients also need to be investigated.

Chemical Stability

Chemical stability can result from a number of different chemical reactions depending on the molecule, conditions, and other components of the dosage form. Typical reactions include oxidation, hydrolysis, and cyclization. Examples of these reactions are shown in Figure 2-14 for moexipril.

Drug molecules can also interact with excipients or other drug molecules in dosage forms.

TABLE 2-9 Physical Stability Conditions for Drug Substances and Products

Study	Storage Conditions	Minimum Time Period
Long-term	25°C/60% RH or 30°C/65% RH	12 months
Accelerated	40°C/75% RH	6 months

Common interactions include the Maillard reaction between amines and reducing sugars, such as lactose, which forms brown pigments (Figure 2-15a), transacylation reactions (Figure 2-15b), and acid-base reactions (Figure 2-15c).

These reactions occur much more readily when the drug is in solution and will be described in chapter 11. When the drug is in the solid state, the two main types are chemical degradation in a surface solution phase and the slower, true solid state chemical degradation. A typical scenario for degradation in the surface solution phase is where water adsorbs to the surface of a drug crystal and a portion of the drug dissolves in that water, creating a surface solution on the drug where drug degradation can readily occur. This type of degradation is the primary reason for testing solids at the high humidities and temperatures of accelerated conditions (Table 2-9).

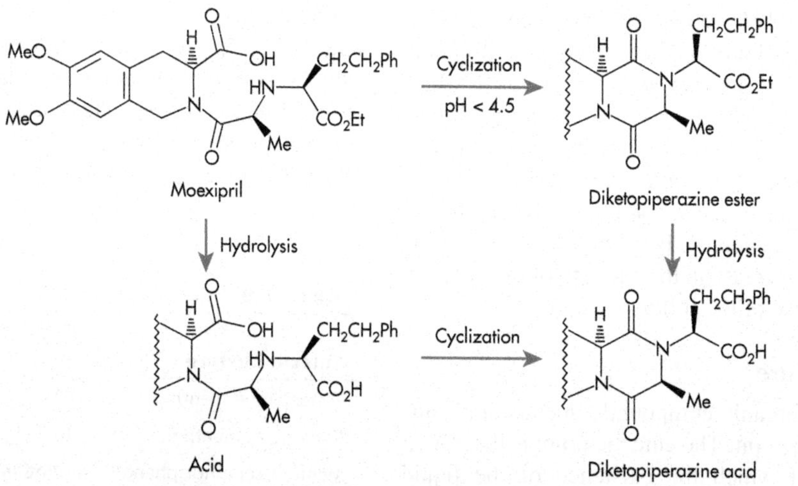

FIGURE 2-14 Degradation (cyclization and hydrolysis) reactions of moexipril. (From **Byrn SR, Xu W, Newman AW**: Chemical reactivity in solid-state pharmaceuticals: Formulation implications. *Drug Deliv Rev* 48(1):115–36. Copyright 2001 Elsevier.)

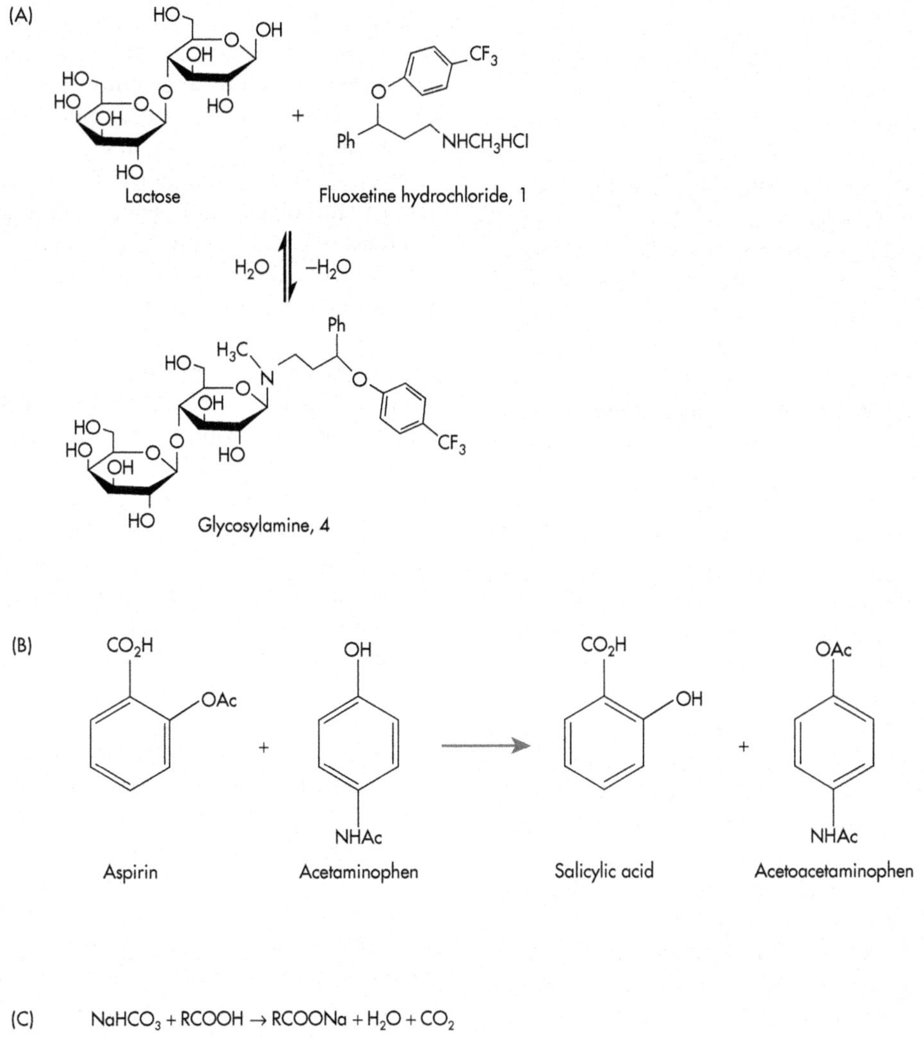

FIGURE 2-15 (a) Maillard reaction between lactose and fluoxetine hydrochloride, (b) transacylation reaction between aspirin and acetaminophen, and (c) acid-base reactions.

True solid state chemical degradation is degradation that occurs to the molecules as they exist in the solid, in the absence of any drug-dissolving medium on the surface. Because of the low molecular mobility of the molecules in a solid, and the lack of a participating medium such as water, true solid state degradation is usually a relatively slow process. It typically occurs in four stages: (1) loosening of molecules at the reaction site, (2) bond breaking and making (i.e., chemical changes), (3) solid solution formation of the degradation product, and (4) separation and crystallization of the degradation product within the parent solid. The four-step process is usually considered to begin at nuclei on the surface, which are weak points (e.g., stress points, imperfections, "hot spots") where individual molecules become exposed. Solid state chemical degradation can be a slow and complex process. It is important to note that amorphous materials will tend to be less chemically stable than crystalline materials, because of the greater molecular mobility in the amorphous state.

Physical Stability

Another type of stability that needs to be addressed is physical stability or the ability of the solid form to resist change upon standing/storage or under stress/processing conditions. Examples of physical form changes and typical conditions that can be encountered are summarized in Table 2-10.

The amorphous form is less stable (metastable) compared to crystalline forms. At elevated temperature or relative humidity conditions, the mobility of the molecules in a solid can increase causing crystallization to an anhydrous or hydrated crystalline form. It is important to determine the temperature and relative humidity where crystallization may occur in order to prevent the crystallization of the amorphous form from occurring. This may include special handling conditions (controlled relative humidity or refrigeration) or special packaging (such as the addition of desiccants or the use of blister packs).

Solid materials will exhibit different thermodynamic stabilities that are related to physical stability changes. Different crystalline polymorphs will also have different stabilities based on free energy. There are two types of systems that describe stability in a polymorphic system called monotropy and enantiotropy. In a monotropic system, one crystalline form will be the most stable form (lower free energy) over the entire temperature range up to the melting point (Figure 2-16a). The lower free energy and more stable form will also exhibit a lower solubility. In an enantiotropic system, there is a transition temperature that describes the stability; below the transition temperature one form is stable and above the transition temperature the other form is stable (Figure 2-16b). In an enantiotropic system it is important to determine the transition temperature to understand the physical stability. The solubility of the two forms will also be determined by the transition temperature. As shown in Figure 2-16b, Form I is more stable below the transition temperature and will be less soluble in this temperature range. Above the transition temperature, Form I becomes less stable and will exhibit a higher solubility compared to Form II in this temperature range. For crystallization experiments, if the transition temperature was 50°C, then Form 1 would crystallize at 40°C, but Form II would crystallize at 65°C. There is a potential to obtain a mixture of both forms at higher temperatures, based on the differences in the kinetics of formation of each form. For example, if Form II is considerably slower to nucleate than Form I, Form I may be present in higher ratios at higher temperatures. The purity of either form is not guaranteed by the temperature. Understanding the relationship between the forms related to temperature is important for processes such as crystallization, drying, and solution formation.

When making solutions, the amorphous material and metastable crystalline forms will have a higher solubility than the stable crystalline form. The most stable (less soluble) crystalline material will commonly precipitate out of these solutions if supersaturation is obtained. It is important to measure the equilibrium solubility of all known forms when producing a solution formulation to ensure that the concentration in the formulated product is below the solubility of the least soluble form. An example of crystallization from a dosage form is ritonavir

TABLE 2-10   Physical Stability Transformations

Initial Form	Final Form	Process	Example Conditions
Amorphous	Crystalline	Crystallizations	Relative humidity, temperature, storage
Crystalline	Amorphous	Amorphization	Processing such as grinding, wet granulation
Crystalline polymorph 1	Crystalline polymorph 2	Transformation	Temperature, storage, processing such as grinding, drying, granulation
Anhydrous	Hydrate/solvate	Hydration	Relative humidity, processing such as wet granulation
Hydrate/solvate	Anhydrous	Drying	Drying at elevated temperature or low relative humidity
Free acid/base	Salt/cocrystal	Reaction	Interaction between excipients and drug
Salt/cocrystal	Free acid/base	Reaction	Relative humidity, temperature, processing such as wet granulation

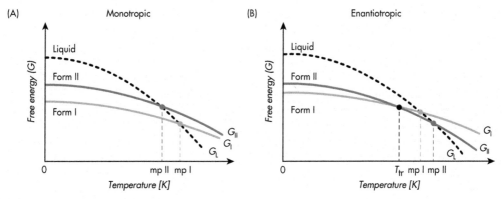

FIGURE 2-16 Thermodynamic relationships between polymorphs: (a) monotropic and (b) enantiotropic systems. Melting points (mp) are where the free energy curve of the solid (Form I or II) intersect with the liquid line. The transition temperature (T_{tr}) is where the thermodynamic stability changes in an enantiotropic system.

dissolved in soft gel capsules (see sidebar on p. 28). The soft gels were developed based on the solubility of Form I, which was significantly higher than Form II. The amount of drug dissolved in the soft gels was supersaturated with respect to Form II and resulted in Form II crystallizing out. Once crystals were present in the soft gel, the drug was no longer bioavailable when taken by the patient, resulting in subtherapeutic levels of the drug.

Reactions can occur in the dosage forms that will change the solid form. Free base or free acid forms of the drug can react with buffers or excipients to form salts in drug products, as illustrated in Figure 2-16c with sodium bicarbonate and a carboxylic acid. Cocrystals can also be produced between neutral molecules. An example is AMG517, which was dissolved in a solution containing 0.1% sorbitol as a preservative. During animal studies, solubility-limited absorption was observed at high doses and was attributed to a new solid form. The solid form was identified as a 1:1 AMG517:sorbic acid cocrystal, and its presence severely affected the performance of the dosage form even in small amounts.[9] It is also possible for salts or cocrystals to dissociate and produce the free acid/base or drug molecule. In the case of the delavirdine mesylate salt, exposure to high relative humidity resulted in poor dissolution of the tablets. When analyzed, it was found that the delavirdine mesylate reacted with the excipient sodium cross-carmellose to produce the delavirdine free base. When the free base was present, the dissolution properties were affected.[10] Special packaging would be needed to prevent the exposure to water and formation of the free base.

ACKNOWLEDGMENT

We would like to thank Dr. Keith Chadwick (Purdue University) for his review and significant contributions to improving this chapter.

PROBLEMS

1. Calculate the weight in grams of propane gas (C_3H_8) in a 5-L steel cylinder at 125°C (257°F) and 100 atm, using the ideal gas equation ($R = 0.082$ 1. atm. deg^{-1} mole^{-1} and $T = 273° + °C$).
2. Given geometric isomers such as *cis*(e) and *trans*(z), which has the higher melting point? The higher heat of fusion, ΔH_f? The stronger intermolecular forces?
3. If atmospheric pressure was considerably lower in one region of the country because of a change of altitude compared to another region, in which region would the P_{CO_2} in the arterial blood of a person be lower?

4. If a noticeable increase in barometric pressure occurs, which biological properties, such as P_{CO_2} and P_{O_2} in arterial blood, vapor pressure of water in the body, and oxygen bound to hemoglobin, will increase?
5. Two organic compounds have the same molecular weight, but compound A has a higher heat of fusion than compound B. What physical properties could you predict would also be higher for compound A? Consider melting point, boiling point, vapor pressure, and solubility.
6. List four structural modifications in this organic molecule that would raise the vapor pressure.

[Structure: Z (trans) isomer]

7. Given that the molecular weights of phenol (94.1) and resorcinol (110) are different by 16 atomic units, how would you expect the melting points, the boiling points, the vapor pressures, and the heats of fusion to differ?
8. The so-called goat acids (C_6–C_{10}) used in the manufacture of esters for artificial fruit flavors are:
   n-capric acid (deconic acid)
   n-caprylic acid (octanoic acid)
   n-caproic acid (hexanoic acid)
   Which of the acids would have the highest boiling point? Which would have the strongest intermolecular forces of attraction and the greatest heat of vaporization?
9. How would the temperature and the volume of a liquid affect its vapor pressure?
10. Ducosanoic acid and the omega-3 fatty acid DHA are both C_{22} carboxylic acids. DHA has six double bonds, all cis(e), while ducosanoic has no double bonds. One is an oily liquid, and the other is a waxy solid. Which one is predicted to be a waxy solid? Which one has the greater heat of fusion? Which has a higher melting point?

ANSWERS

1. Solving the ideal gas equation for moles, $n = pv/RT$, $T = 398°K$, $n = 1.53$ moles, the molecular weight for propane is 44 g/mole; the weight is 674 g.
2. For geometric isomers, the trans(z) isomer can form stronger intermolecular bonds and therefore has a higher heat of fusion and a higher melting point.
3. Arterial blood gases are influenced by altitude. Lower atmospheric pressure occurs at a higher altitude, resulting in lower P_{CO_2}.
4. An increase in barometric pressure will increase the partial pressure of oxygen P_{O_2} and the P_{CO_2} but will have no effect on the vapor pressure of water and probably result in little if any increase in oxygen bound to hemoglobin.
5. Compound A is held together by stronger intermolecular forces; therefore, its melting point tends to be higher, its boiling point is higher, its vapor pressure is lower, and its solubility is decreased.
6. Add double bond(s); remove hydroxyl and remove the CH_3 groups; use cis(e) isomer; shorten the chain by one CH_2 group.
7. Resorcinol would be expected to have stronger intermolecular forces. Resorcinol has a higher melting point, a higher boiling point, and a higher heat of fusion. Phenol has higher vapor pressure.
8. The C_{10} acid, n-capric acid, would have the highest boiling point, the greatest heat of vaporization, and the strongest intermolecular forces.
9. Vapor pressure is influenced by the temperature of a liquid, not by its volume.
10. Ducosanoic acid would be predicted to be a waxy solid, have a greater heat of fusion, and have a higher melting point.

KEY POINTS

- The interactions of molecules are governed by a balance of various attractive and repulsive forces that will determine the likelihood and strength of molecular bonding.
- Among the common attractive forces are van der Waals forces, ion dipole forces, hydrogen bonding, and hydrophobic interactions.
- van der Waals forces include dipole-dipole, dipole-induced dipole, and induced dipole-induced dipole. These are relatively weak forces compared to covalent forces, but they are very important in pharmacy, being significant for solubility, complexation, and other types of physical bonding.
- The very common drug salts are held together by ion dipole forces, where the ionized drug molecule is bound to a counterion, which is often an inorganic salt ion, such is sodium or chloride.
- The hydrogen bond is a particularly strong type of dipole-dipole attraction that results from the interaction between the hydrogen atom and an electronegative atom. It is important for the behavior of many molecules in water.
- Hydrophobic interactions are attractive forces that arise from the inability of nonpolar molecules (or parts of molecules) to interact with water, and are thus driven together by the hydrophobic effect, which is important for a protein's tertiary structure as well as for the formation of lipid bilayers and micelles.
- Solids may be either crystalline [composed of repeating (periodic) arrangement of the individual molecules to create a crystal lattice], or it can be amorphous (i.e., without crystalline form). Most drug solids are crystals.
- A particular drug can crystallize in different shapes, depending on how the molecules within the crystal are arranged. Drugs that can exist in more than one crystalline shape are called polymorphs.
- Because polymorphic crystals have different molecular interactions, they will differ in such properties as melting point and solubility. Amorphous forms generally dissolve more quickly than crystal forms.
- Differences in the solubility of a drug may lead to differences in its bioavailability.
- For polymorphic crystals, the polymorph with weaker molecular interactions is considered relatively metastable.
- Both metastable and amorphous solids may be considered less physically stable, because they can convert to more stable crystalline structures over time.
- Chemical degradation of a drug's molecules that make up the crystal is generally slow, but chemical degradation can be accelerated if the drug on a crystal's surface dissolves in a thin film of water that has adsorbed onto the crystal (e.g., in humid conditions).

CLINICAL QUESTIONS

1. In sickle cell anemia, the gene encoding hemoglobin has a mutation that causes the replacement of a polar amino acid, glutamic acid, with a nonpolar amino acid, valine, on the surface of the protein's beta subunits. Explain how this causes the hemoglobins in sickle cell patients to stick to each other, which will ultimately lead to sickle-shaped red blood cells.
2. Ranitidine is a histamine H_2-receptor antagonist used to reduce gastric pH. It exists in two polymorphic forms. However, the drug product is formulated with ranitidine hydrochloride, in which both forms are *highly soluble*, so its absorption is likely not dissolution-rate limited (described in Chapter 1). Cefuroxime axetil is a *poorly soluble* cephalosporin antibiotic that can exist in amorphous and crystalline forms. The innovator brand utilizes the amorphous form. In approving generics for these two drugs, why might the FDA be more concerned about the polymorphism of cefuroxime axetil than that of ranitidine? In your explanation, utilize the concepts of bioavailability and a rate-limiting step for absorption.

3 Physical Properties of Solutions

Beverly J. Sandmann, Antoine Al-Achi, and Robert Greenwood

Learning Objectives

After completing this chapter, the reader should be able to:

- Define and classify liquid systems on the basis of their physical properties.
- List the identification terms, symbols, and definitions of common concentration expressions.
- Clearly understand each concentration expression.
- Solve problems by utilizing concentration expressions and convert from one concentration unit to another.
- Define and demonstrate an understanding of the colligative properties of nonelectrolyte and electrolyte solutions.
- Calculate the molecular weight of a solute from any of the colligative properties.
- Recognize the significance of L_{iso} and ΔT_f.
- Calculate osmolarity from concentration and vice versa.
- Show familiarity with the physiologic implications of colligative properties.

GENERAL CONSIDERATIONS

A *solution* is defined as a chemically and physically homogeneous mixture of two or more substances. *Homogeneous* is a term used to imply that a mixture is uniform; that is, all the parts are identical. When subjected to routine chemical and physical analysis, the parts test the same. A binary solution is a mixture of only two components. These two components are called the *solute* and the *solvent*. For a solution of a solid material in a liquid such as water, the solute is the solid component and the solvent is the liquid component—water in the case of most pharmaceutical applications. The possibilities for a solute and solvent mixture as a solution for all states of matter are $3 \times 3 = 9$, as shown here:

Solute	Solvent	Example
Gas	Gas	Air
Liquid	Gas	Fog
Solid	Gas	Iodine vapor in air
Gas	Liquid	Carbonated water
Liquid	Liquid	Alcohol in water
Solid	Liquid	Aqueous sodium chloride solution
Gas	Solid	Hydrogen in palladium
Liquid	Solid	Mineral oil in paraffin
Solid	Solid	Gold-silver mixture, mixture of alums

Pharmaceutical solutions are mostly solids dissolved in a liquid, which is usually water or water combined with other liquids. They may be named for their use in the patient as outlined here:

- Define an isotonic solution.
- Perform the calculations for preparing an isotonic solution.
- Use the appropriate tables to find freezing point depression values of solutions, sodium chloride equivalent values, and Sprowls' values.

Classification	Characteristics
Oral solutions (elixir, syrup)	Taken by mouth, inactives for flavor and viscosity, water usual solvent
Topical solutions	Application to skin or mucous membranes, aqueous or alcoholic
Otic solutions (aural)	For the ear; usually not aqueous; glycerin, propylene glycol, and polyethylene glycol
Ophthalmic solutions	For use in the eye, inactives for viscosity, free of particulates, isotonic, sterile
Parenteral solutions	For injection IV, IM, or SQ; few if any inactives; rigid standards for sterility; free of particulates; free of pyrogens; isotonic; immediate effect

The advantages of solutions as a dosage form include faster onset of activity, good for children and the elderly, homogeneous, always uniform, flexible dosing, and given by any route of administration. The disadvantages are bulkiness, leakage from the container, less stability than solid dosage forms, and more pronounced taste.

CONCENTRATION EXPRESSIONS

In the pharmaceutical sciences, liquid homogeneous aqueous systems are termed solutions, having a solute and a solvent. The concentration of solute is expressed in a number of different forms that range from the scientific to the applied. Concentration units are independent of the volume of solution. This is important to remember in problem solving. Table 3-1 shows the different concentration expressions.

Each of these concentration expressions is used in different applications in the pharmaceutical sciences. Molarity is a concentration expression based on moles of solute in a specific volume of solution. To calculate moles from grams, molecular weight (MW) is used. Grams of solute divided by molecular weight converts the weight to moles:

$$\text{Calcium carbonate, } 5\,g$$
$$CaCO_3$$
$$MW = 100.09\,g/mole$$

Converting grams to moles by dimensional analysis:

$$5\,g\ CaCO_3 \times \frac{mole}{100.09\,g} = 0.0499\ mole$$

Rounding the answer to one significant figure: 0.05 mole.

TABLE 3-1   Concentration Expressions

Expression	Symbol	Definition
Molarity (temperature-dependent)	M,c,C,[ ]	Moles (gram molecular weight) of solute in 1 L of solution, also millimoles/mL (mM)
Normality (temperature-dependent)	N	Equivalents (gram equivalent weights) of solute in 1 L of solution, also mEq/mL
Molality (temperature-independent)	m	Moles of solute in 1000 g of solvent
Mole fraction $X_1 = \dfrac{n_1}{n_1+n_2}, X_2 = \dfrac{n_2}{n_1+n_2}$ $X_1 + X_2 = 1$	X X_1 = solvent X_2 = solute	Ratio of the moles of one constituent (e.g., the solute n_2) of a solution to the total moles of all constituents (solute and solvent $n_2 + n_1$)
Mole percent	$X_2 \times 100$	Moles of one constituent in 100 moles of the solution; mole percent is obtained by multiplying mole fraction by 100
Percent by weight (temperature-independent)	% w/w	Grams of solute in 100 g of solution
Percent by volume (temperature-dependent)	% v/v	Milliliters of solute in 100 mL of solution
Percent weight in volume (temperature-dependent)	% w/v	Grams of solute in 100 mL of solution
Milligrams per deciliter (temperature-dependent)	mg/dL	Milligrams of solute in 100 mL of solution
Osmolality (temperature-independent)	mOsmol/kg	The mass of solute that, when dissolved in 1 kg of water, will exert an osmotic pressure equal to that exerted by a gram molecular weight of an ideal unionized substance dissolved in 1 kg of water
Osmolarity (temperature-dependent)	mOsmol/L	The mass of solute that, when dissolved in 1 L of solution, will exert an osmotic pressure equal to that exerted by a gram molecular weight of an ideal unionized substance dissolved in 1 L of solution

Note: For high concentrations (over 1 M or 10%), molality > molarity and mOsmol/kg > mOsmol/L, whereas at low concentrations they are nearly the same.

Molarity is a common unit for concentration. Because it is based on the volume of a solution, there is a slight variation in that volume with a change in temperature. Most solutions in pharmacy practice are kept near room temperature [25°C (77°F)] or refrigerated [4°C (39.2°F)]. This 20-degree difference in temperature would not cause a significant change in volume and therefore is not a serious concern. Normality is a concentration term based on chemical activity. It is gram equivalent weights of solute in 1 L of solution. For solutes with one replaceable hydrogen, molarity and normality are the same. However, for solutes with two replaceable hydrogen ions or hydroxyls, the gram equivalent weight is the molecular weight divided by 2. For 1 M or 1 N of hydrochloric acid, the moles of HCl are equal to the equivalent unit of HCl, and so there is no difference between molarity and normality. However, for the salt, magnesium sulfate ($MgSO_4$), the gram molecular weight is 120.39 g, and the gram equivalent weight is 60.195 g, which is 120.39/2. The difference is that magnesium sulfate has two equivalents per mole with two replaceable hydrogen ions from sulfuric acid and two replaceable hydroxyls from magnesium hydroxide to give $MgSO_4$.

Therefore, a 1 N solution of magnesium sulfate is a 0.5 M solution. Conversely, a 1 M solution of magnesium sulfate is a 2 N solution. When these ions are in solution and are not being titrated or neutralized, the calculation of the concentrations in mmoles/mL (M) and mEq/mL (N) can be done as shown here:

For 1 M sulfuric acid
H_2SO_4, MW = 98.08 g/mole

Ionization is essentially complete:

$$H_2SO_{4(aq)} \rightarrow 2H^+_{(+aq)} + SO^=_{4(aq)}$$

Molar concentration: 2 M 1 M

The hydrogen ion concentration is 2 M because there are two hydrogen ions for each sulfuric acid molecule.

To calculate the normality of this solution, the two hydrogen ions provide two equivalents per mole for the sulfuric acid; therefore,

$$1 \text{ mole } H_2SO_4 \times \frac{2 \text{ EQ}}{\text{mole}} = 2 \text{ equivalents } H_2SO_4$$

Applying that to the concentration gives one molar H_2SO_4, which is two normal H_2SO_4:

$$H_2SO_{4(aq)} \rightarrow 2H^+_{(aq)} + SO^=_{4(aq)}$$

Normal concentration   2 N 2 N

The hydrogen ion concentration in solution is 2 M and 2 N where the sulfate ion concentration is 1 M and 2 N. The normality is twice the molarity for the sulfate ion, which has a charge of -2. In solutions of ions with no chemical reaction occurring, it is generally possible to determine the number of equivalents for more highly charged ions, $+2, -2, +3, -3$, by considering the charge on the ion. Other examples of salts with more highly charged ions are $CaCl_2$, $ZnCl_2$, K_2HPO_4, and Na_2HPO_4.

A comparison of molar to molal at two different concentrations shows that differences show up in the second or third decimal place with a change in sodium chloride concentration from 0.9% w/v to 3% w/v.

For 0.9% NaCl (normal saline):

Density: 1.005 g/mL
MW: 58.45

Conversion of percent concentration to molar using dimensional analysis:

$$0.9 \text{ g}/100 \text{ mL} = 9 \text{ g}/1000 \text{ mL} \times 1 \text{ mole}/58.45 \text{ g}$$
$$= 0.154 \text{ } M$$

Using density gives the weight of water required for molal concentration:

1 L weighs 1005 g (9 g NaCl and 996 g H_2O)

Conversion to molal using a simple proportion:

$$0.154 \text{ mole}/996 \text{ g } H_2O = x/1000 \text{ g}$$
$$x = 0.155 \text{ } M$$

The comparison is between 0.154 M and 0.155 m.
For 3% NaCl:

Density: 1.02 g/mL

Conversion of percent concentration to molar:

$$3 \text{ g}/100 \text{ mL} = 30 \text{ g}/1000 \text{ mL} \times 1 \text{ mole}/58.45 \text{ g}$$
$$= 0.513 \text{ } M$$

Using density gives:

1 L of 3% NaCl weighs 1020 g
(30 g NaCl, 990 g H_2O)

Conversion to molal using a simple proportion:

$$0.513 \text{ mole}/990 \text{ g } H_2O = x/1000 \text{ g}$$
$$x = 0.518 \text{ } m$$

The comparison is between 0.513 M and 0.518 m.

The molar and molal concentrations for most pharmaceutical applications show differences in the third decimal point. Therefore, molar concentration can be used over a range of concentrations and temperatures, can be substituted for molal, and can give values within a percent or so difference.

CLASSIFICATION OF AQUEOUS SOLUTION SYSTEMS

Solutions as Ideal or Real Based on Thermodynamics

Ideal solutions are solutions for which there is no change in the physical properties of the components other than dilution when they are mixed. No heat is given off or taken in, and the volume does not shrink or expand. The final volume is the sum of the component volumes. Few real aqueous mixtures behave like ideal solutions. For the ideal solution, a relationship between the vapor pressure for a liquid and its mole fraction in solution is given by Raoult's law: The vapor pressure of each volatile constituent is equal to the vapor pressure of the pure constituent multiplied by its mole fraction in the solution. For a two-component ideal system of A and B:

$$P_A = p_A°X_A$$

where P is the vapor pressure in solution

$$P_B = p_B°X_B$$

X is the mole fraction $p°$ in the vapor pressure of the pure constituent.

This is a linear relationship passing through the origin, as shown in Figure 3-1.

One solution that is not ideal is a mixture of alcohol and water. The mixture gets warm, heat is given off, and the final volume of the solution is less than that of the sum of the two components. Raoult's law is not linear for mixtures of alcohol and water.

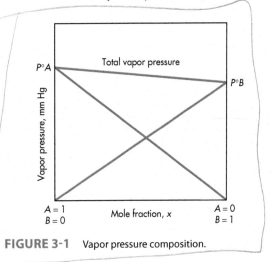

FIGURE 3-1 Vapor pressure composition.

Solutions for which changes in the physical properties of the solution occur when the components are mixed are called *real solutions*. Real solutions do not show linear relationships for Raoult's law. The lines in the diagram curve up or down, depending on whether there are stronger cohesive or stronger adhesive forces of attraction between the solute and the solvent. Most solutions in pharmacy are aqueous with low percentages of volatile solutes, with cosolvents such as alcohol, and therefore tend to obey Raoult's law and behave nearly like ideal solutions when they are mixed.

Solutions Based on Type of Solute

Solutes can be considered on the basis of how they behave in water. This classification includes nonelectrolytes, strong electrolytes, and weak electrolytes. Table 3-2 shows the characteristics of each type of solute.

TABLE 3-2 Characteristics of Each Type of Solute

Solutions of Nonelectrolytes	Solutions of Strong Electrolytes	Solutions of Weak Electrolytes
In nonaqueous media behave as ideal solution	In nonaqueous media for ion pairs	In nonaqueous media behave as ideal solution
In aqueous media:	In aqueous media:	In aqueous media:
Do not conduct current	Conduct strong current	Weak current
No ions present	Totally ionized	Partially ionized
Exhibit regular colligative properties	Exhibit irregular colligative properties (based on number of ions)	Colligative properties based on ionization
Sugars, alcohols, acetamide, acetone, glycerin	Mineral acids, strong bases, all salts: HCl, NaOH, NaCl, KOH, KCl, KAc	Weak acids, weak bases, HOAc, NH_4OH, benzoic acid, atropine

PHYSICAL PROPERTIES OF SOLUTIONS

The physical properties of solutions vary according to the concentration of the dissolved solute. Four important physical properties, called *colligative properties*, are bound together through a common source. Each of the properties depends only on the number of solute particles (ions or molecules) dissolved in the solution. The four colligative properties are:

1. Vapor pressure lowering
2. Boiling point elevation
3. Freezing point depression
4. Osmotic pressure

Several assumptions are operating in the application of colligative properties to pharmaceutical systems:

- Solute is nonvolatile, and solvent is water.
- For nonelectrolytes, values of colligative properties are the same for different solutes at the same molar concentration.
- For strong electrolytes, values of colligative properties depend on the number of ions.
- For weak electrolytes, values of colligative properties depend on ionization.
- Solutions are dilute.

For water as the solvent, the physical properties of the pure liquid are as follows: vapor pressure at room temperature equals 23.77 mmHg, boiling point is 100°C (212°F), melting point is 0°C (32°F), and osmotic pressure does not exist. When a nonvolatile solute is dissolved in water to create a solution, the properties of water are changed when the solution is formed. For the variation of the vapor pressure of water with the addition of solute, there is vapor pressure lowering.

The vapor pressure of a solution is less than that of the pure solvent. This is expressed by Raoult's law: $P_1 = X_1 \, p_1°$. When P_1 is the vapor pressure of the solvent above the solution, X is the solvent mole fraction and $p_1°$ is the vapor pressure of pure solvent. For $X_1 + X_2 = 1$, $X_1 = (1 - X_2)$ can be used, where X_2 is the mole fraction of solute.

Substitution into Raoult's law gives $P_1 = (1 - X_2) \, p_1°$. Rearrangement gives $p_1° - P_1 = \Delta P = X_2 \, p_1°$, where ΔP, the lowering of vapor pressure from pure solvent to solution, is directly proportional to the mole fraction of solute. The lowering of vapor pressure is a colligative property. However, it is not convenient to measure vapor pressure in pharmacy practice, and vapor pressure lowering is not used in calculations for colligative properties.

Boiling point elevation is the increase in the boiling point of a solution over that of the pure liquid solvent. The change in boiling point is small and is subject to barometric pressure. Therefore, measurement of this property is not practical.

Osmotic pressure is the pressure that results from osmosis. All aqueous solutions of nonvolatile solutes exert an osmotic pressure. *Osmosis* is the diffusion of solvent through a semipermeable membrane that allows only solvent to move through it. Osmosis always takes place in the direction that will equalize the concentration of all the components on both sides of the membrane. When osmotic pressure is equal on both sides of the membrane, the system is isoosmotic. In biological fluids, this equilibration is termed isotonic.

The freezing point depression is the difference in the melting points of pure water and a solution. There is always a decrease in the temperature for the freezing point of the solution compared to that for pure water. Adding salt to pure water causes it to freeze at a temperature below 0°C (32°F).

The equation that relates the freezing point depression to the concentration of solute is the van't Hoff equation:

$$\Delta T = iK_f m$$

where ΔT = absolute value of the decrease in melting point in degrees
  i = ionization factor
  m = molal concentration of solute, which is moles solute per kilogram of water
  k_f = molal depression constant, which may be derived from Raoult's law and the Clapeyron equation; for water, it is 1.86

A modification of the van't Hoff equation uses the L_{iso} value for the iK_f term:

$$\Delta T_f = L_{iso} m$$

TABLE 3-3 Approximate L_{iso} Values

Compound Type by Valence	L_{iso} Value
Nonelectrolyte	1.86
Weak electrolyte	2.0
Di-divalent electrolyte (divalent cation and divalent anion)	2.0
Uni-univalent electrolyte (single charge on each)	3.4
Uni-divalent electrolyte	4.3
Di-univalent electrolyte	4.8

TABLE 3-4 Colligative Properties of Biological Fluids

Osmotic pressure: $\Pi = 6.5$ atm (0°C), 7.6 atm (38°C)
Vapor pressure lowering: $\Delta p = 0.12$ atm at 100°C
Freezing point depression: $\Delta T_f = 0.52°$ (older value is 0.56)
Boiling point elevation: $\Delta T_b = 0.15°$ (100°C)
Serum osmolality (mOsmol/kg water): 290–310 mOsmol/kg
Calculated osmolarity (mOsmol/L water): 308 mOsmol/L*

*Calculated from NaCl only.

The L_{iso} value is determined by the ionization properties of the solute. Values for L_{iso} based on the valence of the compound are given in Table 3-3. For example, one can calculate the freezing point depression for 1% sodium propionate, a uni-univalent electrolyte. The first *uni* is the monovalent cation sodium; the second *uni* is for the monovalent anion propionate.

Sodium propionate:

$$MW = 96$$

$$L_{iso} = 3.4$$

$1\% = 0.104\ M$ where molal $\cong$ molar in dilute solutions

Using $\Delta T = L_{iso}m$, substitution gives:

$$\Delta T = 3.4(0.104) = 0.35\ \text{degree}$$

The freezing point depression properties can be used to determine freezing point depressions for nearly all solutes in dilute solution. In concentrated solutions, the L_{iso} values would be different from those in the table. The actual temperature for the freezing point for 1% sodium propionate is 272.7°K, or −35°C.

PHYSIOLOGIC APPLICATIONS OF COLLIGATIVE PROPERTIES

General Considerations

A healthy human body maintains a delicate balance of water and electrolytes that are highly regulated by osmotic pressure and ion concentration. The normal values for the colligative properties of biological fluids are given in Table 3-4.

The water content of an adult is about 60% (42 L for a 70-kg person) of total body weight. About 40% (28 L) is intracellular, 15% (10.5 L) is intercellular, and 5% is extracellular (plasma). Plasma volume is about 3.5 L, or close to 1 gallon.

The distribution of fluids in the body may be modeled by using compartments, where the compartments are plasma separated from intercellular fluid. The compartments are bound by semipermeable cellular membranes, as shown in the following diagram:

```
plasma    //    intercellular    //    intracellular
       capillary              cell
        wall                membrane
```

These membranes allow the passage of water and some ions. Water flows freely between the compartments. The compartmental balance is maintained by blood pressure and osmosis. The approximate distribution of electrolytes is given in Table 3-5.

The units used for the ions are mEq/L. This is not normality or mEq/mL but a different concentration unit similar in magnitude to that used for osmolar concentrations.

Physical factors determine the direction of the flow of water across the capillary membrane coming from the artery end to the venous end (Figure 3-2). Blood pressure essentially forces liquid into the extracellular compartment. That pressure basically is a positive 10 mmHg pressure forcing the water from the plasma into the extracellular fluid, where 32 mmHg is counter to 22 mmHg

TABLE 3-5 Approximate Distribution of Electrolytes

Plasma				Extracellular Fluid				Intracellular Fluid			
Cations		Anions		Cations		Anions		Cations		Anions	
Na^+	142	Cl^-	101	Na^+	143	Cl^-	114	Na^+	10	Cl^-	3
K^+	4	HCO_3^-	27	K^+	4	HCO_3^-	31	K^+	160	HCO_3^-	10
Ca^{2+}	5	HPO_4^{2-}	2	Ca^{2+}	5	HPO_4^{2-}	2	Ca^{2+}	2	HPO_4^{2-}	100
Mg^{2+}	2	SO_4^{2-}	1	Mg^{2+}	2	SO_4^{2-}	1	Mg^{2+}	26	SO_4^{2-}	20
		Organic acids	6			Organic acids	7				
		Protein	16			Protein	1			Protein	65
153 mEq/L		153 mEq/L		156 mEq/L		156 mEq/L		198 mEq/L		198 mEq/L	

as a result of colloidal oncotic pressure (osmotic pressure including colloidal proteins). At the venous end of the capillary there is lower pressure, down 20 mmHg. Water comes out of the extracellular space into the capillary, and the difference is a negative 10 mmHg. The colloidal oncotic pressure (COP) is still 22 mmHg.

Body osmolality varies from about 290 to about 310 mOsmol, with the normal figure being 300 mOsmol. As osmolality increases by another 10 mOsmol to over 310 mOsmol, a person actually starts to feel some effects, such as sweating, weakness, and stress. These effects are summarized in Table 3-6.

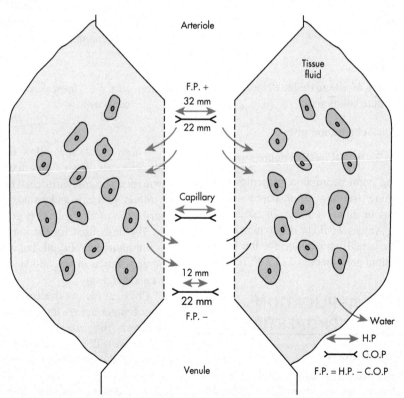

FIGURE 3-2 Direction of water flow across the capillary membrane coming from the arterial end to the venous end.

TABLE 3-6 Physiologic Effects Associated with Increases in Osmolality

Serum Osmolality, mOsmol	Physiologic Changes
330	Fainting, central nervous system changes
320	Weakness
300 ± 10	Normal
250	Weakness
233	Seizures, coma

Fluid and electrolyte therapy for patients usually involves the administration of intravenous fluids to treat dehydration and electrolyte and pH imbalance as well as for the administration of medicinal agents. Infusion of intravenous solutions that are not isotonic can result in substantial osmotic differences, which, depending on the magnitude to which the solution deviates from isotonicity, may be severe. Solutions that are not close to being isotonic will cause pain at the site of injection and postinfusion phlebitis. Infiltration of nonisotonic solutions can cause tissue damage and necrosis. Red blood cells are affected by the tonicity of intravenous solutions. Solutions that are more dilute than isotonic are called hypotonic, and those that are more concentrated are called hypertonic. The following diagram shows the effect of hypertonic and hypotonic solutions on cells:

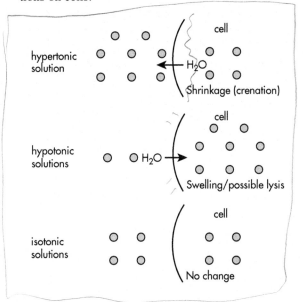

Osmolarity Calculations

1. For nonelectrolytes:

$$\frac{g/L}{MW} \times 1000 = mOsmol/L$$

2. For strong electrolytes:

$$\frac{g/L}{MW} \times (\text{no. ions}) \times 1000 = mOsmol/L$$

3. For any ion

$$\frac{g \text{ ion}/L}{\text{Ionic weight}} \times 1000 = mOsmol/L$$

Osmolarity Calculations: From % w/v, use

$$mOsmol/L = \frac{g/L}{MW} \times \text{number of ions} \times 1000$$

Calculate the osmolarity for 0.9% NaCl (MW 58.45):

$$mOsmol/L = 9 \text{ g}/58.45 \times 2 \times 1000$$
$$= 308 \text{ mOsmol/L}$$

Calculate the osmolarity for 5% dextrose (D5W). Dextrose (MW 180) is dextrose monohydrate (MW 198.17):

$$mOsmol/L = 50 \text{ g}/198.17 \times 1 \times 1000$$
$$= 252 \text{ mOsmol/L}$$

Electrolyte Calculations

EXAMPLE 1 ▶▶▶

How much $CaCl_2$ should be added to 200 mL of D5W injection if the $CaCl_2$ stock solution in the pharmacy contains 2 mEq/mL and the physician wishes to run 20 mEq/L over 10 h?

Solution
The concentration to run is 20 mEq/L.
Determine the mEq of $CaCl_2$ in the 200 mL. Use a simple proportion. If

$$\frac{20 \text{ mEq}}{1000 \text{ mL}} = \frac{x}{200}$$

$$x = 4 \text{ mEq } CaCl_2 \text{ required}$$

For calculation of the volume of stock solution needed to give the $CaCl_2$, use dimensional analysis:

$$4 \text{ mEq } CaCl_2 \frac{mL}{2 \text{ mEq } CaCl_2}$$

$$= 2 \text{ mL stock solution}$$

Note: The addition of 2 mL to 200 mL is considered insufficient to adjust the volume to 202 mL, particularly for diluting injection fluids.

EXAMPLE 2 ▶▶▶

What is the concentration of calcium ion and chloride ion in mEq/mL and mmole/mL for 10% $CaCl_2$ (anhydrous)?

Solution
Calcium chloride MW 110.
Calculate mmoles/mL concentration from %.

$$\text{Using } 10\% = \frac{10 \text{ g}}{100 \text{ mL}}, \text{ convert to liters.}$$

$$\text{Multiply by 10 to give } \frac{100 \text{ g}}{1000 \text{ mL}}$$

Calculate moles using dimensional analysis:

$$\frac{100 \text{ g}}{L} \times \frac{\text{mole}}{110 \text{ g}} \times \frac{1000 \text{ mg}}{g} \times \frac{1}{1000 \text{ mL}}$$

$$= 0.9 \frac{\text{mmole } CaCl_2}{mL}$$

Consider the dissociation of $CaCl_2$:

$$CaCl_2 \rightarrow Ca^{2+} + 2Cl^-$$

1 mole $CaCl_2$ yields 1 mole Ca^{2+} and moles Cl^-. Therefore, we have 0.9 mmoles/mL Ca^{2+} and 1.8 mmoles/mL Cl^-.

Calculate mEq/mL concentration. Use 2 mEq/mL for $CaCl_2$ to convert moles to equivalents:

$$\frac{100 \text{ g}}{1} \times \frac{\text{mole}}{110 \text{ g}} \times \frac{2 \text{ mEq}}{\text{mole}} \times \frac{1000 \text{ mg}}{g} \times \frac{L}{1000 \text{ mL}}$$

$$= \frac{1.8 \text{ mEq } CaCl_2}{mL}$$

Consider the dissociation of $CaCl_2$:

$$CaCl_2 \rightarrow Ca^{2+} + 2Cl^-$$

1 mEq $CaCl_2$ yields 1 mEq Ca^{2+} and 1 mEq Cl^-

Therefore, the concentration of calcium ion and chloride ion is 1.8 mEq/mL.

ISOTONIC SOLUTIONS

Solutions prepared to be isotonic have colligative properties similar to those of body fluids. In addition, these solutions contain solutes that are not capable of diffusing through biological membranes. If the biological membrane is permeable to the solutes in a solution with the same osmotic pressure as blood, this solution may be labeled isoosmotic but not isotonic. Isotonic implies a biological compatibility, whereas isoosmotic implies similarity of chemical and/or physical composition. The osmolality of serum is approximately 285 mOsmol/kg. The osmolality of some other biological fluids is presented in Table 3-7.[2,3]

Solutions that are labeled either hypotonic or hypertonic contain concentrations of solutes that are lower or higher than those found in an isotonic solution, respectively. A typical standard isotonic solution employed in pharmacy is sodium chloride

TABLE 3-7 The Osmolality of Some Biological Fluids

Fluid	Osmolality, mOsmol/kg (mean ± SD)
Bile, gallbladder	349 ± 40
Bile, hepatic	284 ± 17
Gastric	282 ± 22
Feces	357 ± 37
Ileal	292 ± 12
Jejunal	285 ± 10
Pancreatic	320 ± 10
Parotid saliva	127 ± 35

Source: Originally published in **Dickerson RN, Melnik G:** Osmolality of oral drug solutions and suspensions. *Am J Hosp Pharm* 45:832, 1988. American Society of Hospital Pharmacists, Inc. All rights reserved. Reprinted with permission. (R1403)

TABLE 3-8 The Osmolality (mOsmol/kg) of Some Medicinal Preparations as Determined by the Vapor Pressure Method

Drug	Manufacturer	Concentration	Osmolality (mOsmol/kg)
Acetaminophen drops	McNeil	80 mg/0.8 mL	10,535
Acetaminophen elixir	Roxane	65 mg/mL	5400
Amantadine HCl solution	DuPont	10 mg/mL	3900
Amoxicillin suspension	Squibb	50 mg/mL	2250
Calcium glubionate syrup	Sandoz	360 mg/mL	2550
Chloral hydrate syrup	Pharmaceutical Associates	50 mg/mL	4400
Cimetidine solution	Smith Kline & French	60 mg/mL	5550
Digoxin elixir	Glaxo-Wellcome	0.05 mg/mL	2630
Docusate sodium syrup	Roxane	3.3 mg/mL	3900
Erythromycin ethylsuccinate suspension	Abbott	40 mg/mL	1750
Ferrous sulfate liquid	Mead Johnson	75 mg/0.6 mL	4445
Furosemide solution	Hoechst-Roussel	10 mg/mL	3335
Haloperidol concentrate	McNeil	2 mg/mL	500
Lactulose syrup	Roerig	670 mg/mL	3600
Lithium citrate syrup	Roxane	1.6 mEq/mL	6850
Methyldopa suspension	Merck Sharp & Dohme	50 mg/mL	2050
Nystatin suspension	Squibb	100,000 units/mL	3300
Phenytoin sodium suspension	Parke-Davis	25 mg/mL	1500
Theophylline solution	Laser	80 mg/15 mL	360
Vitamin D_2 solution	Winthrop	8000 IU/mL	13,565
Vitamin E drops	USV	15 IU/0.3 mL	3620

Modified with permission from *Pediatrics*, **Ernst JA, Williams JM, Glick MR, et al:** Osmolality of substances used in the intensive care nursery. *Pediatrics* 72(3):347, 1983; **Dickerson RN, Melnik G:** Osmolality of oral drug solutions and suspensions. *Am J Hosp Pharm* 45:832, 1988. Originally published copyright © 1988, American Society of Hospital Pharmacists, Inc. All rights reserved. Reprinted with permission (R9958).

solution 0.9% w/v.[4] This solution, as do all other aqueous isotonic solutions, has a freezing point depression (Δt_f) of 0.52°C. A hypotonic aqueous solution has a freezing point (t'_f) between 0°C and −0.52°C, whereas a hypertonic aqueous solution has a t'_f below −0.52°C. When a hypotonic solution comes in contact with living cells, it will cause swelling and bursting of the cells; this phenomenon is known as *hemolysis* when the living tissue is the blood. In contrast, a hypertonic solution causes shrinkage of the living cells when it is in contact with them.[5-7] Solutions that are isoosmotic but not isotonic may behave in the same manner as a hypotonic solution as a result of the ability of the solute molecules to enter the cells and cause swelling.[1] A classic example of an isoosmotic solution that causes hemolysis to red blood cells is boric acid solution (1.9% w/v). It is interesting to note that the same boric acid solution is isotonic with regard to the mucous tissues covering the eye.[8] Thus, boric acid solution 1.9% w/v may be used safely in the eye but not in the blood. The osmolality of some selected medicinal substances is presented in Table 3-8.[3]

It is the intention of pharmacists to prepare isotonic solutions for the comfort and safety of their patients. Situations may arise in which an isotonic buffer solution is needed; Table 3-9 lists some examples of such isotonic buffer solutions. There are situations, however, in which hypertonic or hypotonic solutions may be given to patients for a specific

TABLE 3-9 Some Isotonic Buffer Solutions

Buffer Solution	pH	NaH$_2$PO$_4 \cdot$H$_2$O	Na$_2$HPO$_4$	KH$_2$PO$_4$	K$_2$HPO$_4$
Sodium phosphate	6.6	11.44	8.73		
Sodium phosphate	7.4	3.06	14.70		
Potassium phosphate	6.6			11.03	10.46
Potassium phosphate	7.4			3.07	18.36

Source: **Turner RH, Mehta CS, Benet LZ:** Apparent directional permeability coefficients for drug ions: In vitro intestinal perfusion studies. *J Pharm Sci* 59(5):590, 1970. Reprinted with permission of Wiley-Liss, Inc., a subsidiary of John Wiley & Sons, Inc., copyright © 1970.

application. For example, the administration of a hypertonic solution of mannitol is warranted because of its diuretic effect.[9]

SODIUM CHLORIDE EQUIVALENT VALUES (E)

Drugs and other solutes, when present in solutions, contribute a weight equivalent in their osmotic pressure to a specified weight of sodium chloride. As was mentioned earlier, a sodium chloride solution in a concentration of 0.9% w/v is isotonic. By convention, 1 g of any drug that can replace the entire amount of sodium chloride in this solution will have a sodium chloride equivalent value of 0.9 ($E = 0.9$). The E value of any drug can be estimated from knowledge of L_{iso} and the drug's molecular weight:

$$E \cong 17(L_{iso}/MW)$$

EXAMPLE 3 ▶▶▶

Estimate the E value for a solute with an MW of 340 g/mole and an L_{iso} value of 3.4.

Solution
Since the ratio of L_{iso} to MW is 0.01, the E value of the drug is 0.17. This means that 1 g of the drug is osmotically equivalent to 0.17 g of sodium chloride.

EXAMPLE 4 ▶▶▶

A pharmacist prepared 100 mL of a 1% solution of a drug ($E = 0.17$). How much boric acid is needed to prepare 100 mL of solution with the same osmotic pressure as the drug solution? The E value for boric acid is 0.50.

Solution
The drug solution contains 1 g of drug in every 100 mL of solution, which is equivalent to 0.17 g of sodium chloride. A solution of boric acid with the same osmotic pressure as the drug solution will have (0.17/0.50) g, or 0.34 g, of boric acid in every 100 mL of solution.

Preparation of Isotonic Solutions
Freezing Point Depression Method

The freezing point for an aqueous isotonic solution is −0.52°C. Thus, to make a hypotonic solution isotonic, the freezing point of the hypotonic solution must be lowered to −0.52°C. The lowering of the freezing point is brought about by the addition of an inert solute (such as sodium chloride) to the hypotonic solution. As was mentioned earlier, 0.9% sodium chloride solution is isotonic with biological fluids and has a freezing point depression of −0.52°C. To calculate how much sodium chloride is needed to make the solution isotonic, the following equation can be used:[1]

$$\text{Amount of NaCl needed in 100 mL of solution} = [0.9 \times (0.52 - t'_f)]/0.52$$

EXAMPLE 5 ▶▶▶

The freezing point of a drug solution is –0.18°C. Is this solution hypotonic? How much farther does its freezing point need to be lowered?

Solution
Since the freezing point is above –0.52°C, this solution is hypotonic. The solution's freezing point has to be lowered by (0.52 – 0.18) or 0.34°C.

EXAMPLE 6 ▶▶▶

How much sodium chloride is needed to make 100 mL of the solution in Example 5 isotonic?

Solution
Amount of NaCl needed in 100 mL solution
= [0.9 × (0.52 – t'_f)]/0.52
= [0.9 × 0.34]/0.52
= 0.59 g

Sodium Chloride Equivalent Method

From the definition of the E value presented earlier, one can estimate the amount of sodium chloride or another inert substance needed to render a hypotonic solution isotonic with body fluids. The first step is to convert the amount of drug in the solution to its equivalent in sodium chloride, and then compare this amount with that present in a 0.9% sodium chloride solution. The difference between the latter amount and the equivalent amount is the amount needed for tonicity adjustment:[1]

Step 1: E × (amount of drug in solution) = X
Step 2: (0.9 g/100 mL) × (volume of solution in mL) = Y
Step 3: $Y – X$ = amount of NaCl needed

EXAMPLE 7 ▶▶▶

A 500-mL solution of drug (E = 0.21) contains 3 g of the drug. How much sodium chloride is needed to render the solution isotonic?

Solution
Step 1: 0.21 × 3 g = 0.63 g
Step 2: (0.9 g/100 mL) × (500 mL) = 4.5 g (equivalent grams of sodium chloride needed if making 500 mL of isotonic NaCl alone)
Step 3: 4.5 – 0.63 = 3.87 g (amount of NaCl needed for 500 mL of solution)

White-Vincent Method

White and Vincent published a paper in the *Journal of the American Pharmaceutical Association* that described a method for adjusting the tonicity of a solution. This method calculates the volume of an isotonic solution of a drug that can be prepared by dissolving the drug in a proper amount of water. The following equation, named after its investigators, calculates that volume (V):[10]

$$V = \left[\sum(W_i \times E_i)\right] x\, v$$

where W_i = weight in grams of the ith solute in the formulation
E_i = sodium chloride equivalent of the ith solute in the formula
v = volume of the sodium chloride solution (0.9%) that contains 1 g of NaCl (this volume is 111.1 mL)

To prepare the total volume requested, another isotonic solution, such as sodium chloride 0.9%, can be used to add to the isotonic drug solution.

EXAMPLE 8 ▶▶▶

Chloramphenicol (MW = 323.13 g/mole, L_{iso} = 2) 1 g
Sterile water for injection, q.s. ad 100 mL
M. ft. isotonic solution

Solution
The E value for chloramphenicol can be calculated readily by

$$E \cong 17\,(L_{iso}/MW)$$
$$E \cong 17\,(2/323.13) = 0.10$$

Applying the White-Vincent equation:

$$V = 1\,g \times 0.10 \times 111.1$$
$$= 11.1\,mL$$

Thus, to prepare this solution, dissolve 1 g of chloramphenicol in enough sterile water for injection to make 11.1 mL of solution and add enough sodium chloride injection USP to make the volume measure 100 mL.

Sprowls' Method

Sprowls proposed this method in the *Journal of the American Pharmaceutical Association* in 1949. That author relied on the White-Vincent equation to calculate the volume of an isotonic solution that could be prepared by dissolving 0.3 g of a drug in enough water. The weight 0.3 g was chosen because it represents the amount of drug in a fluid ounce of a 1% solution.[11] The Sprowls' volumes also can be used to calculate the amount of solute needed to prepare isotonic solutions.

EXAMPLE 9 ▶▶▶

Glycerin has a Sprowls' value of 11.7. Estimate the E value of glycerin.

Solution
Since the Sprowls' value is the V in the White-Vincent method,

$$11.7 = 0.3\,g \times E \times 111.1$$
$$E = 0.35$$

EXAMPLE 10 ▶▶▶

Estimate the freezing point depression of a 1% solution of glycerin.

Solution
Since the E value of glycerin is 0.35, 1 g of glycerin is osmotically equivalent to 0.35 g of sodium chloride. Thus, a 1% solution of glycerin behaves as does a 0.35% solution of sodium chloride. Since a 0.9% solution of sodium chloride has a freezing point depression of 0.52°C, the freezing point depression of a 0.35% sodium chloride solution (or a 1% glycerin solution) is approximately 0.202°C.

EXAMPLE 11 ▶▶▶

An isotonic silver nitrate solution prepared by the Sprowls' method had a volume of 13.0 mL. How much silver nitrate is needed to prepare 100 mL of isotonic solution?

Solution
Based on the Sprowls' method, there is 0.3 g of silver nitrate in the 13-mL isotonic solution. Therefore, by proportion,

$$(0.3\,g)(100\,mL)/(13\,mL) = 2.3\,g \text{ of silver nitrate}$$

PROBLEMS

1. **a.** Calculate the numbers of milliliters of glacial acetic acid (specific gravity 1.05) required to prepare 1 L of 25% w/w acetic acid (specific gravity 1.05).
   b. What is the molarity of the solution that is prepared?
2. **a.** How many grams of citric acid monohydrate is required to prepare 1 L of a 0.25 M citric acid solution?
   b. What is the percentage of only the citric acid in the solution prepared?
   c. What is the mole percentage of citric acid? Assume a density of 1.
3. What is the L_{iso} value for sodium chloride in a 0.9% solution of sodium chloride?
4. What is the freezing point depression of a 1% solution of sodium proprionate? Of a 10% solution? Assume a density of 1.

5. Calculate the mole fraction of urea in water at 25°C for a solution containing 20 moles of urea/kg of water.
6. How many milliequivalents of Na^+ and HCO_3^- are in 50 mL of a 7.5% solution of $NaHCO_3$?
7. How many milliequivalents of sodium are in 500 mL of normal saline (normal saline = 0.9% NaCl)?
8. Two grams of sodium sulfate contains how many mEq each of sodium and sulfate? How many mmoles of each?
9. How many milligrams of drug X are in 10 mL of a 1:200 solution of X?
10. a. What is the percentage concentration of a 1:10,000 solution of epinephrine HCl (MW 219.20)?
    b. If 50 mL of this solution was prepared, what would be the molarity?
    c. What is the milligram per deciliter concentration?
11. How many milligrams of 10% w/v KCl must be used to make a liter of solution containing 60 mEq/L?
12. How many grams of manganese sulfate quadrahydrate must be added to make 1 L of solution containing 0.72 mEq/mL of manganese ion?
13. How many mEq of calcium and of iodide are contained in 5 mL of Calcidrine Syrup? Calcidrine contains 152 mg of CaI_2/mL.
14. If a parenteral feeding order specified 10 mEq calcium, 10 mEq potassium, 9 mEq sodium, and 29 mEq chloride per liter, how many mEq/mL of NaCl, KCl, and $CaCl_2$ would be needed?
15. Given the following prescription, calculate the molarity of potassium ion.

    Rx
    Sodium sulfacetamide   0.5%
    Monobasic potassium phosphate   1%
    Purified water qs ad   60 mL
    M solution

16. A 145-lb man has a normal potassium content of 32 mEq/kg. If his laboratory reports showed serum potassium to be 2.7 mEq/L and serum pH is 7.5, determine from Figure 3-3 the approximate percentage depletion of potassium and the total body potassium depletion.

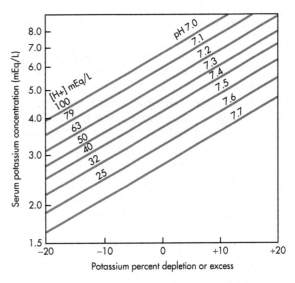

FIGURE 3-3 Percent abnormality in potassium content: Capacity as a function of both serum pH and serum potassium.

17. What percentage of NaCl must a solution contain to have an osmolarity of 300 mOsmol/L? What is the osmolarity of a solution containing 0.45% NaCl?
18. a. If a solution had an osmolarity of 100 mOsmol/L, what percentage of NaCl would be present?
    b. What percentage of dextrose monohydrate would be present?
19. Determine the sodium chloride equivalent and freezing point depression for 1% silver acetate.
20. Calculate the freezing point depression of a 5% solution of ephedrine HCl. Is the solution hypo-, hyper-, or isotonic?
21. What is the osmolarity of a solution containing 0.45% NaCl and 10% dextrose?
22. What is the molecular weight of cefazolin sodium if a 2% aqueous solution has an osmolarity of 83.9 mOsmol/L?
23. If there is 1 g KH_2PO_4 and 1 g K_2HPO_4 in 100 mL of solution, what is the mEq/mL of potassium?
24. Baby calves very quickly become acidotic. Dehydration and electrolyte balance are secondary to the acidosis. Treatment consists of the addition of baking soda to sterile water for injection or normal saline given intraperitoneally as a liter of fluid. One vial of sodium

bicarbonate at 84 mg (1mEq)/mL is available. What dilution of sterile water is required so that the solution is isotonic and can be diluted with sterile normal saline to make a liter of fluid?

25. Suppose that a pharmaceutical company is developing a new antiviral drug for use in the treatment of AIDS. A 1% solution of the drug has a freezing point of 0.053°C.
   a. You are asked to prepare 100 mL of an isotonic solution containing 1% of the drug in water for injection. Show your calculation.
   b. Estimate the E value for this antiviral drug if its L_{iso} is 4.8.
   c. Calculate the Sprowls' value for this drug.

26. Cogentin (benztropine mesylate, Merck Sharp & Dohme; MW = 403.54 g/mole, valence = 1) contains 1 mg of benztropine mesylate in 1 mL of sterile solution for injection. The solution also contains sodium chloride for tonicity adjustment. The drug is an anticholinergic and antihistaminic that is used as an adjunct in therapy for Parkinson's disease.
   a. Calculate the amount of NaCl needed for 1 mL of the sterile solution. The E value for benztropine mesylate is 0.26.
   b. Express the concentration of benztropine mesylate in the solution as mg%.

27. Micro-K (potassium chloride, AH Robins) is used as a potassium supplement in patients with a potassium deficiency. When taking Micro-K, patients are advised to take one capsule (8 mEq of K^+) with a full glass of water (240 mL) after meals.
   a. Calculate the amount of KCl in one Micro-K capsule (in milligrams). The ionic weight of potassium is 39.09 g/mole.
   b. If 8 mEq of KCl is added to enough water to make 240 mL of solution, how much potassium phosphate monobasic (MW = 136.09 g/mole) is needed to render the solution isotonic with body fluids?
   c. Calculate the concentration of potassium ion (mEq/L) in the isotonic solution in b.

28. Sodium formate (MW = 68.02 g/mole) is an astringent. The L_{iso} of the salt is 3.4.
   a. One gram of sodium formate is osmotically equivalent to how many grams of NaCl?
   b. Calculate the freezing point depression of a 1% aqueous solution of the salt.
   c. How much dextrose is needed to render a 1% aqueous solution of the salt isotonic? The E value of dextrose is 0.16.

29. A newly discovered weak acid was found to diffuse freely through the red blood cell membrane. A 2.3% solution of this weak acid has the same freezing point as blood.
   a. Is the 2.3% solution isotonic, isoosmotic, or both isoosmotic and isotonic?
   b. Is the normal freezing point of blood 0°C, −0.052°C, −0.52°C, or −0.90°C?
   c. When a volume of the 2.3% solution in a is mixed with blood, red blood cells will retain their normal size and shape, shrink and become wrinkled, or swell and finally burst?
   d. When two isotonic solutions are mixed (assuming no reaction), the resulting solution is also isotonic. True or false?

30. Consider the following data:

   Drug A: MW = 356 g/mole, L_{iso} = 4.3
   Drug B: MW = 179 g/mole, E = 0.18
   Drug C: $\Delta t_f^{1\%}$ = 0.17°C

   a. Calculate $\Delta t_f^{1\%}$ of drug A.
   b. What is the L_{iso} value for drug B?
   c. A 1% solution of drug C was prepared. How much sodium chloride is needed for tonicity adjustment?
   d. According to the Sprowls' method, what is the value of V for drug B?

31. Fluorouracil   1 g
    Sterile water qs ad   100 mL
    M. Ft. topical solution

   Fluorouracil (MW = 130.08 g/mole) is an anticancer agent. However, this formulation is intended to be applied on precancerous lesions of the skin. The E value for the drug is 0.13.
   a. Calculate the amount of sodium chloride needed to render the solution isotonic.
   b. If boric acid replaced sodium chloride in the solution in a, how much boric acid would be required? The E value of boric acid is 0.5.
   c. Boric acid should never be used for adjusting topical solutions, intravenous solutions, or oral solutions?

32. Acyclovir sodium (MW = 248.21 g/mole) 0.5 g
    Sterile water for injection qs ad 10 mL
    M. Ft. solution

    Acyclovir sodium is an antiviral drug used in the management of herpes simplex. The L_{iso} for this salt is 3.4.
    a. Estimate the freezing point depression of the solution.
    b. This solution is hypotonic, isotonic, or hypertonic?
    c. According to the Sprowls' method, what is the volume of an isotonic solution that can be prepared by means of the addition of enough water to acyclovir sodium?

33. A solution of a newly discovered drug (20 g/L) has a freezing point in water of –0.52°C. The drug has a L_{iso} value of 1.9.
    a. This solution is hyposmotic, isoosmotic, or hyperosmotic?
    b. According to the Sprowls' method, what is the weight of the drug needed to prepare an isotonic solution?

34. The freezing point depression of an aqueous solution of carbachol is 0.205°C.
    a. Estimate the osmotic pressure (atm) of the solution at room temperature. K_f = 1.86, R = 0.0821 atm/deg mole
    b. An isotonic solution of carbachol contains 2.82 g of the drug in every 100 mL of solution. What is the sodium chloride equivalent value for carbachol?
    c. How much sodium chloride is needed to render the solution isotonic?
    d. An aqueous hypotonic solution of carbachol has a boiling point below 100°C. True or false?
    e. If 500 mL of 0.9% NaCl solution is mixed with 100 mL of 2.82% carbachol solution, the resulting solution will be hypotonic, isotonic, or hypertonic?

ANSWERS

1. a. 250 mL; b. 4.38 M
2. a. 52.5 g; b. 4.8% w/v; c. 0.45%
3. L_{iso} = 3.4
4. 1%, 0.35 deg, 10%, 3.9 deg
5. 0.27
6. 44.6 mEq of each
7. 77 mEq
8. 28 mEq Na and SO_4, 28 mM Na and 14 mM SO_4
9. 50 mg
10. a. 0.01%; b. 4.56 × 10^{-4} M; c. 10 mg/dL
11. 45 mL
12. 80.3 g
13. 5.2 mEq
14. 0.009 mEq/mL NaCl, 0.01 mEq/mL KCl, 0.01 mEq/mL $CaCl_2$
15. 0.073 M
16. 12% depletion, 253 mEq low
17. 0.87%, 155 mOsmol/L
18. a. 0.29%; b. 1.98%
19. 0.17 deg, E = 0.28
20. 0.85 deg, hypertonic
21. 659 mOsmol/L
22. 476.8 g/mole
23. 0.19 mEq/mL
24. Dilute 1 mL $NaHCO_3$ with 5.5 mL sterile water and add to 1 L normal saline.
25. a. Since the freezing point depression of a 1% solution of this drug is 0.053°C, this solution is hypotonic. To adjust for tonicity, its freezing point needs to be brought down to –0.053°C. Therefore, sodium chloride will be added to make up the difference in the freezing point (0.52 – 0.053 = 0.467°C). Since a 0.9% NaCl solution has a freezing point depression of 0.52°C, by proportion, a reduction of 0.467°C in the freezing point requires 0.81 g of NaCl in every 100-mL preparation. The resulting solution is isotonic.
    b. From the definition of the E value, since a 100-mL isotonic solution of the drug contains 1 g of drug and 0.82 g of NaCl, and comparing this solution to 0.9% NaCl solution, (0.9 – 0.81 = 0.09 g), or 0.09 g of NaCl is osmotically equivalent to 1 g of the drug. By definition, this is the E value of this drug, or 0.09.

c. The Sprowls' value is calculated by using the White-Vincent method:

$$V = W \times E \times 111.1$$
$$= (0.3 \text{ g})(0.09)(111.1)$$
$$= 3.0 \text{ mL}$$

26. **a.** $E \times W = (0.26) \times (0.001 \text{ g}) = 0.00026$ g
    (0.9 g NaCl/100 mL) (1 mL) = 0.009 g NaCl
    0.009 − 0.00026 = 0.0087 g, or 8.7 mg of NaCl needed per 1 mL of solution
    b. Since there is 1 mg of drug in 1 mL of solution, or 100 mg in 100 mL of solution, the %mg is 100%.

27. **a.** Since there is 8 mEq of potassium ion per capsule, we should have 8 mEq of KCl per capsule. One mEq of KCl is 74.55 mg, and 8 mEq is (8 × 74.55 mg = 596.4 mg).
    b. $W \times E = 596.4 \times 0.76 = 453.3$ mg
    (0.9 g NaCl/100 mL) × (240 mL) = 2.16 g = 2160 mg
    2160 − 453.3 = 1706.7 mg NaCl
    However, since we are using monobasic potassium phosphate to adjust for tonicity, (1706.7 mg)/0.44 = 3878.9 mg of monobasic potassium phosphate, or 3.9 g of this salt in each 240 mL of solution.
    c. Potassium ion is present in this solution from two sources: potassium chloride and monobasic potassium phosphate. Potassium chloride contributes 8 mEq and monobasic potassium phosphate adds another 28.5 mEq of potassium in 240 mL of solution. By proportion, in 1 L of solution there is 152.1 mEq of potassium ion.

28. **a.** $E \cong 17 (L_{iso}/MW) = 17(3.4/68.02) = 0.85$
    One gram of sodium formate is osmotically equivalent to 0.85 g of sodium chloride.
    b. $\Delta t_f^{1\%} = 3.4 [(1/68.02)/0.1] = 0.5°C$
    c. (0.85)(1) = 0.85 g
    (0.9 g NaCl/100 mL) (100 mL) = 0.9 g
    0.9 − 0.85 = 0.05 g NaCl
    (0.05)/(0.16) = 0.3 g dextrose to adjust for tonicity

29. **a.** Isoosmotic
    b. −0.52°C
    c. Swell and finally burst
    d. True

30. **a.** $\Delta t_f^{1\%} = 4.3 \times (10/356) = 0.12°C$
    b. $0.18 = 17 (L_{iso}/179)$
    $L_{iso} = 1.9$
    c. 0.52 − 0.17 = 0.35°C reduction in the freezing point
    (0.9 × 0.35)/0.52 = 0.60% NaCl needed, or 0.60 g of NaCl in every 100 mL of preparation
    d. $V = 0.18 \times 0.3 \times 111.1 = 6.0$ mL

31. **a.** 0.18×1 g = 0.18 g
    (0.9 g NaCl/100 mL) (100 mL) = 0.9 g
    0.9 − 0.18 = 0.72 g of NaCl
    b. 0.72/0.5 = 1.44 g of boric acid
    c. Intravenous solutions

32. **a.** $\Delta t_f^{5\%} = 3.4 \times (50/248.21) = 0.68°C$
    b. Hypertonic
    c. $E \cong 17 (3.4/248.21) = 0.23$
    $V = 0.23 \times 0.3 \times 111.1 = 7.8$ mL

33. **a.** Isoosmotic
    b. 0.3 g

34. **a.** $\Delta t_f = K_f m$
    $m = 0.205/1.86 = 0.11$ molal
    $\pi = i m R T = (1) (0.11) (0.082) (25 + 273.16) = 2.7$ atm
    b. Since 2.82 g of the drug is equivalent to 0.9 g of sodium chloride, by proportion, 1 g of the drug is osmotically equivalent to 0.32 g of NaCl. Thus, the E value for carbachol is 0.32.
    c. 0.52 − 0.205 = 0.315°C
    (0.315) × (0.9% NaCl)/0.52 = 0.54% of NaCl needed, or 0.54 g of NaCl in every 100 mL of solution
    d. False
    e. Isotonic

KEY POINTS

- A solution is a single phase, homogeneous mixture of two or more substances, where a solute is molecularly dispersed (dissolved) within a solvent.
- Pharmacists must be able to convert between different expressions of solution concentrations, including percentages, molarity, normality, and osmolarity.
- Two important categories of solutions relevant to pharmacy practice are solutions of electrolytes, such as sodium chloride, and nonelectrolytes, such as dextrose. Drugs that are weak acids or weak bases, are weak electrolytes. Other drugs are nonelectrolytes.
- Among the important properties of solutions are colligative properties, which are a function of the concentration of solutes.
- Among the most pharmaceutically relevant colligative properties, caused by the presence of solutes, are vapor pressure and freezing point reduction, and osmolarity, which has widespread physiological and pharmaceutical importance.
- For certain drug products that are administered to sensitive areas, such as parenterals and ophthalmics, it is important to consider their relative osmolarity to body fluids, in order to avoid untoward consequences, such as cell and tissue damage and patient discomfort.

CLINICAL QUESTIONS

Extravasation is the unintended leakage of fluid into the tissues surrounding a vein that occurs during the administration of certain fluids. The process can lead to necrosis (cell death) of the affected tissues. Among the fluids that can cause extravasation and necrosis are hypertonic fluids such as 10% dextrose solution, particularly when administered into small veins such as in the arm and hand. Explain how the intravenous infusion of a hypertonic fluid can lead to extravasation and subsequent necrosis, based on the possible effects of fluid on the encountered cells, beginning with those in the venous endothelium.

Intravenous mannitol solutions are indicated for several purposes including the reduction of intracranial pressure. When infused it remains extracellular and is readily filtered through the glomeruli with very low tubular reabsorption. Consequently, it increases diuresis. Explain how the increased diuresis is due to osmotic effects, and how enhanced diuresis ultimately reduces intracranial pressure.

4 Ionic Equilibria and Buffers

Beverly J. Sandmann
Alekha K. Dash
Antoine Al-Achi
Robert Greenwood

Learning Objectives

After completing this chapter, the reader should be able to:

- Define a strong electrolyte, a weak electrolyte, and a nonelectrolyte.
- Differentiate between a cation and an anion.
- Define an acid, a base, and the different types of salts.
- Recognize and give examples of pharmaceutically important acids, bases, and salts.
- Describe the ionization equilibria of strong and weak electrolytes.
- Calculate and interpret the ionization constants of weak acids and bases (K_a, K_b, pK_a, and pK_b).
- Calculate the pH and pOH and the concentrations of the components of aqueous solutions of weak electrolytes.
- Calculate the pH and pOH of mixtures of weak acids and bases, diprotic acids, and ampholyte solutions.
- Calculate the percent ionization of weak electrolytes as a function of pH.

ELECTROLYTES VERSUS NONELECTROLYTES

An *electrolyte* is a substance (an acid, a base, or a salt) that in an aqueous solution ionizes to positive ions (cations) and negative ions (anions). A salt such as sodium chloride (NaCl) is an example of an electrolyte. In an aqueous solution, the ions formed are Na^+ (a cation) and Cl^- (an anion). Therefore, for each molecule, two independent entities are formed in solution. In an aqueous solution, electrolytes exhibit the following important properties that differentiate them from nonelectrolytes:

- They exhibit anomalous colligative properties compared with nonelectrolytes (see Chapter 3).
- They can conduct an electrical current.
- They tend to show rapid chemical reactions compared with nonelectrolyte solutions.

Electrolytes can be classified into two groups—strong and weak—depending on their ability to ionize in an aqueous solution. Strong electrolytes are substances that are completely ionized in water. Hydrochloric acid, for example, is a strong electrolyte, and its degree of ionization is pH-independent. The weak electrolytes, in contrast, ionize only partly in water. Atropine, phenobarbital, and sulfadiazine, and most other drugs, are examples of weak electrolytes. Nonelectrolytes are substances that do not ionize in water at all and therefore do not conduct an electric current in solution. Examples of nonelectrolytes include sucrose, fructose, urea, and glycerol.

IMPORTANCE OF IONIZATION IN PHARMACY

Many drugs and pharmaceutical ingredients are organic weak electrolytes. Their degree of ionization is an important physicochemical property and has extensive physiologic implications. Depending on the pH of the environment in which these organic weak

- Describe the physiologic implications of percentage of ionization of weak electrolytes in the absorption, transport, and excretion of drugs (pH partition hypothesis).
- Define a buffer solution and list its components.
- Calculate the pH of a buffer solution.
- Define and calculate the buffer capacity of a buffer solution.
- Calculate the quantity of ingredients needed to prepare a buffer solution.
- Recognize the various types of buffer solutions and their methods of preparation.
- Define ionic strength and understand its effect on ionization.

electrolytes are present, they can exist in either an ionized form or an unionized form. This degree of ionization can affect the absorption, transport, and excretion of drugs. In general, the ionized forms are more water-soluble but much less soluble in lipid, while the unionized forms have greater lipid solubility and are more permeable to lipid membranes. The effect of the degree of ionization of organic weak electrolytes on drug absorption leads to the development of the important *pH-partition hypothesis*. According to this hypothesis, the absorption of a weak electrolyte is determined mainly by the extent to which the drug exists in its unionized form at the site of absorption. As an example, the gastrointestinal-blood membrane is impermeable to the ionized form of a drug because of the poor lipid solubility of the drug in this form. (The pH-partition hypothesis is discussed in Chapters 5 and 6.)

A small change in pH can cause a large change in the ratio of ionized to unionized molecules and a correspondingly large change in properties such as solubility, dissolution rate, and partition coefficient. Hence, the degree of ionization can have a profound effect on the absorption, distribution, and elimination of drugs and also can affect factors such as the antimicrobial effectiveness of preservatives, chromatographic separation, and the physicochemical stability of a drug product.

ACIDS AND BASES

Arrhenius Classification

The first acid-base theory was proposed by Arrhenius in 1923. According to this theory, acids are substances that yield hydrogen ions in an aqueous solution, whereas bases yield hydroxyl ions in an aqueous solution:

$$HA \rightleftharpoons H^+ + A^-$$
(Acid)

$$BOH \rightleftharpoons B^+ + OH^-$$
(Base)

Brønsted-Lowry Concept

The major drawback of the Arrhenius definition of acids and bases is its lack of generality: It does not account for the acid-base relation in solutions other than water. In 1923, Brønsted-Lowry defined an acid as a species with a tendency to lose a proton, whereas a base has a tendency to accept a proton. Thus, an acid is a proton donor and a base is a proton acceptor. The greater the tendency of a substance to lose a proton, the stronger that substance is as an acid. In contrast, the strength of a base is measured by its tendency to accept a proton. The complementary relationship between an acid and a base (conjugate acid-base pairs) is evident from the following equilibriums:

$$HA + H_2O \rightleftharpoons A^- + H_3O^+$$
$$\text{(Acid)} \qquad \text{(Base)}$$

$$B + H_2O \rightleftharpoons BH^+ + OH^-$$
$$\text{(Base)} \qquad \text{(Acid)}$$

In these cases, the anion (A^-) acts as a base because it accepts a proton. The cation (BH^+) acts as a Brønsted acid because it can donate a proton. Ordinarily, free protons are not released by acids. Protons are transferred from one substance (an ion, atom, or molecule) to another. Therefore, actual acid-base behavior involves an interaction between two sets of acid-base pairs, as shown here for water:

$$A_1 + B_2 = B_1 + A_2$$
$$\text{acid}_1 \quad \text{base}_2 \quad \text{base}_1 \quad \text{acid}_2$$

$$HCl + H_2O = Cl^- + H_3O^+$$
$$\text{acid}_1 \quad \text{base}_2 \quad \text{base}_1 \quad \text{acid}_2$$

$$CH_3COOH + H_2O = CH_3COO^- + H_3O^+$$
$$\text{acid}_1 \qquad \text{base}_2 \qquad \text{base}_1 \qquad \text{acid}$$

$$NH_4^+ + H_2O = NH_3 + H_3O^+$$
$$\text{acid}_1 \quad \text{base}_2 \quad \text{base}_1 \quad \text{acid}_2$$

A_1 and B_1, HCl and Cl^-, and CH_3COOH and CH_3COO^- constitute three conjugate acid-base pairs; similarly, H_3O^+ and H_2O constitute the other conjugate acid-base pair. The acetate ion (the anion) is the conjugate base of the weak acid, whereas the ammonium ion (the cation) is the conjugate acid of the weak base ammonia. The stronger an acid, the weaker its conjugate base, and vice versa (Table 4-1).

TABLE 4-1 Examples of Conjugate Acid-Base Pairs

Conjugate Acid	Conjugate Base
HA molecular acid	A^- anion base
HCl molecular acid	Cl^- anion base
HOCl molecular acid	OCl^- anion base
H_2O molecular acid	OH^- anion base
NH_4^+ cation acid	NH_3 molecular base
BH^+ cation acid	B molecular base

Lewis Electronic Theory

In 1923, G. N. Lewis introduced a new theory of acids and bases. According to this theory, an acid is a molecule or ion that accepts an electron pair to form a covalent bond. A base is defined as a substance that provides the pair of unshared electrons by which the base coordinates with an acid. According to this definition, some species that do not contain a hydrogen can be considered acids. The reaction between boron trichloride (considered a Lewis acid) and ammonia (a Lewis base) is an example of a Lewis acid-base pair:

$$\begin{array}{c}Cl\\|\\Cl-B\\|\\Cl\end{array} + \begin{array}{c}H\\|\\:N-H\\|\\H\end{array} = \begin{array}{c}Cl\ H\\|\ \ |\\Cl-B:N-H\\|\ \ |\\Cl\ H\end{array}$$

$$\text{acid} \qquad \text{base}$$

IONIZATION OF WATER

Ionization Constants (pK_a)

Ionization constants are used to measure the strength of acids and bases and to calculate the extent of ionization of a weak acid or base at a given pH. Water can behave as an acid as well as a base because it is capable of losing as well as accepting a proton. Water ionizes to an extremely small extent into hydronium (solvated hydrogen ions) and hydroxyl ions:

$$H_2O + H_2O \underset{K_r}{\overset{K_f}{\rightleftharpoons}} H_3O^+ + OH^-$$
$$\text{acid} \quad \text{base} \qquad \text{acid} \quad \text{base}$$

where K_f = rate constant for forward reaction

K_r = rate constant for reverse reaction

The rate of forward reaction (R_f) ∝ concentration of reactants is as follows:

$$R_f \propto [H_2O][H_2O]$$
$$R_f \propto [H_2O]^2$$
$$R_f = K_f[H_2O]^2$$

Similarly, the rate of reverse reaction $R_r = K_r[H_3O^+][OH]$.

At equilibrium, the rate of forward reaction equals the rate of reverse reaction. Therefore:

$$K_r[H_2O]^2 = K_r[H_3O^+][OH]$$

$$\frac{K_f}{K_r} = K_{H_2O} = \frac{[H_3O^+][OH^-]}{[H_2O]^2}$$

Water ionizes to an extremely small extent. Therefore, $[H_2O] \gg [H^+]$ and $[H_2O] \gg [OH]$; $[H_2O] =$ constant.

$$K_{H_2O} = \frac{[H_3O^+][OH^-]}{[H_2O]^2}$$

$$K_{H_2O}[H_2O]^2 = [H_3O^+][OH^-]$$

$$K_{H_2O}[H_2O]^2 = K_w = \text{ionic product of water}$$

$$K_w = [H_3O^+][OH]$$

or $\qquad K_w = [H_3O^+][OH] \qquad (1)$

where K_w is the ionic product of water and, for simplicity of representation, H^+ is used as a symbol for the solvated hydronium ion. Here the symbol H^+ does not refer only to a proton.

Water ionizes to give 10^{-7} mol L^{-1} H$^+$ and 10^{-7} mol L^{-1} OH$^-$ at about 25°C (77°F):

$$K_w = [H^+][OH]$$
$$K_w = 10^{-7} \times 10^{-7}$$
$$K_w = 10^{-14} (\text{mole/L}^{-1})^2 (\text{units usually omitted})$$
$$pH = -\log[H^+]$$
$$pH = -\log[1 \times 10^{-7}] = 7$$
$$pOH = -\log[OH] = 7$$
$$K_w = [H^+][OH]$$

Multiplying both sides by $(-\log)$:

$$-\log K_w = -\log[H^+] - \log[OH]$$
$$pK_w = pH + pOH \qquad (2)$$
$$pK_w = 7 + 7 = 14 \text{ at } 25°C (77°F)$$

EXAMPLE 1 ▶▶▶

For a 0.001 M solution of HCl, what are (a) the H$^+$ concentration, (b) the pH, (c) the OH$^-$ concentration, and (d) pOH?

HCl is a "strong" inorganic acid (i.e., it is 100% ionized in dilute solution).

Therefore, when 0.001 mole of HCl is introduced into 1 L of H$_2$O, it immediately dissociates to produce 0.001 M H$^+$ and 0.001 M Cl$^-$.

Note: In dealing with strong acids, the H$^+$ contribution from the ionization of water is neglected. Concentration in moles per liter (molar) is indicated with brackets [ ].

Solution

a. H$^+$ = 0.001 M

b. pH = $-\log[H^+]$
$= -\log[0.001] = -\log 10^{-3}$
$= -(-3) = 3$
pH = 3

c. $[H^+][OH^-] = K_w$

or $[OH] = \dfrac{K_w}{[H^+]}$

$[OH] = \dfrac{1 \times 10^{-14}}{1 \times 10^{-3}}$

$[OH] = 1 \times 10^{-11} M$

d. pOH = $-\log[OH]$
$= -\log(10^{-11})$
pOH = $-(-11) = 11$

pOH also can be determined as follows:

pH + pOH = 14
pOH = 14 − pH = 14 − 3 = 11
pOH = 11

EXAMPLE 2 ▶▶▶

What is the molar concentration of HNO$_3$ in a solution that has a pH of 3.4?

Solution

pH = $-\log[H^+]$ = 3.4

or

$$\log \frac{1}{[H^+]} = 3.4$$

$$\frac{1}{[H^+]} = \text{antilog } (3.4)$$

$$[H^+] = 3.98 \times 10^{-4} M$$

The concentration of HNO_3 is equal to the concentration of $[H^+] = 3.98 \times 10^{-4}$ M.

Relationship Between pK_a and pK_b

The use of acidic constants for the ionization of bases was introduced in 1923 by Brønsted, who realized the advantages of expressing the ionization of acids and bases on the same scale, just as a pH value can denote both alkalinity and acidity.

Some textbooks continue to use K_b and pK_b for bases. This increases the number of equations that must be remembered or derived. It is more convenient if both acids and bases are related in terms of a single quantity, K_a and pK_a. The stronger a base is, the higher its pK_a is. The pK_a of substances of pharmaceutical interest is provided in many textbooks[1,2] and the pharmaceutical literature.[3-5]

If only K_b or pK_b values are available, they can be converted to K_a or pK_a by using:

$$K_w = K_a \cdot K_b \quad (3)$$

Taking $(-\log)$ on both sides of equation (3) yields:

$$pK_w = pK_a + pK_b$$
$$pK_a = pK_w + pK_b \quad (4)$$

EXAMPLE 3 ▶▶▶

The K_b for triethanolamine at 25°C (77°F) is 5.3×10^{-5}. Calculate (a) its K_a and (b) its pK_a.

Solution

a. $K_a = \dfrac{K_w}{K_b} = \dfrac{1 \times 10^{-14}}{5.3 \times 10^{-5}} = 1.9 \times 10^{-10}$

b. $pK_a = -\log(K_a) = -\log(1.9 \times 10^{-10}) = 9.72$

IONIZATION OF ELECTROLYTES

Strong Acids and Strong Bases

For a strong acid, hydrochloric acid:

$$\underset{\text{acid}}{HCl} + \underset{\text{base}}{H_2O} \longrightarrow \underset{\text{acid}}{H_3O^+} + \underset{\text{base}}{Cl^-}$$

Hydrochloric acid in water, for example, is considered a strong acid because it has a strong tendency to ionize, and the reverse reaction occurs only to a very small extent in more concentrated solutions. Therefore, its conjugate base Cl^- is very weak (has a weak tendency to accept a proton). Equilibrium shifts completely to the right in a dilute solution. Strong acids and bases are defined as those that are completely ionized at all pH values. In other words, their extent of ionization is pH-independent. For weak acids:

$$\underset{\text{acid}}{C_6H_5OH} + \underset{\text{base}}{H_2O} \rightleftharpoons \underset{\text{base}}{C_6H_5O^-} + \underset{\text{acid}}{H_3O^+}$$

Phenol in the presence of water is an example of this type. It has a weak tendency to ionize, and the conjugate base, phenoxide ion, is moderately strong. Therefore, the equilibrium shifts considerably to the left. Weak acids and bases are incompletely ionized at some pH values. Their extent of ionization is pH-dependent.

Organic Weak Acids

The organic weak acid HA dissolves in water with the following ionization equilibrium:

$$\underset{\text{acid}}{HA} + \underset{\text{base}}{H_2O} = \underset{\text{acid}}{H_3O^+} + \underset{\text{base}}{A^-}$$

The concentration of water generally is omitted from this equilibrium equation for simplicity.

$$HA \underset{K_2}{\overset{K_1}{\rightleftharpoons}} H^+ + A^- \quad (5)$$

Rate 1 (the forward reaction) = $K_1[HA]$

Rate 2 (the reverse reaction) = $K_2[H^+][A]$

At equilibrium, rate 1 = rate 2, or:

$$K_1[HA] = K_2[H^+][A]$$

$$\frac{K_1}{K_2} = K_a = \frac{[H^+][A^+]}{[H]} \quad (6)$$

Thus, the product of the concentration of ions derived from the ionization of the acid always has a fixed ratio to the concentration of the unionized molecules. This ratio is called the acidic ionization constant or, more simply, the ionization constant K_a (units: moles/L^{-1} usually are omitted). Equation (6) is only an approximation because the law of mass action is based on the concept of thermodynamic activity as opposed to molar concentration. Hence, K_a varies slightly with concentration. However, equation (6) and other equations derived from it are accurate enough for pharmaceutical purposes:

$$K_a = \frac{[H^+][A^-]}{[HA]}$$

For a weak acid, the ionization in water can be written as:

$$H_a = H^+ + A^-$$

$$K_a = \frac{[H^+][A^-]}{[HA]}$$

(See Table 4-2.)

In the absence of any common ion in solution, one should expect the [H$^+$] to be equal to [A$^-$]. Therefore, the ionization constant equation can be expressed as:

$$K_a = \frac{[H^+]^2}{[HA]}$$

or

$$[H^+]^2 = K_a[HA]$$

$$[H^+] = \sqrt{K_a[HA]}$$

TABLE 4-2 Ionization Constants for Weak Acids at 25°C (77°F)

Weak Acids	Molecular Weight	K_a	pK_a
Acetaminophen	151.16	3×10^{-10}	9.5
Acetic	60.05	1.75×10^{-5}	4.76
Acetylsalicylic	180.15	3.27×10^{-4}	3.49
Ampicillin	349.42	$K_1\ 3 \times 10^{-3}$	2.65
		$K_2\ 5.6 \times 10^{-8}$	7.2
Ascorbic acid	176.12	$K_1\ 6.3 \times 10^{-5}$	4.2
		$K_2\ 2.5 \times 10^{-12}$	11.6
Barbituric	128.09	1.05×10^{-4}	3.98
Benzoic	122.12	6.30×10^{-5}	4.20
Benzyl penicillin (G)	344.38	1.74×10^{-3}	2.76
Boric	61.84	$K_1\ 5.8 \times 10^{-10}$	9.24
Carbonic	44.01	$K_1\ 4.31 \times 10^{-7}$	6.37
		$K_2\ 4.7 \times 10^{-11}$	10.33
Cefazolin	454.51	7.9×10^{-3}	2.1
Cephalothin	395.4	5.2×10^{-3}	2.5
Cephradine	349.41	$K_1\ 2.3 \times 10^{-3}$	2.63
		$K_2\ 5.4 \times 10^{-8}$	7.27
Citric (1 H$_2$O)	210.14	$K_1\ 7.0 \times 10^{-4}$	3.15
		$K_2\ 1.8 \times 10^{-5}$	4.78
		$K_3\ 4.0 \times 10^{-7}$	6.40
Folic acid	441.40	3.16×10^{-3}	2.5
Formic	43.02	1.77×10^{-4}	3.75
Furosemide	330.77	2.24×10^{-4}	3.65
α-D-Glucose	180.16	8.6×10^{-13}	12.1
Glycine	75.07	$K_1\ 4.5 \times 10^{-3}$	2.35
		$K_2\ 1.7 \times 10^{-10}$	9.78
Indomethacin	357.81	3.2×10^{-5}	4.5
Lactic	90.08	1.39×10^{-4}	3.86
Mandelic	152.14	4.29×10^{-4}	3.37
Methicillin	380.42	9.77×10^{-4}	3.01
Nafcillin	414.46	2.24×10^{-3}	2.65
Nicotinic	123.11	1.3×10^{-5}	4.87
Nitrofurantoin	238.16	6.31×10^{-8}	7.2
Oxacillin	401.44	1.3×10^{-3}	2.88
Oxalic (2 H$_2$O)	126.07	$K_1\ 5.5 \times 10^{-2}$	1.26
		$K_2\ 5.3 \times 10^{-5}$	4.28
Pentobarbital	248.26	7.94×10^{-9}	8.1
Phenethicillin	363.53	1.5×10^{-3}	2.82
Phenobarbital	232.23	3.9×10^{-8}	7.41
Phenol	95.12	1×10^{-10}	10.0
Phenophthalein	318.31	2×10^{-10}	9.7

Weak Acids	Molecular Weight	K_a	pK_a
Phenytoin	252.26	4.8×10^{-9}	8.32
Phosphoric	98.00	$K_1\ 7.5 \times 10^{-3}$	2.12
		$K_2\ 6.2 \times 10^{-8}$	7.21
		$K_3\ 2.1 \times 10^{-13}$	12.67
Propionic	74.08	1.34×10^{-5}	4.87
Saccharin	183.18	3.5×10^{-2}	1.60
Salicylic	138.12	1.06×10^{-3}	2.97
Stearic	284.47	1.8×10^{-6}	5.75
Succinic	118.09	$K_1\ 6.4 \times 10^{-5}$	4.19
		$K_2\ 2.3 \times 10^{-6}$	5.63
Sulfadiazine	250.28	3.3×10^{-7}	6.48
Sulfamerazine	264.30	8.7×10^{-8}	7.06
Sulfapyridine	249.29	3.6×10^{-9}	8.44
Sulfathiazole	255.32	7.6×10^{-8}	7.12
Tartaric	150.09	$K_1\ 9.6 \times 10^{-4}$	3.02
		$K_2\ 4.4 \times 10^{-5}$	4.36
Trichloroacetic	163.40	1.3×10^{-1}	0.89
Uric acid	168.11	$K_1\ 1.78 \times 10^{-6}$	5.75
		$K_2\ 5 \times 10^{-11}$	10.3
Warfarin	308.32	8.01×10^{-6}	5.05

If the ionization of the acid is very low, one can substitute the total concentration of the acid (C) equal to [HA]:

$$[H^+] = \sqrt{K_a C}$$

Taking ($-\log$) of both sides:

$$-\log[H^+] = \tfrac{1}{2}(-\log K_a - \log C)$$
$$pH = \tfrac{1}{2}(-\log K_a - \log C)$$
$$= \tfrac{1}{2} pK_a - \log C \qquad (7)$$

This equation is useful for calculating the pH of a weak acid in water if one knows the molar concentration (C) and the pK_a of the weak acid.

EXAMPLE 4 ▶▶▶

Calculate the pH of a 0.1 M solution of a weak acid at 25°C (77°F). The pK_a of the acid is 4.76 at 25°C (77°F).

Solution

Because this is a weak acid, one cannot assume 100% ionization of the acid. Therefore, the pH of the solution can be determined by using the following equation:

$$pH = \tfrac{1}{2} pK_a - \tfrac{1}{2}\log C$$
$$= \tfrac{1}{2}(4.76) - \tfrac{1}{2}\log(0.1)$$
$$= 2.38 + 0.5$$
$$= 2.88$$

The ratio of the ionized form to the unionized form of the drug in fluids such as physiologic fluids may be calculated by taking the $-\log$ of equation (6):

$$K_a = \frac{[H^+][A^-]}{[HA]}$$
$$-\log K_a = -\log[H^+] - \log[A^-] + \log[HA]$$
$$pK_a = pH + \log[HA] - \log[A^-]$$

where A^- is the conjugate base.

Rearrangement gives:

$$pH = pK_a + \log\frac{[A^-]}{[HA]} \qquad (8)$$

(*Note:* This is the Henderson-Hasselbalch equation for weak acids, which will be described further in the section on buffers.)

EXAMPLE 5 ▶▶▶

Salicylic acid is an organic weak acid with a pK_a of 3.0. Calculate (*a*) the ratio of the ionized form to the unionized form of this drug in the stomach with pH 1.2 and (*b*) the ratio of the unionized form to the ionized form and the percent of the unionized form of the drug in the stomach.

Solution

a. For a weakly acidic drug, use equation (8):

$$pH = pK_a + \log\frac{[A^-]}{[HA]}$$
$$1.2 = 3.0 + \log\left\{\frac{[\text{ionized}]}{[\text{unionized}]}\right\}$$
$$-1.8 = \log\left\{\frac{[\text{ionized}]}{[\text{unionized}]}\right\}$$

Therefore, the ratio of ionized to unionized salicylic acid will be 0.016/1.

b. The ratio of unionized to ionized salicylic acid will be the reciprocal of the ratio of ionized to unionized.

$$1/0.016$$

or

$$6.31/1$$

Percent unionized salicylic acid will be:

$$\frac{[unionized]}{([ionized]+[unionized])} \times 100$$

$$63.1/(1+63.1) \times 100$$

$$98.44\%$$

Organic Weak Bases

When an organic weak base (B) is added to water, it takes up hydrogen ions from the water to form a conjugate acid BH^+, and the resulting solution contains hydroxyl ions and becomes alkaline:

$$\underset{\text{base}}{B} + \underset{\text{acid}}{H_2O} \rightleftharpoons \underset{\text{acid}}{BH^+} + \underset{\text{base}}{OH^-}$$

$$K_b = \frac{[BH^+][OH^-]}{[B]}$$

$$K_b = \frac{[OH^-]^2}{[B]}$$

$$[OH^-] = \sqrt{K_b[B]} = \sqrt{K_b C}$$

To change the ionization expression for bases to K_a, the equilibrium is written as follows. Values for K_b, pK_b, and pK_a for a weak base are given in Table 4-3.

$$BH^+ \underset{K_2}{\overset{K_1}{\rightleftharpoons}} B + H^+$$

$$\text{Rate 1} = K_1[BH^+]$$

$$\text{Rate 2} = K_2[B][H^+]$$

At equilibrium, rate 1 = rate 2, or $K_1[BH^+] = K_2[B][H^+]$:

$$\frac{K_1}{K_2} = K_a = \frac{[B][H^+]}{[BH^+]} \qquad (9)$$

$$K_a = \frac{[B][H^+]}{[BH^+]}$$

TABLE 4-3 Ionization Constants for Weak Bases at 25°C (77°F)

Weak Bases	Molecular Weight	K_b	pK_b	pK_a
Acetanilide	135.6	(40°) 4.1×10^{-14}	13.39	(0.14)0.3
Ammonia	35.05	1.74×10^{-5}	4.76	9.24
Apomorphine	267.31	1.0×10^{-7}	7.00	7.00
Atropine	289.4	4.5×10^{-5}	4.35	9.65
Benzocaine	165.19	6.0×10^{-12}	11.22	2.78
Caffeine	194.19	(40°) 4.1×10^{-14}	13.39	(0.14)0.61
Chlorcyclizine	300.85	1.4×10^{-6}	5.85	8.15
Chlordiazepoxide	299.76	5.75×10^{-10}	9.24	4.76
Chlorpheniramine	390.87	1.6×10^{-5}	4.8	9.2
Chlorpromazine	318.88	2.1×10^{-5}	4.68	9.32
Clindamycin	461.44	4×10^{-7}	6.4	7.6
Cocaine	303.35	2.6×10^{-6}	5.59	8.41
Codeine	299.36	9×10^{-7}	6.1	7.9
Diazepam	284.75	2.5×10^{-11}	10.6	3.4
Diphenhydramine	255.35	9.54×10^{-6}	5.02	8.98
Ephedrine	165.23	2.3×10^{-5}	4.64	9.36
Erythromycin	733.92	6.3×10^{-6}	5.2	8.8
Gentamicin complex	463.57	1.58×10^{-6}	5.8	8.2
Glycine	75.07	2.3×10^{-12}	11.65	2.35
Hydroquinone	110.11	4.7×10^{-6}	5.33	8.67
Isoproterenol	211.24	4.36×10^{-6}	5.36	8.64
Kanamycin	484.58	1.58×10^{-7}	6.8	7.2
Lidocaine	234.33	7.41×10^{-7}	6.13	7.87
Lincomycin	406.5	3.16×10^{-7}	6.5	7.5
Meperidine	283.79	5.01×10^{-6}	5.3	8.7
Methapyrilene	261.38	7.07×10^{-6}	5.15	8.85
Morphine	285.33	7.4×10^{-7}	6.13	7.87
Oxytetracycline	460.44	$K_3\ 3.16 \times 10^{-11}$	10.5	3.5
		$K_2\ 4 \times 10^{-7}$	6.4	7.6
		$K_1\ 1.6 \times 10^{-5}$	4.8	9.2
Papaverine	339.38	8×10^{-9}	8.1	5.9
Phenylephrine	203.67	$K_1\ 5.9 \times 10^{-6}$	5.23	8.77

Weak Bases	Molecular Weight	K_b	pK_b	pK_a
		$K_2\ 6.3 \times 10^{-5}$ (phenolic)	4.2	9.8
Phenylpropa-nolamine	151.21	2.5×10^{-5}	4.6	9.4
Physostigmine	275.34	$K_1\ 7.6 \times 10^{-7}$	6.12	7.88
		$K_2\ 5.7 \times 10^{-13}$	12.24	1.76
Pilocarpine	208.25	$K_1\ 7 \times 10^{-8}$	7.2	6.8
		$K_2\ 2 \times 10^{-13}$	12.7	1.30
Polymyxin B	(1200)	1×10^{-6}	6	8
Procaine	236.30	7×10^{-6}	5.2	8.8
Promazine	284.41	2.5×10^{-5}	4.6	9.4
Promethazine	284.41	1.2×10^{-5}	4.92	9.08
Pyridine	79.10	1.4×10^{-9}	8.85	5.15
Quinidine	324.4	$K_1\ 3.7 \times 10^{-6}$	5.40	8.6
		$K_2\ 1 \times 10^{-10}$	10.0	4.0
Quinine	324.41	$K_1\ 1.0 \times 10^{-6}$	6.00	8.0
		$K_2\ 1.3 \times 10^{-10}$	9.89	4.11
Reserpine	608	4×10	7.4	6.6
Strychnine	334.40	$K_1\ 1 \times 10^{-8}$	6.0	8.0
		$K_2\ 2 \times 10^{-12}$	11.7	2.3
Tetracycline	444.43	2×10^{-6}	5.7	8.3
Theobromine	180.17	(40°) 4.8×10^{-14}	13.32 (0.21)	0.68
Theophylline (base)	180.17	5×10^{-14}	13.3	0.7
Thiourea	76.12	1.1×10^{-15}	14.96	
Urea	60.06	1.5×10^{-14}	13.83	0.17

For weak bases, K_a is only an approximation based on molar concentration rather than thermodynamic activity. The K_a for bases varies only very slightly with concentration. When weak bases dissolve in water, the concentration of $[B] \ll [BH^+]$. Taking the log of both sides of the previous equation gives:

$$\log K_a = \log[H^+] + \log[B] - \log[BH^+]$$

Equating $[H^+]$ to $[B]$ and rearranging gives:

$$-2\log[H^+] = -\log K_a - \log[BH^+]$$
$$pH = \tfrac{1}{2} pK_a - \tfrac{1}{2} \log[BH^+]$$

The pH of the protonated organic weak base can be calculated from this formula.

The pH of the organic weak base (free base) in water can be calculated by using pK_a, pK_w, and the concentration of the base $[B]$, or C, as given here:

$$pH = \tfrac{1}{2} pK_w + \tfrac{1}{2} pK_a + \tfrac{1}{2} \log[B]$$

This equation is useful in calculating the pH of an organic weak base in water.

EXAMPLE 6 ▶▶▶

Calculate the pH of a 0.1 M solution of a weak base (trimethylamine) at 25°C (77°F).

Solution
Since this is a weak base, the pH of the solution can be determined by using the following equation:

$$pH = \tfrac{1}{2} pK_w + \tfrac{1}{2} pK_a + \tfrac{1}{2} \log C$$
$$= 7 + \tfrac{1}{2}(9.72) + \tfrac{1}{2} \log(0.1)$$
$$= 11.36$$

in a solution that is pH-adjusted to provide $[B] \cong [BH^+]$.

Connecting equation (9) by using log forms yields:

$$pK_a = pH + \log[BH^+] - \log[B] \qquad (10)$$

Rearrangement gives:

$$pH = pK_a + \log\left(\frac{[B]}{[BH^+]}\right) \qquad (11)$$

(*Note:* This is the Henderson-Hasselbalch equation for weak bases, which will be described further in the section on buffers.)

EXAMPLE 7 ▶▶▶

Calculate the approximate percentage of codeine present as free base in the small intestine (pH 6.5). pK_a codeine = 7.9.

Solution

$$pH = pK_a + \log\left(\frac{\text{free base}}{\text{ionized}}\right)$$

$$6.5 = 7.9 + \log\left(\frac{\text{free base}}{\text{ionized}}\right)$$

$$6.5 = 7.9 + \log\left(\frac{\text{free base}}{\text{ionized}}\right)$$

$$\left(\frac{\text{free base}}{\text{conjugate acid}}\right) = 0.398$$

or

$$\left(\frac{\text{free base}}{\text{conjugate acid}}\right) = \frac{0.398}{1}$$

Total codeine concentration in parts = $[B] + [BH^+]$ = $0.0398 + 1 = 1.0398$:

Percentage free base in the intestine

$$= \frac{[B]}{([B]+[BH^+])} \times 100$$

$$= \frac{0.0398}{1.0398} \times 100 = 3.8\%$$

Theoretically, for a 0.1% solution of codeine in water, using $pK_a = 7.9$, the pH equation for the free base gives:

$$pH = \tfrac{1}{2} pK_w + \tfrac{1}{2}(7.9) + \tfrac{1}{2}\log 0.1$$

$$= 7 + 3.95 - \tfrac{1}{2}$$

$$= 10.45$$

IONIZATION OF SALTS

A salt is formed by an acid-base reaction involving either a proton donation or a proton acceptance. When a salt is added to water, the solution can be neutral, acidic, or basic, depending on the salt. Interaction of the ions of the salts with the ions of water is called *hydrolysis*. Salts can be classified into the following four categories:

1. Salts of strong acids and strong bases
2. Salts of weak acids and strong bases
3. Salts of strong acids and weak bases
4. Salts of weak acids and weak bases

Salts of Strong Acids and Strong Bases

Salts of this class do not undergo hydrolysis; therefore, the concentrations of hydrogen and hydroxyl ions remain unchanged. The salt solution in water therefore shows a neutral reaction. Sodium chloride is an example of this class.

Salts of weak acids and strong bases have a basic pH in water. Salts of this category completely ionize in aqueous solution, and the hydrolysis reaction results in a basic solution:

$$NaA \rightarrow Na^+ + A^-$$
complete ionization

The conjugate anion A^- interacts with water to form the molecular acid and hydroxide ion, resulting in an alkaline solution:

$$H_2O + A^- \rightleftharpoons HA + OH^-$$
acid base acid base

The acid formed (HA) is a molecular unionized or free acid. Sodium acetate is a salt of this class. The combined reaction is given by:

$$NaA \rightarrow A^- + Na^+$$
$$\Updownarrow$$
$$HA + OH^-$$

The equilibrium expression for K_b is:

$$K_b = \frac{[HA]_{eq}[OH^-]_{eq}}{[A^-]_0} = K_b = \frac{K_w}{K_a}$$

As a result of the hydrolysis reaction, when dissolved in water, this salt will give an alkaline pH and act as a base. The hydrolysis reaction is like a K_b that can be converted to K_a as was done previously for a weak base:

$$pH = \tfrac{1}{2} pK_w + \tfrac{1}{2} pK_a + \tfrac{1}{2} \log [A^-]$$

Salts of Weak Acids and Strong Bases

To calculate the pH of sodium acetate in water, the pK_a of acetic acid is 4.76 and the concentration of the salt is 0.1 M. For calculating the pH of the salt of a strong base and a weak acid that acts as a base in water, the pH is alkaline. Substituting in the values for pK_a and C where C is the concentration of the salt in water gives:

$$pH = 7 + \tfrac{1}{2}(4.76) + \tfrac{1}{2} \log(0.1)$$

$$= 7 + 2.38 - 0.5$$

$$= 8.9$$

For the reaction of an organic weak acid with calcium or magnesium hydroxide to form a salt, it takes 1 mole of a strong base for each 2 moles of weak acid. An example of such a salt is magnesium stearate (MgS_2):

$$\text{ionization}$$
$$MgS_2 \rightarrow Mg^{2+} + 2S^-$$
$$\text{Hydrolysis} \updownarrow H_2O$$
$$2HS + 2OH^-$$

The concentration of the salt to use in the pH formula for a weak acid is two times the salt concentration C:

$$[S^-] = 2C$$

EXAMPLE 8 ▶▶▶

To calculate the pH of 0.01 M magnesium stearate in water, for stearic acid, the pK_a is 5.75:

$$pH = \tfrac{1}{2}pK_w + \tfrac{1}{2}pK_w + \tfrac{1}{2}\log 2C$$
$$= 7 + \tfrac{1}{2}(5.75) + \tfrac{1}{2}\log 0.02$$
$$= 7 + 2.875 - 0.85$$
$$= 9.025$$

3. Salts of Strong Acids and Weak Bases

Salts of weak bases and strong acids have an acidic pH in water. When a salt of a weak base and a strong acid is added to water, it is completely ionized in the aqueous solution, as follows:

$$BHCl \rightarrow BH^+ + Cl^-$$
$$\text{complete} \quad \text{ionization}$$

However, the conjugate cation acid, BH^+, can react with water to form the molecular base (also unionized or free base), and the resulting solution is therefore acidic in nature:

$$BH^+ + H_2O \rightleftharpoons B: + H^+$$
$$\text{ionization} \quad \text{expression}$$

Ephedrine hydrochloride is a salt of this class. It is important to note that very many weak acid and weak base drugs are formulated as the salt form. The main purpose of formulating a drug in its salt form is that, being already ionized, the salt form has greater water solubility than its free acid or free base counterpart.

The conjugate cation (BH^+) interacts with water to form the free base and hydrogen ion, resulting in an acid solution:

$$H_2O + BH^+ \rightleftharpoons B: + H^+$$
$$\text{hydrolysis} \quad \text{expression}$$

The equilibrium expression for K_h is:

$$K_h = \frac{[B]_{eq}[H^+]_{eq}}{[BH^+]_0} = K_a = \frac{K_w}{K_b}$$

This salt in water will give an acidic pH and act as an acid. The hydrolysis reaction is like a K_a:

pH expression: $pH = \tfrac{1}{2}pK_a - \tfrac{1}{2}\log[BH^+]$

EXAMPLE 9 ▶▶▶

To calculate the pH of 0.1 M ephedrine hydrochloride in water, $pK_a = 9.36$:

$$pH = \tfrac{1}{2}pK_a - \tfrac{1}{2}\log C$$
$$= \tfrac{1}{2}(9.36) - \tfrac{1}{2}\log(0.1)$$
$$= 4.68 + 0.5$$
$$= 5.2$$

For the reaction of an organic weak base with sulfuric acid, a strong acid, to form a salt, it takes 2 moles of weak base for each mole of sulfuric acid. An example of such a salt is ephedrine sulfate:

$$\text{ionization}$$
$$(EpH)_2 SO_4 \rightarrow 2 EpH^+ + SO_4^-$$
$$\text{Hydrolysis} \updownarrow H_2O$$
$$2 EP + 2H^+$$

The concentration of the salt to use in the pH formula for a weak base is two times the salt concentration (C):

$$[EpH^+] = 2C$$

To calculate the pH of 0.1 M ephedrine sulfate, for ephedrine, the pK_a is 9.36:

$$pH = \tfrac{1}{2}pK_a - \tfrac{1}{2}\log(2C)$$
$$= \tfrac{1}{2}(9.36) - \tfrac{1}{2}\log(0.2)$$
$$= 4.68 + 0.35$$
$$= 5.03$$

BUFFERS

A solution containing either a weak acid with its conjugate base or a weak base with its conjugate acid has the capacity to function as a buffer. The term *buffer* implies protection or shielding. In a sense, that is exactly what a buffer does to a pharmaceutical preparation: It protects the formulation from a sudden change in pH. Table 4-4 lists examples of pharmaceutical preparations and their pH.[6] To keep the hydronium ion concentration constant during the life of a preparation, a buffer system may be included in the formulation. Also, buffer solutions are necessary in many experiments conducted in pharmaceutical or biomedical research. For example, drug stability studies,[7–12] diffusion-dissolution studies,[13–16] and partitioning studies[17] are just a few of the applications of buffers in research.

TABLE 4-4 The pH of Some Liquids as Determined by a pH Meter

Product	pH at Room Temperature
Acetaminophen liquid (Harber)	5.12
Alkalol (Alkalol Co.)	8.63
Apple juice (Mott's Co.)	3.70
Burows solution, USP	4.62
Club soda (Schweppes)	5.07
Coffee, instant (Nestles)	4.74
Cough syrup (Harber Pharmacal)	2.74
Debrox ear drops (Dow, Inc.)	7.27
Diphenhydramine syrup (Harber)	5.03
Distilled vinegar (Heinz Co.)	2.58
Ferrous sulfate solution, USP	2.54
Gatorade (Gatorade Co.)	2.99
Hydrogen peroxide, USP	4.22
Iodine tincture (Drug Guild)	5.13
Ipecac syrup, USP	1.71
Kolex nasal spray (Drug Guild)	5.51
Lemon juice (Borden, Inc.)	2.51
Listerine (Warner-Lambert Co.)	3.87
Magnesium citrate solution, USP	3.94
Mercurochrome (Drug Guild)	7.32
Normal saline solution, USP	6.83
Pepsi (Pepsico Co.)	2.57
Povidone-iodine solution, USP	3.60
Ready-to-use enema (Fleet)	5.52
Robitussin DM (A.H. Robins)	2.53
Tea (Lipton Co.)	5.14
Visine eye drops (Pfizer, Inc.)	6.32
White wine (Holland House)	2.67

Source: Reprinted by permission from **Zuckerman CL, Nash RA:** The pH value of common household products using meters, papers and sticks. *IJPC* 2(2):137, 1998. Copyright *International Journal of Pharmaceutical Compounding*. Corning Model 220 pH meter; Corning, Inc., Elmira, NY.

A buffer acts by neutralizing any hydrogen ions or hydroxyl ions added to it.[18] The weak acid (HA) present in a buffer can react with the added hydroxyl ions, and the weak base (A$^-$) can react with the added proton (i.e., hydrogen ions). The net result is a slight change in the concentration of the weak acid and base present in the solution with a corresponding small change in the pH:

$$HA + OH^- \rightarrow A^- + H_2O \qquad (12)$$

$$A^- + H_3O^+ \rightarrow HA + OH^- \qquad (13)$$

EXAMPLE 10 ▶▶▶

A solution contains phenobarbital and sodium phenobarbital. Describe what will happen if acetic acid is added to the solution.

Solution
The solution containing phenobarbital and its sodium salt is a buffer system. Since acetic acid is going to yield protons in the solution, sodium phenobarbital will react with the added protons and produce more phenobarbital according to equation (13).

The Buffer Equation

The Henderson-Hasselbalch equation,[19,20] or the *buffer equation*, also can be used to calculate the pH of a buffer solution:

$$\text{For weak acids: pH} = pK_a + \log\frac{[A^-]}{[HA]}$$

$$\text{For weak bases: pH} = pK_a + \log\left(\frac{[B]}{[BH^+]}\right)$$

EXAMPLE 11 ▶▶▶

Calculate the pH of the buffer solution described in Example 10 if the molar concentration of phenobarbital is 0.03 M and that of sodium phenobarbital is 0.02 M. The pK_a for phenobarbital is 7.4.

Solution
According to the Henderson-Hasselbalch equation:

pH = 7.4 + log(0.02)/(0.03)
 = 7.2

EXAMPLE 12 ▶▶▶

If the addition of 0.1 mole of HCl [molecular weight (MW) = 36.5 g/mole] to 1 L of a buffer solution caused a drop in the pH by 0.4 unit, calculate the capacity of the buffer.

Solution
Since one equivalent weight of HCl is 36.5 g,

$$\beta = 0.1/0.4 = 0.25$$

EXAMPLE 13 ▶▶▶

The buffer capacity of a solution is 0.02. How much 1 N solution of NaOH (MW = 40 g/mole) must be added to 1 L of this solution to change its pH by 1 unit?

Solution
Since β is 0.02 and pH is 1, β is 0.02. Thus, 20 mL of the NaOH solution is needed to bring about this change (assume that the addition of 20 mL of solution to 1 L will not cause a great change in the volume or concentration).

The Buffer Capacity (b)

Buffers are able to protect preparations from large swings in pH. However, every buffer will reach a point where it no longer can protect the preparation from outside insults. In 1922, Van Slyke published a paper defining the protective strength of a buffer.[18] According to Van Slyke, the strength of the buffer is measured by its *buffer capacity*. Other terms used to describe this strength are *buffer effect* and *buffer value*. The buffer capacity is a function of the ionic dissociation constant, the hydronium ion concentration, and the total molar concentration (C) of the components (i.e., the weak acid and the conjugate base):

$$\beta = 2.3\, C\{(K_a \times [H_3O^+])/(K_a + [H_3O^+])^2\}$$

It also is defined approximately as the ratio of the change in the gram equivalent weight per liter (Eq/L) of an acid or a base (B) needed to produce a particular change in the pH:[18, 21]

$$\beta = B/pH$$

When β values are expressed, no units usually are stated; however, the number represents the Eq/L for a unit pH change.

The value of β reaches its maximum when the hydronium ion concentration in the solution is equal to the dissociation constant for the weak acid. The maximum buffer capacity is defined as:

$$\beta_{max} = 0.575\, C$$

Preparation of Buffers

Later in this chapter we describe many USP and other buffer systems, along with guidelines for their preparation. Here are general steps for pharmaceutical buffer preparation:[22]

1. Decide for what pH the buffer is needed.
2. Choose a conjugate acid-base pair so that the ionization constant for the weak acid is as close as possible to the hydronium ion concentration desired (from step 1). This will guarantee that β will be at its maximum value.
3. Apply the Henderson-Hasselbalch equation and calculate the ratio of proton acceptor to proton donor.
4. Choose the concentration of the proton acceptor and the proton donor so that β is within the range of 0.01 to 0.1.
5. Verify the pH of the resulting buffer by using a pH meter.

> **EXAMPLE 14** ▶▶▶
>
> Prepare a pharmaceutical buffer with pH of 4.76.
>
> **Solution**
> Acetic acid has $pK_a = 4.76$. An acetic acid–sodium acetate buffer will be prepared. According to the Henderson-Hasselbalch equation, the ratio of salt to acid is 1. Choose the salt and acid concentration that makes β between 0.01 and 0.1.
>
> $0.05 = 2.3\,C\{(1.75 \times 10^{-5}) \times 10^{-4.67} / [(1.75 \times 10^{-5}) + 10^{-4.67}]^2\}$
>
> $C = 0.088\,M$ (or $0.044\,M$ acetic acid and $0.044\,M$ sodium acetate, since the ratio of the salt to acid is 1)

> **EXAMPLE 15** ▶▶▶
>
> For a buffer solution containing $0.1\,M$ of a weak base and $0.06\,M$ of its conjugate acid, calculate its buffer capacity if the pH of the solution equals the pK_a of the conjugate acid.
>
> **Solution**
> Since $pK_a = pH$, β is at its maximum value:
>
> $\beta_{max} = 0.575\,C$
> $\quad\quad = 0.575(0.1 + 0.06)$
> $\quad\quad = 0.092$

Physiologic Buffer Systems

The blood contains a variety of buffer systems that maintain pH at 7.4. These are the hemoglobin, bicarbonate, phosphate, and plasma protein buffers.[21] At physiologic pH, hemoglobin has the highest buffering capacity.[21] The normal buffer capacity of the blood is in the range of 0.0385, and it remains relatively unchanged over the pH range of 6.6 to 7.4.[21] Values of blood pH that differ significantly from 7.4 are considered incompatible with life. Eye fluid also contains buffers that maintain pH at the physiologic value of 7.4.[22] The proteins, carbonic acid, and organic acids in tears act as buffer systems in the eye.[23] The pH of tears may be slightly less acidic because of the loss of CO_2 that dissolves in it during the collection process.[22] Solutions instilled in the eye may have a pH significantly different from 7.40; such solutions cause little discomfort to the patient because of the buffer systems present in the eye.[23] It seems that the buffer capacity of the solution instilled in the eye has more effect on irritation than the pH.[22,23] Generally, tissue irritation is correlated with the differences in the pH between the physiologic fluid and the solution being administered, the volume of the solution being used relative to the physiologic fluid's volume, and the buffer capacity of the drug solution and the physiologic fluid. The greater the departure from the physiologic fluid by the drug solution in those parameters, the greater the discomfort felt by the patient.[24]

IONIZATION OF AMPHOTERIC ELECTROLYTES

Of great biological importance are organic molecules, which contain both acidic and basic functional groups. Under certain conditions these molecules will act as an acid and under other conditions behave as a base. Such electrolytes are called *amphoteric electrolytes* or simply *ampholytes*. The amino acids and proteins have both carboxylic acid and amine functional groups. From an acid solution in which the carboxylic acid and the amine functions are protonated to higher pH with the successive dissociation of the carboxylic acid and the amine function, the amino acid passes through the *zwitterion* form (a molecule with positive and negative charge) providing a classic example of biological ampholytes. For the simplest amino acid, glycine, the zwitterionic form is $NH_3^+ - CH_2 - COO^-$. The ionization equations for glycine can be given as:

$$K_{a_1} = \frac{[H^+][NH_3^+ - CH_2 - COO^-]}{[NH_3^+ - CH_2 - COOH]} = 4.5 \times 10^{-3}$$

$$K_{a_2} = \frac{[H^+][NH_2 - CH_2 - COO^-]}{[NH_3^+ - CH_2 - COO^-]} = 1.7 \times 10^{-10}$$

The isoelectric point, the pH of the zwitterions formation, is the pH where net charge equals zero. Experimentally, this pH has been determined to be 6.1 for glycine. At this isoelectric pH, there is no

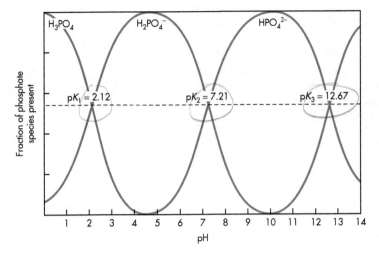

FIGURE 4-1 Distribution diagram for (ortho) phosphoric acid.

migration of protein in an applied electric field during electrophoresis.

IONIZATION OF POLYPROTIC ACIDS

A polyprotic acid is an acid containing two or more ionizable hydrogens. These acids always ionize in successive steps. Ionization of phosphoric acid (a triprotic acid) is as follows:

$H_3PO_4 \rightleftharpoons H^+ + H_2PO_4^-$ (primary ionization)

$H_2PO_4^- \rightleftharpoons H^+ + HPO_4^{2-}$ (secondary ionization)

$HPO_4^{2-} \rightleftharpoons H^+ + PO_4^{3-}$ (tertiary ionization)

$$K_1 = \frac{[H^+][H_2PO_4^-]}{[H_3PO_4]}$$

$$K_2 = \frac{[H^+][HPO_4^{2-}]}{[H_2PO_4^-]}$$

$$K_3 = \frac{[H^+][PO_4^{3-}]}{[HPO_4^{2-}]}$$

The ionization constants for these three reactions can be derived as follows:

$$\frac{[H^+][PO_4^{3-}]}{[H_3PO_4]} = K_1 \times K_2 \times K_3 = K = 2.23 \times 10^{-22}$$

at 25°C (77°F)

A distribution diagram for (ortho) phosphoric acid is given in Figure 4-1.

ACTIVITY AND ACTIVITY COEFFICIENT

In the case of nonideal solutions and solutions at a higher concentration, the effective concentration of ions in solution is lower than the actual concentration. This effective concentration of the ion is called the *activity of the ion*. Therefore, activity related to the concentration is:

$$a = \gamma C$$

where a = activity (moles/L)

γ = activity coefficient

C = concentration (moles/L)

For an infinitely dilute solution, γ approaches unity and concentration is equal to activity. For an electrolyte, the activity of the individual ions is generally unequal. Therefore, mean activity of the ions generally is considered as follows:

$$a_\pm = \sqrt{a_+ a_-}$$

The activity coefficient of electrolytes is also expressed as the mean activity coefficient:

$$\gamma_\pm = \sqrt{\gamma_+ \gamma_-}$$

Debye and Huckel Equation of Ionic Strength

The ionic strength of a solution is a measure of the concentration of ions in that solution. An empirical equation was developed by Debye and Huckel to determine the activity coefficient of an ion without any experimental data. At a very low concentration, the limiting form of this equation is given by:

$$\log \gamma = -Kz^2 \sqrt{v}$$

where y = activity coefficient of the ion
z = valence of the ion (neglecting its sign)
v = ionic strength of the solution
K = a constant

$$v = \sum \frac{1}{2} cz^2$$

The Σ symbol in this equation represents a summation of the term ($1/2\ cz^2$) for each type of ion present in solution.

EXAMPLE 16 ▶▶▶

A solution contains 0.001 M sodium chloride and 0.003 M calcium chloride. Calculate (a) the ionic strength of this solution and (b) the activity of sodium ion in the solution if the K value at the specified temperature is 0.51.

$$v = \sum \frac{1}{2} cz^2$$

Solution

a. The value of $\frac{1}{2}\ cz^2$ for each of the ions in solution should be calculated as follows:

$$\text{Sodium ion} = \frac{1}{2} \times 0.001 \times 1^2$$
$$= 0.0005$$
$$\text{Chloride ion from sodium chloride} = \frac{1}{2} \times 0.001 \times 1^2$$
$$= 0.0005$$
$$\text{Chloride ion from calcium chloride} = \frac{1}{2} \times 0.006 \times 1^2$$
$$= 0.003$$
$$\text{Calcium ion} = \frac{1}{2} \times 0.003 \times 4^2$$
$$= 0.024$$
$$\text{Ionic strength} = 0.028$$

b. $\log \gamma N_{a+} = -0.51 \times 1^2 \times \sqrt{0.028} = -0.085339$
$\gamma N_{a+} = 0.8216$
$a_{Na^+} = \gamma_{Na^+}$ (Na$^+$ concentration)
$= 0.8216 \times 0.001 = 0.0008216\ M$

ACID-BASE TITRATION AND TITRATION CURVE

Figure 4-2 shows the volume of alkali added and the pH for a monoprotic acid. A titration curve is obtained experimentally by adding a small increment of a standard base to a solution of weak acid, or vice versa, and determining the pH of the solution after each addition of the standard base. A solvent correction curve is obtained by adding the same volume of base to the solvent alone. Finally, the solvent correction curve is subtracted from the titration curve for the acid to give the true titration curve for the acid. In this case, the volume of the alkali added (y coordinate) is plotted against the pH (x coordinate), and this plot is called a titration curve. A hypothetical titration curve for a polyprotic acid is shown in Figure 4-3.

The sigmoidal nature of the curve indicates that the pH does not change at a constant rate when equal increments of alkali are added to the acid. There is an initial rapid change in the pH, followed by a very small change in the pH and finally by rapid change again in the pH.

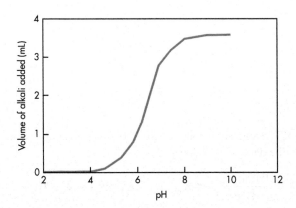

FIGURE 4-2 Graph showing the volume of alkali added and pH for a monoprotic acid.

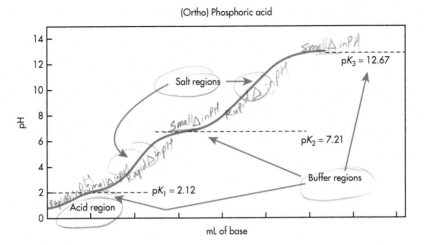

FIGURE 4-3 A hypothetical titration curve.

CASE STUDIES

Case 1

A 32-year-old male was on sulfadiazine for toxoplasmosis therapy.[25] The patient was receiving 2 g of sulfadiazine four times a day for two weeks. The patient developed dysuria and problems voiding, including the presence of red urine, red stones, and severe urethral pain. The pH of the urine was found to be acidic (5.5). Based on your knowledge of ionic equilibriums and the pK_a of this weakly acidic drug (6.48), describe the cause of this clinical situation and suggest possible remedies for this patient.

Case 2

The bacteriostatic property of benzoic acid has been shown to be due to the presence of the undissociated moiety rather than the ionized benzoate ion.[26] This is believed to be due to the high penetration of the unionized molecule into the living membrane, as opposed to the ionized species. If the pK_a of benzoic acid is 4.2, explain the ineffectiveness of this preservative at a biological pH (7.4).

Case 3

The bioavailability of benzoic acid ($pK_a = 4.2$) and sodium benzoate from the fatty suppository base in humans was evaluated.[27] Compared with the sodium salt, benzoic acid was absorbed from the fatty base more rapidly and with less interindividual variation. Justify this result in light of the pH partition hypothesis.

Case 4

The equilibrium distribution of a weakly acidic drug (salicylic acid) with $pK_a = 3$ between gastric fluid (pH 1.2) and blood (pH 7.4) was studied. Using the dissociation equation for a weak acid, justify the validity of the following statement: The total equilibrium concentration of a weakly acidic drug is approximately 25,000 times greater in the blood than in the stomach.

Case 5

According to pH partition hypothesis, when administered orally, a weakly basic drug is poorly absorbed from the stomach. Assuming that the pK_a of the basic drug is 5, the pH of the stomach is 1.2, and the pH of blood is 7.4, justify this statement.

Case 6

The absorption of ketoconazole has been shown to be impaired in patients with drug-induced achlorhydria.[28] The drug absorption was increased significantly to approximately 65% when the drug was taken with an acidic beverage such as Coca-Cola (pH 2.5). Ketoconazole is a synthetic azole antifungal agent and a weakly dibasic compound with pK_a of 2.9 and 6.5. Using the information provided and the pH partition hypothesis, justify the use of Coca-Cola to improve the absorption of this antifungal agent.

Case 7

A depressed patient was admitted to the emergency room and was found to have consumed excessive amobarbital sodium tablets (pK_a 8), and the pH of the patient's urine was found to be 6. After consultation with the clinical pharmacist, the physician in charge recommended the intravenous administration of sodium bicarbonate to increase the patient's urine pH. Justify the use of this alkalizing agent in this emergency situation.

Case 8

A postmenopausal woman experienced severe systemic side effects after treating her vaginal yeast infection with an over-the-counter miconazole nitrate (pK_a 6.9) topical antifungal agent. The determination of vaginal pH of the patient was 7.0. Using your knowledge of ionic equilibriums and the pH partition hypothesis, explain this adverse drug reaction.

PROBLEMS

1. **a.** What is the pH of a 5 g per 100 mL solution of phenol?
   b. What is the hydroxyl ion concentration of the solution?
   c. What is the pH of sodium phenolate?
2. What is the pH of a 1% solution of lidocaine HCl?
3. Calculate the pH of a 1% w/v solution of ephedrine sulfate.
4. **a.** What is the pH of a solution containing both 0.1 mole of ephedrine and 0.01 mole of ephedrine hydrochloride per liter of solution?
   b. What is the pH of each alone?
5. Calculate the pH of the following solution of a monocarboxylic acid and its sodium salt, which was intended to serve as a medium for an ophthalmic drug. The K_a of the acid is 6.3×10^{-5}.

Acid (MW 122)	0.3 g
Na salt (MW 144)	1.7 g
Dist. H$_2$O qs ad	100 mL

6. Taking the pK_a for acetic acid from the pK_a ionization constant for weak acids (see Table 4-2), calculate how much solid sodium acetate (in moles) must be added to a liter of 0.1 N acetic acid to produce a buffer solution of pH 4.50 (ignore the volume change).
7. **a.** If 0.1 mole of NaOH is added to a liter of 2% boric acid solution, what is the pH of the resulting solution?
   b. What was the original pH of the boric acid solution?
8. One desires to adjust a solution to pH 8.8 by using a boric acid–sodium borate buffer. What approximate ratio of acid and salt is required?
9. What is the pH of a phosphate buffer containing 150 mL of 0.2 M sodium biphosphate and 100 mL of 0.1 M sodium phosphate dibasic?
10. If plasma H$_2$CO$_3$ is mostly in the form of dissolved CO$_2$ gas, what will the actual concentration (mEq/L) of bicarbonate ion be when the PCO$_2$ is 44 mmHg for normal plasma pH?
11. Plot the pH titration curve for the neutralization of 0.1 N barbituric acid by 0.1 N NaOH. What is the pH of the solution at the equivalence point? What is the percent ionization at pH 4 and pH 6? The volumes of solution may be selected.
12. At the pH of tears (7.4), what proportion of pilocarpine would be present as the free base?
13. **a.** What is the mole percent of free phenobarbital in solution at pH 8.00?
    b. What is the mole percent of free cocaine in solution at pH 8.00?

ANSWERS

1. **a.** 5 g/100 mL = 50 g/1000 mL
   50 g/95.12 = 0.53 mole
   pH = 1/2 pK_a – 1/2 log C
   = 5 – 1/2 log 0.53
   = 5.13
   b. pOH = 8.87
   [OH$^-$] = 1.3 × 10^{-9}
   c. 5 g NaP/100 mL = 50 g/L
   50 g/116 g = 0.43
   pH = 1/2 pK_w + 1/2 pK_a + log C
   = 7 + 5 – .183
   = 11.82

2. 1% = 10 g/L = .037 M
   MW = 270.79
   pH = 1/2 pK_a – 1/2 log C
   = 7.87/2 + 0.716
   = 4.65

3. Ephedrine sulfate is uni-divalent.
   (BH$^+$)$_2$SO$_4^{2-}$/2BH$^+$ + SO$_4$
   [BH$^+$] = 2 conc. salt.
   1% = 1 g/100 mL × mole/428.6 g × 1000 mL
   = 0.023 M salt = .046 M BH$^+$ BH$^+$/B + H$^+$
   K_h = [B][H$^+$]/[BH$^+$] = K_a = K_w/K_b
   [H$^+$]^2 = (0.46)(4.3 × 10^{-10}) or pH = 1/2 pK_a – 1/2 log (2C)
   = 4.45 × 10^{-6} = 9.36/2 – 1/2 log (.046)
   pH = 5.35

4. pH = pK_a + log B/S   0.1 M ephedrine
                              0.01 M ephedrine HCl
   = 9.36 + log 0.1/0.01 pH = 1/2 pK_w + 1/2 log .1 pH = 1/2 pK_a – 1/2 log C
   a. pH = 10.36   **b.** pH = 11.2, pH = 5.2

5. K_a = 6.3 × 10^{-5}, pK_a = 4.2
   pH = pK_a + log S/A
   [salt] = 1.7 g/100 mL × mole/144 g × 1000 mL
   = 0.12 M
   [acid] 3 g/100 mL × mole/122 g × 1000 mL = 0.025 M
   pH = 4.2 + log 0.12/0.025 = 4.88

6. pH = pK_a + log S/A
   4.50 = 4.76 + log S/A
   S/A = 0.6

   A = 0.1
   S = 0.06 M or 0.06 mole

7. 2% B.A = 2 g/100 mL × 1000 mL × mole/62 g = 0.322 M
   H$_3$B.O$_3$ + NaOH → NaH$_2$BO$_3$ + H$_2$O
   init:.322   0   ~0
   equil.: .322. – 1   .1   .1
   pH = pK_a + log S/A
   = 9.24 + log .1/.222 = 8.89
   original pH
   H$_3$BO$_3$ = H$^+$ + H$_2$BO$_3^-$
   K_a = [H$^+$][H$_2$BO$_3^-$]/[H$_3$BO$_3$] = [H$^+$]^2/0.322 = 5.8 × 10^{-10}
   [H$^+$] = 1.36 × 10^{-5} or   pH = 1/2 pK_a – 1/2 log C
                                = 4.62 + .25 = 4.87

8. H$_3$BO$_3$, pK_a = 9.24
   pH = pK_a + log S/A
   8.8 = 9.24 + log S/A
   S/A = 0.364/L

9. pKa_2 = 7.21
   150 mL × 0.2 mole/1000 mL = .03 mole
   .03 mole/250 mL = 0.12 M
   100 mL × 0.1 mole/1000 mL = .01 mole
   .01/250 = 0.04 M
   pH = 7.21 + log 0.04/0.12
   = 6.73

10. 40 × 0.03 = 1.2 mEq/L
    pH = pK_a + log [HCO$_3^-$]/1.2
    log [HCO$_3^-$]/1.2 = 1.3
    [HCO$_3^-$] = 23.94 = 24 mEq/L

11. at pH 4% I = 51%
    at pH 6% I = 99.05%
    At equivalence pt. pH = pH of salt
    [salt] = .1 N = .1 M

12. pOH = pK_b + log S/B
    or
    pH = pK_a + log B/S for a base
    7.4 = 6.8 + log B/S
    B/S = 3.98   solve as simultaneous equations
    B + S = 100
    80%

13. a. Mole % phenobarbital at pH 8, $pK_a = 7.41$
$pH = pK_a + \log S/A$
$8 = 7.41 + \log S/A$
$S/A = 3.89$ $\;\;$ | solve as simultaneous
$S + A = 100\%$ | equations
$A = 20.4\%$

b. Mole % cocaine at pH 8, $pK_a = 8.41$
$pH = pK_a + \log B/S$
$8.0 = 8.41 + \log B/S$
$B/S = .3876$
$S + B = 100\%$
$B = 27.9\%$

KEY POINTS

▶ A solution is a single-phase, homogeneous mixture of two or more substances, where a solute is molecularly dispersed (dissolved) within a solvent.

▶ Pharmacists must be able to convert between different expressions of solution concentrations, including percentages, molarity, normality, and osmolarity.

▶ Two important categories of solutions relevant to pharmacy practice are solutions of electrolytes, such as sodium chloride, and nonelectrolytes, such as dextrose. Drugs that are weak acids or weak bases, are weak electrolytes. Other drugs are nonelectrolytes.

▶ Among the important properties of solutions are colligative properties, which are a function of the concentration of solutes.

▶ Among the most pharmaceutically relevant colligative properties, caused by the presence of solutes, are vapor pressure and freezing point reduction, and osmolarity, which has widespread physiological and pharmaceutical importance.

▶ For certain drug products that are administered to sensitive areas, such as parenterals and ophthalmics, it is important to consider their relative osmolarity to body fluids to avoid untoward consequences, such as cell and tissue damage and patient discomfort.

APPENDIX

USP Buffer Systems

The United States Pharmacopeia (USP)[29] recognizes special types of solutions as buffers. According to the USP, standard buffer solutions should be prepared with CO_2-free water and should be stored in type I glass tight containers. A table in the USP gives buffers for pH ranges between 1.2 and 10.0. The six USP buffer systems are hydrochloric acid (pH 1.2–2.2), acid phthalate (pH 2.2–4.0), neutralized phthalate (pH 4.2–5.8), phosphate (pH 5.8–8.0), alkaline borate (pH 8.0–10.0), and acetate (pH 4.1-5.5). For 200 mL of hydrochloric acid buffer solution, 50 mL of potassium chloride solution (0.2 M) is mixed with an appropriate volume of 0.2 M HCl to provide the desired pH. A 200-mL acid phthalate buffer is prepared by mixing 50 mL of potassium biphthalate solution (0.2 M) with a given volume of 0.2 M HCl. For the other three buffer systems, 200 mL of the buffer is made by mixing a specified volume of NaOH 0.2 M solution with 50 mL of potassium biphthalate solution 0.2 M (for the neutralized phthalate buffer), 50 mL of the monobasic potassium phosphate solution 0.2 M (for the phosphate buffer), or 50 mL of boric acid and potassium chloride solution 0.2 M (for the alkaline borate buffer). Table 4-5 lists the preparation formulas for USP Acetate Buffer over its covered pH range.

Other Buffer Systems

Among the most common buffers used in pharmaceutical formulations are phosphate, citrate, and acetate, histidine, tromethamine (Tris), and bicarbonate. Table 4-6 lists several of the common buffers, along with their structures, pKa's, useful pH ranges, and drug product examples.[30,31]

Ionic Equilibria and Buffers

TABLE 4-5 Place the specified amount of sodium acetate trihydrate in a 1000 mL volumetric flask, add the specified volume of 2N acetic acid, then add water to 1000 mL, and mix.

Sodium Acetate Trihydrate, g	2N Acetic Acid, mL	pH
1.5	19.5	4.1
1.99	17.7	4.3
2.99	14.0	4.5
3.59	11.8	4.7
4.34	9.1	4.9
5.08	6.3	5.1
5.23	5.8	5.2
5.61	4.4	5.3
5.76	3.8	5.4
5.98	3.0	5.5

TABLE 4-6 Common Buffers Used in Pharmaceutical Products

Buffer	Structure	pKa (s)	pH range	Drug Product Examples
Citrate		3.1, 4.8, 6.4	3-7	Procrit (epoetin alfa) Injection, Tenormin (atenolol) Injection
Acetate		4.75	4-6	Pegasys (peginterferon alfa-2a) Injection, Neupogen (filgrastim) Injection
Histidine		1.8, 6.2, 9.2	5-6.5	Xolair (omalizumab) Injection, Simponi (golimumab) Injection
Maleate		1.9, 6.2	5.2-6.8	Carnitor (levocarnitine) Oral Solution, Vumon (teniposide) Injection
Phosphate		2.2, 7.2, 12.4	6-8	Enbrel (etanercept) Injection, Alamast (pemirolast potassium) Ophthalmic Solution
Bicarbonate		6.3, 10.3	4-9	Cefotan (cefotetan) Injection, Primaxin (imipenem and cilastatin) For Injection
Tris (Tromethamine)		8	7-9	Efudex (fluorouracil) Topical Solution, Apidra (insulin glulisine) Injection

TABLE 4-7 Mixing x mL of Veronal Sodium with (10 − x) mL of 0.1 N HCl Solution Yields the Desired pH

Veronal Sodium, mL	pH	Veronal Sodium, mL	pH	Veronal Sodium, mL	pH
5.36	7.00	7.16	8.00	9.36	9.00
5.54	7.20	7.69	8.20	9.52	9.20
5.81	7.40	8.23	8.40	9.74	9.40
6.15	7.60	8.71	8.60		
6.62	7.80	9.08	8.80		

Data from **Michaelis L:** Diethylbarbiturate buffer. *J Biol Chem* 87(1):33, 1930. The American Society for Biochemistry & Molecular Biology ©1930; **Gennaro AR (ed.):** Solutions and phase equilibria. In *Remington's Pharmaceutical Science*, 17th ed., pp 207–29, Easton, PA: Mack Publishing Company.

Veronal Buffers

Veronal sodium (sodium diethylbarbiturate) is used in these solutions, where 10.30 g of this salt is added to enough carbon dioxide–free water to make 500 mL of solution. To prepare a buffer solution of this type with a given pH, x mL of this solution is mixed with $(10 - x)$ mL of 0.1 N HCl. A pH in the range of 7.0 to 9.4 at room temperature is obtained (Table 4-7).[32]

Sodium Carbonate-Bicarbonate Buffers

Dissolve 10.599 g of Na_2CO_3 anhydrous in enough water to make 1 L (solution A). Prepare a second solution containing 8.4 g of $NaHCO_3$ in 1 L aqueous solution (solution B). When specific volumes from solutions A and B are mixed together, the desired pH is obtained. Table 4-8 lists the pH values obtained at 20°C (68°F).[33]

Disodium Phosphate and Citric Acid Buffers

Table 4-9 summarizes the pH values obtained by mixing specified volumes of 0.1 M citric acid solution and 0.2 M disodium phosphate solution at room temperature. The final volume of the buffer is 20 mL.[34]

Borate Buffers

These systems give pHs in the range of 7.4 to 9.0. A volume of 0.05 M borax is mixed with a solution of 0.2 M boric acid to give the appropriate pH at room temperature. Table 4-10 provides the volumes of the two solutions to prepare 10 mL of a buffer.[35]

TABLE 4-8 The pH of Solutions Prepared by Mixing Together Sodium Carbonate and Sodium Bicarbonate Solutions at 20°C

Bicarbonate Sol., mL	Sodium Carbonate Sol., mL	pH	Bicarbonate Sol., mL	Sodium Carbonate Sol., mL	pH
9	1	9.16	4	6	10.14
8	2	9.40	3	7	10.28
7	3	9.51	2	8	10.53
6	4	9.78	1	9	10.83
5	5	9.90			

Data from **Delory GE, King EJ:** A sodium carbonate-bicarbonate buffer for alkaline phosphatases. *Biochem J* 39:245, 1945.

TABLE 4-9 The pH of Buffer Solutions Obtained by Mixing a Volume of Disodium Phosphate Solution (0.2 M) with a Volume of Citric Acid Solution (0.1 M)

Disodium Phosphate Sol., mL	Citric Acid Sol., mL	pH	Disodium Phosphate Sol., mL	Citric Acid Sol., mL	pH
0.40	19.60	2.2	10.72	9.28	5.2
1.24	18.76	2.4	11.15	8.85	5.4
2.18	17.82	2.6	11.60	8.40	5.6
3.17	16.83	2.8	12.09	7.91	5.8
4.11	15.89	3.0	12.63	7.37	6.0
4.94	15.06	3.2	13.22	6.78	6.2
5.70	14.30	3.4	13.85	6.15	6.4
6.44	13.56	3.6	14.55	5.45	6.6
7.10	12.90	3.8	15.45	4.55	6.8
7.71	12.29	4.0	16.47	3.53	7.0
8.28	11.72	4.2	17.39	2.61	7.2
8.82	11.18	4.4	18.17	1.83	7.4
9.35	10.65	4.6	18.73	1.27	7.6
9.86	10.14	4.8	19.15	0.85	7.8
10.30	9.70	5.0	19.45	0.55	8.0

Data from **McIlvaine TC:** A buffer solution for colorimetric comparison. *J Biol Chem* 49:183, 1921. The American Society for Biochemistry & Molecular Biology ©1921; **Gennaro AR (ed.):** Solutions and phase equilibria. In *Remington's Pharmaceutical Science*, 17th ed., pp 207–29, Easton, PA: Mack Publishing Company.

Potassium Chloride–Hydrochloric Acid Buffers

Solutions of this type provide pH values in the range of 1.0 to 2.2. To make these buffers, two solutions

TABLE 4-10 Volumes of Sodium Borate (Borax) Solution (0.05 M) and Boric Acid Solution (0.2 M) Needed to Prepare Buffer Solutions at Room Temperature

Boric Acid Solution, mL	Borax Solution, mL	pH
9.0	1.0	7.4
8.5	1.5	7.6
8.0	2.0	7.8
7.0	3.0	8.0
6.5	3.5	8.2
5.5	4.5	8.4
4.0	6.0	8.7
2.0	8.0	9.0

Source: Reprinted by permission from **Holmes W:** Silver staining of nerve axons in paraffin sections. *Anat Rec* 86:157, 1943. Wiley-Liss, Inc., a subsidiary of John Wiley & Sons, Inc. copyright © 1943.

are needed: 5 M potassium chloride and 5 M hydrochloric acid. Table 4-11 shows the volumes of each of the solutions that when mixed together provide the desired pH after dilution with enough water to make 200 mL of solution.[24]

Phthalate-Hydrochloric Acid Buffers

Systems such as these are prepared by mixing specified volumes of acid potassium phthalate solution (5 M) with hydrochloric acid solution (5 M), and the final volume (200 mL) is reached through the addition of water. Table 4-12 shows the volumes needed

TABLE 4-11 Volumes of 5 M Hydrochloric Acid and Potassium Chloride Solution Needed to Prepare Buffer Systems with pH 1.0 to 2.2

KCl Solution (5 M), mL	HCl Solution (5 M), mL	pH
50.0	97.0	1.0
50.0	64.5	1.2
50.0	41.5	1.4
50.0	26.3	1.6
50.0	16.6	1.8
50.0	10.6	2.0
50.0	6.7	2.2

Data from **Clark WM, Lubs HA:** The colorimetric determination of hydrogen ion concentration and its applications in bacteriology. *J Bacteriol* 2:1, 1917. The American Society for Microbiology ©1917.

TABLE 4-12 Volumes of 5 M Hydrochloric Acid and Acid Potassium Phthalate Solution Needed to Prepare Buffer Systems with pH 2.2 to 3.8

Acid Potassium Phthalate Solution (5 M), mL	Hydrochloric Acid Solution (5 M), mL	pH
50.0	46.70	2.2
50.0	39.60	2.4
50.0	32.95	2.6
50.0	26.42	2.8
50.0	20.32	3.0
50.0	14.70	3.2
50.0	9.90	3.4
50.0	5.97	3.6
50.0	2.63	3.8

Data from **Clark WM, Lubs HA:** The colorimetric determination of hydrogen ion concentration and its applications in bacteriology. *J Bacteriol* 2:1, 1917. The American Society for Microbiology ©1917.

from each solution to prepare buffers with a pH between 2.2 and 3.8.[36]

Phthalate-Sodium Hydroxide Buffers

These solutions are similar to the phthalate–hydrochloric acid buffers except that the hydrochloric acid solution is replaced with sodium hydroxide solution (5 M). The pH range of these buffers is 4.0 to 6.2 (Table 4-13).[36]

Phosphate-Sodium Hydroxide Buffers

Sodium hydroxide solution (5 M) is mixed with a solution of acid potassium phosphate, and enough water is added to prepare 200 mL of solution with the needed pH (5.8–7.0) (Table 4-14).[36]

Boric Acid–Sodium Hydroxide Buffers

Two solutions are needed to prepare these buffers: a sodium hydroxide solution (5 M) and a solution containing potassium chloride (5 M) and boric acid (5 M). Solutions with pH in the range of 7.8 to 10.0 may be prepared by mixing a specified volume from each and diluting with water to make 200 mL (Table 4-15).[36]

Sorensen's Buffer Systems

Solutions such as these can be made by mixing x mL of an acid stock solution (sodium acid phosphate

TABLE 4-13 Volumes of 5 M Sodium Hydroxide and Acid Potassium Phthalate Solution Needed to Prepare Buffer Systems with pH 4.0 to 6.2

Acid Potassium Phthalate Solution (5 M), mL	Sodium Hydroxide Solution (5 M), mL	pH
50.0	0.40	4.0
50.0	3.70	4.2
50.0	7.50	4.4
50.0	12.15	4.6
50.0	17.70	4.8
50.0	23.85	5.0
50.0	29.95	5.2
50.0	35.45	5.4
50.0	39.85	5.6
50.0	43.00	5.8
50.0	45.45	6.0
50.0	47.00	6.2

Data from **Clark WM, Lubs HA:** The colorimetric determination of hydrogen ion concentration and its applications in bacteriology. *J Bacteriol* 2:1, 1917. The American Society for Microbiology ©1917.

TABLE 4-14 Volumes of 5 M Sodium Hydroxide Solution and Acid Potassium Phosphate Solution Needed to Prepare Buffer Systems with pH 5.8 to 7.0

Acid Potassium Phosphate Solution (5 M), mL	Sodium Hydroxide Solution (5 M), mL	pH
50.0	3.72	5.8
50.0	5.70	6.0
50.0	8.60	6.2
50.0	12.60	6.4
50.0	17.80	6.6
50.0	23.65	6.8
50.0	29.63	7.0

Data from **Clark WM, Lubs HA:** The colorimetric determination of hydrogen ion concentration and its applications in bacteriology. *J Bacteriol* 2:1, 1917. The American Society for Microbiology ©1917.

monohydrate 9.208 g in 1 L solution) with $(100-x)$ mL of an alkaline stock solution (sodium phosphate anhydrous 9.407 g in 1 L of solution). Table 4-16 shows the pH of the resulting buffers.[37]

TABLE 4-15 Volumes of 5 M Sodium Hydroxide and Boric Acid–Potassium Chloride Solution Needed to Prepare Buffer Systems with pH 7.8 to 10.0

Boric Acid (5 M)–Potassium Chloride (5 M) Solution, mL	Sodium Hydroxide Solution (5 M), mL	pH
50.0	2.61	7.8
50.0	3.97	8.0
50.0	5.90	8.2
50.0	8.50	8.4
50.0	12.00	8.6
50.0	16.30	8.8
50.0	21.30	9.0
50.0	26.70	9.2
50.0	32.00	9.4
50.0	36.85	9.6
50.0	40.80	9.8
50.0	43.90	10.0

Data from **Clark WM, Lubs HA:** The colorimetric determination of hydrogen ion concentration and its applications in bacteriology. *J Bacteriol* 2:1, 1917. The American Society for Microbiology ©1917.

TABLE 4-16 Sorensen's Buffers at 25°C

Acid Stock Solution, mL	Alkaline Stock Solution, mL	pH
90	10	5.90
80	20	6.25
70	30	6.47
60	40	6.65
50	50	6.81
40	60	6.98
30	70	7.19
20	80	7.42
10	90	7.79
5	95	8.13

Source: Reprinted with permission of Wiley-Liss, Inc., a subsidiary of John Wiley & Sons, Inc., copyright © 1969. **Cutie AJ, Sciarrone BJ:** Re-evaluation of pH and tonicity of pharmaceutical buffers at 37°. *J Pharm Sci* 58(8):990, 1969.

Other Examples of Buffer Solutions

Table 4-17 presents some buffer systems with pH ranging from 2.3 to 9.3.[7]

TABLE 4-17 Examples of Additional Buffer Systems

Buffer Composition	Concentration or Amount	Volume, mL	pH
Disodium citrate	0.1 M	340	
HCl qs ad	0.1 N	1000	2.3
Disodium citrate	0.1 M	600	
HCl qs ad	0.1 N	1000	3.9
Disodium citrate	0.1 M	856	
HCl qs ad	0.1 N	1000	4.8
Potassium phosphate (monobasic)	9.073 g		
Sodium phosphate (dibasic)	0.594 g		
Water qs ad		1000	5.5
Potassium phosphate (monobasic)	0.2 M	250	
Sodium hydroxide	0.2 M	28	
Water qs ad		1000	6.0
Potassium phosphate (monobasic)	0.2 M	250	
Sodium hydroxide	0.2 M	195	
Water qs ad		1000	7.3
Boric acid	0.2 M	250	
Potassium chloride	0.2 M	250	
Sodium hydroxide	0.2 M	30	
Water qs ad		1000	8.2
Boric acid	0.2 M	250	
Potassium chloride	0.2 M	250	
Sodium hydroxide	0.2 M	203	
Water qs ad		1000	9.3

Source: Originally published in **Raffanti EF, King JC:** Effect of pH on the stability of sodium ampicillin solutions. *Am J Hosp Pharm* 31:745, 1974. American Society of Hospital Pharmacists, Inc. All rights reserved. Reprinted with permission (R1403).

SUGGESTED READINGS

Albert A, Serjeant EP: *The Determination of Ionization Constants*, 3d ed. New York: Chapman and Hall, 1984.

Foye WO: *Principles of Medicinal Chemistry*, 3d ed. Philadelphia: Lea & Febiger, 1990.

Logan SR: *Physical Chemistry for the Biomedical Sciences*. Philadelphia: Taylor & Francis, 1998.

Martin A, Bustamante P, Chun AHC: *Physical Pharmacy*, 4th ed. Philadelphia: Lea & Febiger, 1993.

5 Solubility, Dissolution, and Partitioning

Beverly J. Sandmann and Mansoor M. Amiji

Learning Objectives

After completing this chapter, the reader should be able to:

▶ Define and understand the terms and concepts of solubility and miscibility.

▶ Identify the descriptive terms for solubility, their meaning, and their percent value.

▶ Recognize factors that affect solubility.

▶ Calculate the solubility of poorly soluble strong electrolytes by using K_{sp} values.

▶ Calculate the solubility of weak electrolytes as a function of pH and understand what the answer means.

▶ Calculate the pH of precipitation for weak electrolytes in aqueous and mixed solvent systems.

▶ Describe the role of drug dissolution from dosage forms in drug bioavailability.

▶ Relate the parameters of the Noyes-Whitney equation to the variables that affect the dissolution rate of drug particles.

▶ Recognize and appreciate the United States Pharmacopeia (USP) requirements for dissolution studies.

SOLUBILITY

General Solubility Concepts

Solubility is the concentration of a solute when the solvent has dissolved all the solute that it can at a given temperature. A useful definition of solubility is the concentration of solute in a saturated solution at equilibrium. Saturated solution concentration is the solubility of the solute at that temperature. Solubility is a physical property, and values for the solubility of pure substances are found in the literature. It is possible for some substances to dissolve in a higher concentration than could be attained at equilibrium, have increased solubility, and form supersaturated solutions. One compound that readily forms supersaturated solutions is caffeine. The term *miscible* is used to refer to the solute when it is a liquid and will form a solution with a solvent over any concentration range.

The solubility of most substances varies with temperature, having a nearly exponential relationship, as seen in one of the solubility curves for two inorganic potassium salts shown in Figure 5-1.

Solubility Expressions

The solubility of solid solutes in liquid solvents, particularly in water, is an important consideration in the preparation, storage, and use of liquid pharmaceutical formulations and products.

Descriptive terms for solubility are found in the *United States Pharmacopeia/National Formularly (USP/NF)* and in *Remington's Pharmaceutical Sciences*. An interpretation of these terms is given in Table 5-1. The descriptive terms are general solubilities and cover a range of values. For example, a substance described as "soluble" requires between 10 and 30 parts (mL) of solvent to dissolve 1 part (g) of solute.

Exact and approximate values of solubility are given in the *Merck Index*, the USP-NF, and *Remington*. Solubility values for solutes reported in the literature often are given as per weight of solvent or per volume of solvent. Expressions such as grams of solute in 100 grams of solvent and 1 g of solute in a volume (mL) of solvent are common in pharmaceutical references. These values

- Discuss how various physicochemical, formulation, and physiological factors can influence dissolution behavior.
- Define and understand the partitioning law, partition coefficients, and apparent partition coefficients.
- Calculate the partition coefficients for different types of solutes in aqueous/organic solvent systems.
- Determine the effect of pH on the partition coefficient of weak electrolytes and on their excretion in different body fluids.

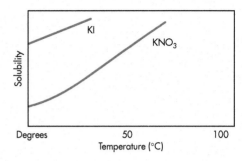

FIGURE 5-1 Solubility curves.

TABLE 5-1 USP Chart of Descriptive Terms

Term	Parts Solvent (mL) Required to Dissolve 1 Part Solute (g)
Very soluble	Less than 1
Freely soluble	1–10
Soluble	10–30
Sparingly soluble	30–100
Slightly soluble	100–1000
Very slightly soluble	1000–10,000
Practically insoluble, insoluble	More than 10,000

do not convert directly to percent concentration or molar solubility in moles per liter, which is represented as S. S indicates saturated solution concentration (solubility) in molar concentration units.

Examples of common medicinal agents and their descriptive solubility terms and solubility values are given in Table 5-2.

Weak electrolytes such as phenobarbital, atropine, and aspirin are very slightly soluble to slightly soluble, whereas their salts, which are strong electrolytes, have much higher solubilities in water. Salts of weak acids and weak bases are completely ionized in water and are very soluble.

The miscibility expression for liquid solute in liquid solvents is found for liquid solutions that mix together at any concentration (Table 5-3). There are no degrees of miscibility; only the word *miscible* is used.

Factors Affecting Solubility

As stated, *solubility* is a term used to express the concentration of a solution at saturation. It is an equilibrium value that is dependent on a given temperature and pressure. Variation of pressure in most pharmaceutical applications is not important. Therefore,

TABLE 5-2 Solubilities in Water at 25°C

Solid	Descriptive Term	Solubility Value
Phenobarbital	Very slightly soluble	1 g in 1000 mL
Phenobarbital sodium	Very soluble	1 g in 1 mL
Atropine	Slightly soluble	1 g in 466 mL
Atropine sulfate	Very soluble	1 g in 0.4 mL
Codeine	Slightly soluble	1 g in 120 mL
Morphine	Very slightly soluble	1 g in 5000 mL
Sulfadiazine	Practically insoluble	1 g in 13,000 mL
Silver chloride	Insoluble	1.93 mg in 1000 mL
Aspirin	Slightly soluble	1 g in 300 mL
Phenol	Soluble	1 g in 15 mL
Camphor	Slightly soluble	1 g in 800 mL
Sodium carbonate	Freely soluble	1 g in 3.5 mL

solubilities are reported at a given temperature. Solubility does not depend on how fast a substance dissolves in the solvent. How fast a solution is formed is not the solubility value.

Temperature

As has been indicated, a principal factor that influences the equilibrium solubility is the temperature. The nearly exponential rise or fall of temperature with solubility is related to the exchange of energy that occurs during solution formation. As a solid solute dissolves, it takes in energy to break apart the crystal lattice structure and separate it into ions or molecules to get to the individual ion or molecule necessary to form a solution. The solvent—water—takes in energy to break apart its hydrogen bonds to allow the solute to come in contact with adjacent molecules of water. When the solute and the solvent come together, energy is released. The overall exchange of energy can be positive or negative. The overall energy exchanged is called the *heat of solution*. The exponential proportionality for the effect of temperature on solubility is given by:

$$S \alpha e^{\frac{-\Delta H_{soln}}{RT}}$$

where S is solubility in moles per liter.

The equation to predict the effect of temperature on solubility is:

$$\log \frac{S_{(T_2)}}{S_{(T_1)}} = \frac{\Delta H_{soln}}{2.303\,R} \left(\frac{T_2 - T_1}{T_1 T_2} \right)$$

The heat of solution for some representative compounds is shown in Table 5-4.

If ΔH_{soln} is negative, increasing the temperature of the solvent will decrease the solubility of the solute. A negative heat of solution indicates that heat must be released for the solute to dissolve. Heat will be given off as the solution is formed. Increasing temperature adds more heat; therefore, less solute will dissolve at the higher temperature. Conversely, if the ΔH_{soln} is positive, a solvent at a higher temperature will dissolve more solute than it will at a lower temperature. An example of a slight temperature effect on solubility is sodium chloride, NaCl, with a ΔH_{soln} of only 1 kcal/mole. There is essentially no difference in solubility for sodium chloride at a higher temperature or a lower temperature.

TABLE 5-3 Miscibilities in Water at 25°C

Liquid	Descriptive Term
Alcohol	Miscible
Glycerin	Miscible
Acetic acid	Miscible
Light mineral oil	Immiscible

TABLE 5-4 Heat of Solution

Solute	ΔH_{soln} (25°C) kcal/mole
HCl (g)	−17.96
KOH$_{(c)}$	−13.7
NaOH$_{(c)}$	−10.6
CH$_3$COOH$_{(l)}$	−.36
KMnO4$_{(c)}$	+10.4
C$_2$H$_2$OH$_{(l)}$	−3.1
KI$_{(c)}$	+4.3
NaCl$_{(c)}$	+1.0
LiCl$_{(c)}$	−8.8
AgCl$_{(c)}$	+15.0
Ca(OH)$_2$	−2.79
Methylcellulose$_{(c)}$	+
Mannitol$_{(c)}$	+5.26
Phenol$_{(c)}$	+
Sodium carbenicillin$_{(c)}$	−
Salts with water of hydration (crystals)	+
Anhydrous salts (crystals)	−

(c) = crystalline.

TABLE 5-5 Dielectric Constants of Some Solvents at 25°C

Solvent	Dielectric Constant
Water	78.5
Glycerin	40.1
N,N-Dimethylacetamide	37.8
Propylene glycol	32.01 (30°)
Methanol	31.5
Ethanol	24.3
N-Propanol	20.1
Acetone	19.1
Benzyl alcohol	13.1
Polyethylene glycol 400	12.5
Phenol	9.7
Cottonseed oil	6.4
Ether	4.3
Ethyl acetate	3.0

Polarity and Hydrogen Bonding

The chemical structure influences solubility through the dipole moment, dielectric properties, and hydrogen bonding (H bonding).

A dipole molecule has a concentration of a positive charge at one end and a negative charge at the opposite end. A substance with a strong dipole effect is said to be polar. Something that is polar has charges that are separated permanently. Generally, molecules that have a high dipole moment are more soluble in polar solvents such as water because of the charge distribution and orientation. *Dielectric properties* relate to the ability to store charge (Table 5-5). This can influence how a substance interacts with solvents such as water, where the oxygen and hydrogen arrangement allows the electronegative oxygen to pull electrons from the hydrogen molecules to create a partial positive charge on the hydrogen molecules and a partial negative charge on the oxygen molecule. Water has a fairly high dielectric constant (~80) at 20°C (68°F) compared to alcohol (~25) at 20°C (68°F) even though both are considered polar solvents. Some inorganic salts with their ionic charges interact with the partial charges on water to have good solubility. The same salts may not be very soluble in alcohol because alcohol does not have the ionic character.

Hydrogen bonding is an important component of solubility because of the formation of intermolecular bonds that hold a substance in solution. Any compound with functional groups such as OH, NH, and SH can hydrogen bond and be attracted to water, and can therefore increase a drug molecule's water solubility.

Particle Size

Above a certain size, solubility is not influenced by the particle size of the solid solute. Dissolving a big particle or a small particle may determine how fast the substance dissolves (discussed later in this chapter), but the difference does not influence the concentration at saturation. However, the reduction in the size of particles beyond normal particle sizes to micronization can influence solubility. A micronized drug may have an increase in solubility over a drug that is not micronized. Micronization increases solubility if the process breaks down the crystal lattice of the solid. Crystal lattice energy holds the solute together and increases the energy required to separate the ions or molecules for solution formation. Breaking the bonds in the crystal lattice by micronization will reduce the energy required to separate the solute from itself and, consequently, increase solubility.

General Solubility Rules

The following are some general solubility rules that can be useful in many cases to help predict water solubility of organic drug molecules:

1. Like dissolves like. The greater the similarity between the solute and the solvent (similar physical-chemical properties), the greater the solubility.
2. Solubility in water is increased by increasing the capacity of the solute for H bonding with polar groups (e.g., OH, NH_2, SO_3H, COOH).
3. Solubility in water is decreased with an increase in the number of carbon atoms in the solute (i.e., an increase in molecular weight without increasing polarity). For example, polymers with a high molecular weight are insoluble.
4. For many organic molecules, a high melting point means low water solubility.
5. cis (z) Isomer is more soluble than trans (e) isomer; cis has a lower melting point.
6. Increasing unsaturation increases solubility in polar solvents.
7. Anhydrous solutes are more soluble than are those that are crystalline.

Solubility Product

The "real" solution solubility of poorly soluble strong electrolytes in water is calculated by using the *solubility product* constant (K_{sp}), which is a constant that is equal to the product of the equilibrium concentrations of dissolved ions of a salt, with each ion raised to its stoichiometric coefficient. This formula works best when there is no complex formation of the ions and is suitable for inorganic salts and hydroxides. Mole concentration is used for solubility, with saturated solution concentration in moles per liter as S.

One can calculate the solubility (S) of calcium carbonate in water by using the solubility product constant K_{sp} from Table 5-6, $K_{sp} = 9 \times 10^{-9}$ (25°C), where the dissociation equation at saturation is:

$$CaCO_3 \text{(solid)} \rightarrow Ca^{+2}_{aq} + CO_{3\ aq}$$

and the stoichiometry gives $[Ca^{+2}] = [CO_3^{-2}]$:

$$K_{sp} = [Ca^{2+}] = [CO_3^{2-}]$$
$$\text{Set } [Cs^{2+}] = S$$
$$K_{sp} = S^2$$
$$S = \sqrt{K_{sp}}$$
$$S = 9.5 \times 10^{-5} \text{ moles/L}$$

The solubility of calcium carbonate in water at 25°C is 9.5×10^{-5} M, or 0.95 mg/100 mL, which is described as practically insoluble. For salts and hydroxides with unequal stoichiometry, this calculation is more involved.

For magnesium hydroxide, $Mg(OH)_2$, use the K_{sp} from Table 5-6:

$$K_{sp} = 1.2 \times 10^{-11}$$

The equation for saturation is:

$$Mg(OH)_2 \text{ solid} \rightarrow Mg^{2+}(_{aq}) + 2OH^-$$

Now the stoichiometry is:

$$[Mg^{2+}] = 2[OH^-]$$

Set $S = [Mg^{2+}]$. Then $2S = [OH^-]$, indicating that the concentration of hydroxide is twice the concentration of magnesium:

$$K_{sp} = [Mg^{2+}][OH^-]^2$$

where concentration is raised to its stoichiometric coefficient.

Substitution gives:

$$K_{sp} = (S)(2S)^2$$
$$K_{sp} = 4S^3$$
$$4S^3 = 1.2 \times 10^{-11}$$
$$S = 1.44 \times 10^{-4} \text{ moles/L}$$

TABLE 5-6 Solubility Products of Some Common Inorganic Compounds

Compound	Formula	Solubility Product (at Temperature Given)
Aluminum hydroxide	$Al(OH)_3$	4×10^{-15} (25°C)
Barium carbonate	$BaCO_3$	8×10^{-9} (25°C)
Barium oxalate	$BaC_2O_4 - 3\frac{1}{2}H_2O$	1.6×10^{-7} (18°C)
Calcium carbonate	$CaCO_3$	9×10^{-9} (25°C)
Calcium fluoride	CaF_2	4×10^{-11} (26°C)
Calcium oxalate	$CaC_2O_4 - H_2O$	2.6×10^{-9} (25°C)
Calcium phosphate	$Ca_3(PO_4)_2$	1.0×10^{-26} (25°C)
Calcium sulfate	$CaSO_4$	6.1×10^{-5} (10°C)
Cupric sulfide	CuS	9×10^{-45} (18°C)
Ferric hydroxide	$Fe(OH)_3$	1×10^{-36} (18°C)
Ferrous hydroxide	$Fe(OH)_2$	1.6×10^{-14} (18°C)
Ferrous sulfide	FeS	3.7×10^{-19} (18°C)
Lead iodate	$Pb(IO_3)_2$	2.5×10^{-12} (25°C)
Lead sulfate	$PbSO_4$	1×10^{-8} (18°C)
Lithium carbonate	Li_2CO_3	1.7×10^{-3} (25°C)
Magnesium carbonate	$MgCO_3$	2.6×10^{-5} (10°C)
Magnesium hydroxide	$Mg(OH)_2$	1.2×10^{-11} (18°C)
Mercuric sulfide	HgS	1×10^{-50} (25°C)
Mercurous chloride	Hg_2Cl_2	2×10^{-18} (25°C)
Silver bromide	$AgBr$	7.7×10^{-28} (25°C)
Silver chloride	$AgCl$	1.56×10^{-10} (25°C)
Silver hydroxide	$AgOH$	1.52×10^{-8} (20°C)
Silver sulfide	Ag_2S	1.6×10^{-49} (18°C)
Zinc hydroxide	$Zn(OH)_2$	6.3×10^{-18} (25°C)
Zinc oxalate	$ZnC_2O_4 \cdot 2H_2O$	1.4×10^{-9} (18°C)
Zinc sulfide	ZnS	1.2×10^{-23} (18°C)

or $\frac{84 \text{ mg}}{100 \text{ mL}}$ (very slightly soluble)

Solubility of Weak Acids and Weak Bases in Water as Influenced by pH

Most drugs are weak acids or weak bases and the more polar ionized form of the drug is the more water-soluble form. Therefore, it is important to have a good understanding of how the drug's pKa(s) can influence its water solubility, depending on the pH of its medium.

Weak Acids

These are organic weak acids (HA) that dissociate partially in water to give hydrogen ions (H^+) and weak acid anions (A^-) in equilibrium, as represented here:

$$HA \rightleftharpoons H^+ + A^-$$

The equlibrium constant K_a is given by:

$$K_a = \frac{[H^+][A^-]}{[HA]}$$

The anion concentration is given by:

$$[A^-] = \frac{K_a[HA]}{[H^+]}$$

If the weak acid is solid and is added to water, it dissolves and continues to dissolve until a saturated solution is formed. The concentration of the saturated solution is the solubility of the weak acid, HA. Because the weak acids dissociate only slightly, the concentration of the acid in water can be assumed to be all HA.

We have $HA_{solid} \rightarrow HA_{sat.soln}$ at saturation. For the solubility of $[HA]_{sat.soln}$, it can be represented as S_0, where S_0 is a symbol for the solubility in moles per liter of the organic acid molecule only, which is referred to also as associated weak acid (S_0 is also noted as K_s). Substitution of S_0 for $[HA]_{sat.soln}$ in the anion concentration expression gives:

$$[A^-] = \frac{K_a S_0}{[H^+]} = \frac{S_0 K_a}{[H^+]}$$

Both the associated acid [HA] and anion [A$^-$] are in solution. The concentration of the anion in pure water can be very low. If the solution is buffered to a higher pH, the concentration of A$^-$ increases. The total saturated concentration (ss) of both the associated weak acid and the anion is given by $S = S_0 + [A^-]_{aq}$, where S is molar *total solubility*. Substitution and rearrangement of the preceding equation calculation of the total solubility is given by:

$$S = S_0 + \frac{S_0 K_a}{[H^+]} \text{ or } S = S_0\left(1 + \frac{K_a}{[H^+]}\right)$$

S_0 values are constants and do not change with pH; K_a is assumed constant. S and [H$^+$] can change. The change in solubility with increasing pH can be seen in the equation. If [H$^+$] is increased, S decreases, and if [H$^+$] is decreased (higher pH), S increases. The calculated molar solubility of a weak acid such as pentobartital (pK_a 8.1) in aqueous buffered solutions is given in Figure 5-2.

The solubility of the associated barbiturate is fairly constant up to pH 7 or so. This is the S_0 value. Above pH = pK_a, the total solubility starts to increase with increasing pH.

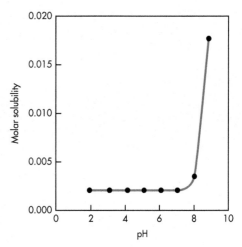

FIGURE 5-2 Calculated molar solubility of pentobarbital as influenced by pH.

Rearrangement of the previous equation and conversion to base 10 logarithms gives:

$$pH = pK_a + \log\frac{S - S_0}{S_0}$$

where now pH = pH$_p$, the pH below which the associated free acid will precipitate out of solution because it has low water solubility. ($S - S_0$) is the anion (salt) molar concentration. S_0 is the solubility of the associated (unionized) weak acid alone.

Weak Bases

These are organic weak bases (B) that react with water to take on a proton and become the associated weak base (BH$^+$) and give hydroxide ion (OH$^-$) in equilibrium, as expressed by:

$$B \Leftrightarrow BH^+ + OH^-$$

The equilibrium constant K_b is given by:

$$K_b = \frac{[BH^+][OH^-]}{[B]}$$

The cation concentration is given by:

$$[BH^+] = \frac{K_b[B]}{OH^-}$$

If the weak base is a solid and is added to water until saturated, the solubility can be given by:

$$B_{solid} \rightarrow B_{sat\ soln}$$

where $[B]_{sat\ soln}$ is given the symbol S_0.

Substitution of S_0 for $[B]_{sat\ soln}$ in the cation equilibria gives:

$$[BH^+] = \frac{K_b S_0}{[OH^-]}$$

The total solubility (S) is given by adding them together:

$$S = S_0 + [BH^+]$$

Substitution, rearrangement, and conversion to K_a of the previous equation allows calculation of the total solubility by:

$$S = S_0 \left(1 + \frac{[H^+]}{K_a}\right)$$

Note the inversion of the terms $[H^+]$ and K_a. For the weak acid, it is K_a over $[H^+]$; for the weak base, it is $[H^+]$ over K_a.

The change in solubility with decreasing pH for a weak base can be shown in the total solubility equation. If the pH is raised, $[H^+]$ decreases and S decreases; if pH is lowered, $[H^+]$ increases and S increases. The calculated molar solubility of the weak base diazepam ($pK_a = 3.4$) in aqueous buffered solutions is shown in Figure 5-3.

Diazepam ($S_0 = 1.05 \times 10^{-2}$ M) is essentially an uncharged basic molecule over the range of physiologic pH. Only at pH below pK_a does the cation increase and therefore increase the amount of drug that can be dissolved. Many other weak bases have a pK_a greater than 5 and are both associated and dissociated in body fluids with solubilities that vary with pH.

Rearrangement of the previous equation and conversion to base 10 logarithms gives:

$$pH = pK_a + \log \frac{S_0}{S - S_0}$$

where now pH = pH_p, the pH above which the molecular base will come out of solution. $(S - S_0)$ is the cation (salt) form in molar concentration, and S_0 is the molar concentration of the (unionized) molecular weak base alone.

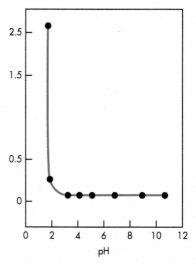

FIGURE 5-3 Calculated molar solubility of diazepam as influenced by pH.

EXAMPLE 1 ▶▶▶

Calculate the pH of precipitation for 1% lidocaine HCl. The solubility of lidocaine (S_0) is 1 g/10,000 mL.

Solution

1% = 1 g/100 mL

1 g/10,000 mL

MW = 234.33 for lidocaine (an organic weak base)

MW = 270.79 for lidocaine HCL without water of hydration

$pK_a = 7.87$

Conversion of 1 g/10,000 mL to molar gives S_0:

$$\frac{1g}{10,000\ mL} = \frac{0.1g}{1000\ mL} \times \frac{mole}{234.33\ g} = 4.3 \times 10^{-3}\ M$$

Conversion of 1 g/100 mL to molar gives S:

$$\frac{1g}{100\ mL} = \frac{10g}{1000\ mL} \times \frac{mole}{270.79\ g} = 3.7 \times 10^{-2}\ M$$

Use

$$pH_p = pK_a + \log \frac{S_0}{S - S_0}$$

Substitution gives

$$pH_p = 7.87 + \log[4.3 \times 10^{-4} / (3.7 \times 10^{-2} - 4.3 \times 10^{-4})]$$

$$pH_p = 5.9$$

Therefore, above pH 5.9 saturated solutions of lidocaine base would become cloudy. Below pH 5.9 solutions of lidocaine would be clear. A solution of 1% lidocaine HCl USP has a calculated pH of 4.7. The 1% solution would remain clear from pH 4.7 to pH 5.9. A 2% solution would remain clear to pH 5.7. The calculated pH values show that changing the pH of aqueous solutions of weak electrolytes may cause precipitation from solution as a result of the poor water solubility of high-molecular-weight organic weak acids and weak bases.

Solubility of Liquids in Liquids

Although some liquids are miscible in all proportions, there are liquid systems that are miscible only within certain limits. When that limit is exceeded, two liquid layers are formed. For liquid-liquid solutions, either component may be regarded as the solvent or the solute, though usually it is the liquid in a greater quantity that is the solvent. Temperature influences the solubility of a liquid in another liquid, and the limit of miscibility changes. Figure 5-4 shows the variation in miscibility with temperature for phenol (MP ~40°C).

The preparation of a 10% phenol in water solution is not possible at 25°C (77°F), but when it is warmed slightly to near body temperature (37°C, or 98.6°F), it becomes a solution. At room temperature (25°C) the tie line shows the range of immiscibility for phenol at 25°C.

Influence of Solvents on Solubility

Strong Electrolytes

Strong acids and bases and their salts are soluble in water. The polar nature of water attracts the ions and forms a solution. Using the inorganic salts potassium iodide and magnesium sulfate as an example of strong electrolytes, Table 5-7 shows the variation in solubility for water, acetone, and alcohol.

Acetone has a low dielectric constant (~19) and no hydrogen bonding, giving the solvent little ability to bond to the ions in solution. Alcohol has a low

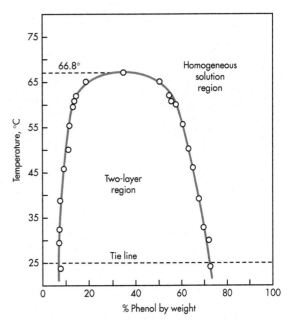

FIGURE 5-4 Variation in miscibility with temperature for phenol.

dielectric constant (~24) and hydrogen bonding but does not solubilize those inorganic ions with a strong crystal lattice in the solid state. The solubility of sodium chloride in alcohol-water mixtures shows the rapid drop-off of dielectric constant and the decrease in solubility with increasing alcohol percentage (Fig. 5-5).

Weak Electrolytes

Weak acids and bases with high molecular weight (MW) are often not soluble in water. Cosolvents such as alcohol, propylene glycol, and polyethylene glycol or mixed solvent systems are required for solubility. Using phenobarbital as an example, Table 5-8 shows the solubility of the weak acid and its sodium salt.

TABLE 5-7 Variation in Solubility for Three Solvents

Solvent	g of KI/100 mL of Solvent	g of MgSO$_4$/100 mL of Solvent
Water	148	36
Acetone	2.9	0.2
Alcohol	1.88	0.3

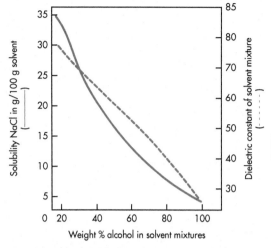

FIGURE 5-5 Solubility and dielectric constant variation with percent alcohol for sodium chloride.

Phenobarbital sodium is soluble in alcohol, a polar hydrogen-bonded solvent; however, there is practically no solubility in the other organic solvents.

Nonelectrolytes

High-molecular-weight organic drugs that do not dissociate or associate in water are generally soluble in organic solvents and have little or no solubility in water. Among the exceptions to this generalization would be sugars such as dextrose. Examples of common nonelectrolyte pharmaceuticals and their solubilities are given in Table 5-9. The list is restricted because many drugs have either carboxyl (acidic) or amino (basic) functional groups that react with water to form salts.

Solvent Effects on the Solubility of Weak Electrolytes in Buffer Solutions

The cosolvent alcohol affects the solubility of a weak electrolyte in a buffer solution in two ways:

TABLE 5-8 Phenobarbital Solubility

Solvent	Free Acid	Na Salt
Water	1 g/1000 mL	1 g/1 mL
Alcohol	1 g/8 mL	1/10 mL
Chloroform	1 g/40 mL	Practically insoluble
Ether	1 g/13 mL	Practically insoluble

TABLE 5-9 Nonelectrolyte Solubility Comparisons (alphabetical order)

Name (MW)	Solubility in Water	Solubility in Alcohol
Camphor (152)	1 g/800 mL	1 g/mL
Cetyl alcohol (242)	Insoluble	Soluble
Mannitol (182)	1 g/6 mL	1 g/83 mL
Methocarbamol (241)	1 g/40 mL	Soluble
Digoxin (781)	Practically insoluble	Slightly soluble

1. The addition of alcohol to a buffered aqueous solution of a weak electrolyte increases the solubility of the uncharged species by adjusting the polarity of the solvent to a more favorable value. The S_0 increases.
2. Being less polar than water, alcohol decreases the dissociation of a weak electrolyte, and the concentration of the drug as the ion in water goes down as the dissociation constant is decreased. Since alcohol decreases the dissociation of a weak acid, the K_a for the drug in the solvent system under consideration must be used for calculation of pH_p. For weak acids, increasing alcohol increases pK_a, as is shown for the weak acid phenobarbital in Figure 5-6.

Alcohol increases the S_0 for phenobarbital:

$$S_0 \text{ in water} = 0.004\ M\ (\text{at } 25°C)$$
$$S_0 \text{ in alcohol}(30\%) = 0.028\ M$$

To calculate the pH_p for 1% sodium phenobarbital in water:

$$S = 1\ g/100\ mL = 0.039\ M$$
$$S_0 = 0.004\ M$$
$$pK_a = 7.41$$

In water:

$$pH_p = pK_a + \log\frac{S - S_0}{S_0} = 7.41 + \log\frac{0.0039 - 0.004}{0.004}$$
$$= 8.35$$

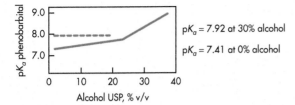

FIGURE 5-6 The influence of alcohol on the ionization of phenobarbital.

The pH of 1% sodium phenobarbital in water is calculated to be 10. The solution may be pH adjusted down to pH = 8.35 before the phenobarbital will precipitate from solution. The pH is rather alkaline, and so lowering the pH_p will make the solution more palatable. Adding a cosolvent such as alcohol will allow a lower pH_p.

To calculate the pH_p for 1% sodium phenobarbital in 30% alcohol:

$$S = 0.039\ M$$
$$S_0 = 0.028\ M$$
$$pK_a\, 30\% = 7.92$$

$$pH_p = pK_a\, 30\% + \log \frac{S - S_0\, 30\%}{S_0\, 30\%}$$

$$= 7.92 + \log \frac{0.039 - 0.028}{0.028} = 7.92 - 0.406$$

$$= 7.51$$

One can formulate 1% sodium phenobarbital at pH 7.51 by using 30% alcohol. This is more desirable than pH 8.35. One can lower the pH by 0.84 unit when using 30% alcohol and still keep the drug in solution.

For weak base solubility in a mixed-solvent buffered solution, the effect of alcohol on K_a is not as dramatic. Increasing alcohol has only a slight effect on K_a because of the nature of the equilibrium:

$$BH^+ \rightleftharpoons B + H^+$$

which has charged ions on both sides of the equilibrium. No generalities can be made about weak bases, as some show an increase in K_a with increasing alcohol and some show a decrease.

DISSOLUTION

Except for solution dosage forms, all other forms need to dissolve in the medium around the site of administration to be absorbed. Dissolution studies therefore are necessary to determine the properties of a drug in the biological environment and correlate them with absorption and bioavailability. *Dissolution* is the process by which a solid solute enters into a solution in the presence of a solvent. An early description of dissolution was provided in a paper by Noyes and Whitney in 1897 in which they described the dissolution characteristics of benzoic acid and lead chloride in water, using a rotating-cylinder method.[1]

In the middle of the twentieth century, Edwards proposed an interesting relationship between the rate of dissolution of the solid dosage form and absorption from the gastrointestinal tract.[2] This proposal was the first to suggest that an *in vitro* test could be used effectively as a predictor of the *in vivo* behavior of a drug formulation. Since that time, dissolution has become an important quality control test for different formulations. The U.S. Food and Drug Administration mandates dissolution testing for the quality control of pharmaceutical products. The United States Pharmacopeia (USP) has established specific guidelines for dissolution studies of different dosage forms. Dissolution testing is also an important tool to evaluate batch-to-batch uniformity of formulations. As more and more patient-specific formulations are compounded in community and clinical pharmacy settings, dissolution studies are performed in some compounding facilities to ensure the quality of the final product.

DISSOLUTION OF PARTICLES

Diffusion Layer Model

Upon their introduction into a beaker or administration into the gastrointestinal tract, most tablets undergo a process of disintegration to produce granules. These granules deaggregate to form a fine powder. Due to an increase in surface area, a large portion of dissolved drug will originate from the fine powder formed due to deaggregation, and once dissolved, the drug can be absorbed across the gastrointestinal tract

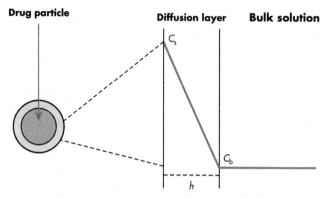

FIGURE 5-7 The dissolution of drug particles according to the diffusion layer model.

into systemic circulation. Dissolution of drug from the powder particle into a large excess bulk medium can be described by the diffusion layer model (Fig. 5-7).

In the diffusion layer model, it is assumed that a *stagnant liquid film* forms on the surface of a particle with a thickness h (in cm). The liquid film constitutes a stationary diffusion layer in which the concentration of the drug is described by C_s. Beyond the thickness of h is the bulk medium, where the concentration of the drug is C_b. The driving force for the dissolution, based on diffusion theory, is the concentration gradient between C_s and C_b ($C_s - C_b$).

Noyes-Whitney Relationship

In 1897, Noyes and Whitney described the quantitative analysis of the amount of drug dissolved from solid particles as a function of time; this has been modified by several other investigators. The *Noyes-Whitney equation* for the rate of dissolution (dM/dt) is described as:[3]

$$\frac{dM}{dt} = (DS/h)(C_s - C_b)$$

where M is the amount of drug dissolved (units = milligrams or mmoles) in time t (units = seconds), D is the diffusion coefficient of the drug (units = cm²/s), S is the surface area of the particles (units = cm²), h is the thickness of the liquid film, and C_s and C_b are the concentrations of the drug at the surface of the particle and the bulk medium, respectively. C_s is considered to be the saturated solubility of the drug at the temperature of the experiment.

EXAMPLE 2 ▶▶▶

Calculate the rate of dissolution (dM/dt) of relatively hydrophobic drug particles with a surface area of 2.5×10^3 cm² and a saturated solubility of 0.35 mg/mL at 25°C (77°F) in water. The diffusion coefficient is 1.75×10^{-7} cm²/s, and the thickness of the diffusion layer is 1.25 μm. The concentration of drug in the bulk solution is 2.1×10^{-4} mg/mL.

Solution

First, convert the units of diffusion layer thickness from micrometers to centimeters. Based on the conversion factor of 1 μm = 10^{-4} cm, the diffusion layer thickness will be 1.25×10^{-4} cm.

Second, the rate of dissolution:

$$\frac{dM}{dt} = \{[(1.75 \times 10^{-7})(2.5 \times 10^3)]/(1.25 \times 10^{-4})\}$$
$$(0.35 - 2.1 \times 10^{-4})$$
$$= 1.22 \text{ mg/s}$$

If the surface area was increased to 4.30×10^4 cm² by micronization of the particles, the new rate of dissolution would be:

$$\frac{dM}{dt} = \{[(1.75 \times 10^{-7})(4.3 \times 10^4)]/(1.25 \times 10^{-4})\}$$
$$(0.35 - 2.1 \times 10^{-4})$$
$$\frac{dM}{dt} = 21.06 \text{ mg/s}$$

For the change in concentration of a dissolved drug as a function of time (dC/dt), a volume term [V (units = cm^3)] is added to the equation:

$$\frac{dC}{dt} = (DS/Vh)(C_s - C_b)$$

In the majority of dissolution cases, the concentration of the drug in the bulk solution is always significantly lower than the saturated solubility ($C_s \gg C_b$). The system therefore represents *sink conditions*, as described in Chapter 6. When the sink conditions assumptions hold, C_b is omitted from the Noyes-Whitney equation and the dissolution rate relationship becomes:

$$\frac{dM}{dt} = (DSC_s/h)$$

Integrating this equation from the limits of zero to infinity, one can obtain the amount of drug dissolved as a function of time:

$$M = (DSC_s/h)t$$

and the concentration of drug dissolved as a function of time:

$$C = (DSC_s/Vh)t$$

Note that the two equations are similar to the *zero-order* transport equations in Chapter 6. As shown in Figure 5-8, the profile of the amount or concentration dissolved under sink conditions will be linear, whereas the profile under nonsink conditions will reach a plateau.

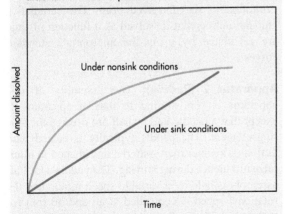

FIGURE 5-8 Relationship of the amount dissolved as a function of time under sink and nonsink conditions.

Hixson-Crowell Cube-Root Relationship

A major assumption in the Noyes-Whitney relationship is that the surface area (S) term in the equation remains constant throughout the dissolution process. This is true for some formulations, such as transdermal patches. However, the size of drug particles from tablets, capsules, and suspensions will decrease as the drug dissolves. This decrease in the size of particles changes the effective area. To take into account the changing surface area, Hixson and Crowell modified the Noyes-Whitney equation to represent the rate of appearance of solute by weight in solution by multiplying both sides by the volume term. Without going into the details of the derivation, the Hixson-Crowell cube-root equation is:

$$M_0^{1/3} - M_t^{1/3} = Kt$$

where M_0 is the amount of drug particles at time zero, M_t is the amount of particles remaining at time t, and K is known as the *cube-root dissolution rate constant*.

USP DISSOLUTION TESTING

The United States Pharmacopeia/National Formulary XXIV[3] monographs provide a detailed description of the *in vitro* dissolution testing procedure for pharmaceutical products. For the tablet dosage form, for instance, USP guidelines are different for uncoated tablets, coated tablets, and enteric-coated tablets. These guidelines have been developed especially for pharmaceutical manufacturers and compounding pharmacists to allow analysis of drug release and quality control in the finished products. Strict adherence to these guidelines is essential for batch-to-batch reproducibility in the analysis. Companies such as VanKel (www.vankel.com), Distek (www.distekinc.com), and Hanson Research (www.hansonresearch.com) are leading U.S. manufacturers of dissolution instruments that meet the USP guidelines. Figure 5-9 shows USP dissolution apparatus 2100B, which is manufactured by Distek, Inc.

Equipment and Procedure

As shown in Figure 5-10, the USP testing procedure varies with the apparatus used for dissolution.

FIGURE 5-9 The Distek 2100B United States Pharmacopeia automated dissolution system. (Image used permission from Distek, Inc.)

Apparatus 1 (Rotating Basket): The most common type of USP dissolution testing equipment, the rotating basket system, consists of a cylindrical stainless-steel basket and shaft assembly. The basket and shaft are made to resist corrosion even in a harsh acidic medium. The standard basket is made with 40-mesh (381-μm) openings. However, many other sizes of openings and basket dimensions are available. Apparatus 1 is used for dissolution studies of capsules, suppositories, and tablets that have low density and tend to float on the dissolution medium or dissolve very slowly. The sample is placed in a basket, and the assembly is inserted into a 1-L glass vessel. The entire system is maintained at a constant temperature of 37°C (98.6°F). A speed-regulating motor controls the rotational speed of the shaft and basket assembly and is selected and maintained at a rate specified in the individual monograph. The most common rotational speed for this system is 100 rpm. The composition and volume of the dissolution medium also are specified in the individual USP monographs. Typically, 900 mL of 0.1 M hydrochloric acid is used to simulate the pH of the gastric fluid. Periodically, small volumes of the dissolution medium are withdrawn and analyzed to find the concentration of the dissolved drug. The cumulative amount and percent dissolved as a function of time are calculated by using the appropriate standard curves.

Apparatus 2 (Paddle): The assembly of this apparatus is very similar to that of apparatus 1, except that a paddle and a shaft are used as the stirring element. The paddle typically is coated with Teflon or another inert material and shaped to minimize turbulence during stirring. The paddle and shaft assembly is held by a variable-speed motor, and the rotational speed is controlled at around 50 rpm for tablets and 25 rpm for suspensions. As in apparatus 1, the assembly is inserted into a large glass vessel

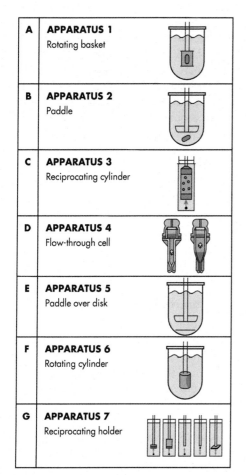

FIGURE 5-10 United States Pharmacopeia–approved dissolution systems.

that holds the dissolution medium and maintains it at 37°C (98.6°F). For effective dissolution studies with apparatus 2, the dosage form must sink to the bottom of the glass vessel and remain at least 2 cm away from the bottom of the paddle. For products, such as capsules, that float on the medium, a *sinker* made of platinum wire is wrapped around the formulation to keep it from floating. It is important to make sure that the sinker does not affect the dissolution properties of the dosage form.

Apparatus 3 (Reciprocating Cylinder): This apparatus consists of a set of flat-bottomed cylindrical glass vessels equipped with a reciprocating cylinder that houses the sample to be tested. Typically, this system is used for drug-containing microparticulate formulations that are designed for sustained release. It also can be used to test the dissolution properties of formulations as a function of the pH or ionic strength of the medium. Six units are tested, and the dissolution medium is maintained at 37°C (98.6°F). The outer tube can hold 100 to 300 mL of the dissolution medium.

Apparatus 4 (Flow-Through Cell): This system consists of a reservoir of dissolution medium that is pumped into a flow cell containing the sample. The flow rate of the dissolution medium can be varied from 4 to 16 mL/min. This apparatus is designed for testing the dissolution properties of sustained- and controlled-release dosage forms where the solubility of the drug in the medium is very low. A leading manufacturer of flow-through cell dissolution apparatus is Sotax, Inc. (www.sotax.com).

Apparatus 5 (Paddle over Disk): This apparatus is especially suitable for the measurement of drug release from topical and transdermal dosage forms such as patches. It consists of a sample holder or disk assembly that sits at the bottom of the glass vessel. For a transdermal patch, the sample is mounted on the disk and the assembly is inserted in the dissolution medium, which is maintained at 32°C (89.6°F). A paddle is installed similarly to the method used for apparatus 2, providing the stirring mechanism, and drug solutions are withdrawn periodically for analysis of the dissolved material. Six units can be tested at once, and the criteria for acceptance are described in the individual USP monographs.

Apparatus 6 (Cylinder Method): This system is modified from apparatus 1 and is designed for testing the dissolution properties of transdermal patches. Instead of a basket, a stainless-steel cylinder is used as a sample holder. The sample is mounted on an inert cellulose membrane, and the entire system is allowed to adhere to the metal cylinder. The cylinder is fixed to a stainless-steel shaft as in apparatus 1, and the assembly is inserted into the dissolution medium, which is maintained at 32°C (89.6°F). The rotational speed of the shaft is described in the

Apparatus 7 (Reciprocating Disk Method): Also used for testing the release of drug from transdermal systems, apparatus 7 consists of a motor drive assembly that reciprocates the system vertically. The samples are placed in a disk holder, using cellulose membrane supports. Drug release studies are carried out at 32°C (89.6°F) when the system is reciprocating at a speed of about 30 cycles per minute. The volume of the dissolution medium can vary from 20 to 275 mL.

Acceptance Criteria

Unless specified differently in the individual drug monograph, the USP guidelines for acceptance or rejection of a drug formulation batch according to dissolution requirements are given in Table 5-10. The term Q is defined as the amount of active ingredient dissolved in the units that are tested and is expressed as a percentage of the labeled amount. The value of Q is specified in the individual drug monograph. The USP advises three stages of testing (S_1, S_2, and S_3) unless the results conform to either S_1 or S_2.

For many products, the passing of Q is set at 75% of the labeled amount in 45 min. Some products dissolve faster, and Q is set at 85% in 30 min; others dissolve more slowly, and Q is set at 75% in 60 min. For a new drug product, the burden of establishing the dissolution guidelines lies with the company that plans to market the product. The guidelines are based on the physicochemical properties of the active ingredient and the formulation factors.

The USP provides dissolution calibrators that can be purchased from its office in Rockville, MD (visit the USP at www.usp.org for more details). Although there are many variables in the dissolution that must be reproduced to obtain meaningful results when one is comparing different batches of a formulation, occasional calibration of the instruments can provide the necessary performance validation. USP calibrators are available as disintegrating (prednisone) tablets and nondisintegrating (salicylic acid) tablets. These calibrators can be used for the rotating basket (apparatus 1) method as well as the paddle (apparatus 2) method.

SIGNIFICANCE OF DISSOLUTION STUDIES

Quality Control Test

According to the USP guidelines, dissolution analysis is an important quality control test for pharmaceutical products. Dissolution studies with solid dosage forms, for instance, are performed with six articles to provide the total time for release of the drug. The USP drug monographs provide detailed guidelines for dissolution studies, including volume, the dissolution medium, pH, ionic strength, and the rotational speed of the basket (apparatus 1) or paddle (apparatus 2). Periodically, samples of the dissolution medium are removed, and the dissolved drug is analyzed according to the assay described in the USP. The cumulative amount and percent of labeled drug are calculated and plotted as a function of time, as shown in Figure 5-11. In the majority of oral dosage forms, the USP requires determination of the time for 75% of the labeled amount of drug to dissolve (i.e., t_{75}).

TABLE 5-10 USP Acceptance Criteria for Dissolution Results

Stage	Number Tested	Acceptance Criteria
S_1	6	Each unit is not less than $Q + 5\%$
S_2	6	Average of 12 units ($S_1 + S_2$) is equal to or greater than Q, and no unit is less than $Q - 15\%$
S_3	12	Average of 24 units ($S_1 + S_2 + S_3$) is equal to or greater than Q, and no more than 2 units are less than $Q - 15\%$

Source: Adapted from United States Pharmacopeia XXIV/National Formulary XIX. United States Pharmacopeial Convention, Rockville, MD, 2000.

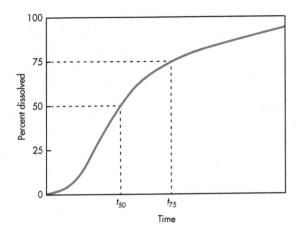

FIGURE 5-11 Relationship of percent drug dissolved as a function of time in a dissolution system for calculation of the time for 50% (t_{50}) and 75% (t_{75}) dissolution.

Dissolution, Absorption, and Bioavailability

When a dosage form reaches its intended site of absorption, the formulation must release the active ingredient to be absorbed into the systemic circulation. For oral dosage forms such as regular tablets, the process begins with disintegration into granules, then deaggregation to form a fine powder, which increases exposure of the drug particles and therefore tends to enhance their dissolution. This process was described earlier in the section entitled "Dissolution of Particles" and is depicted in Figure 5-12. After dissolution, the drug is absorbed primarily from the upper intestine into the systemic circulation through the hepatic portal vein. In these processes of dissolution and absorption, the rate at which the drug reaches the systemic circulation is determined by the slowest step in the sequence. The slowest step in a sequential process is known as the *rate-limiting step*.

For a hydrophilic drug such as many salt forms, dissolution in the aqueous medium is usually very rapid and the absorption process becomes the rate-limiting step. In some instances, the drug in solution may accumulate at the absorption site or may be eliminated prematurely from the body, because the absorption is slower than dissolution. For a hydrophobic drug, in contrast, the dissolution process is usually the rate-limiting step. The *bioavailability* (fraction of the dose administered that is available in the systemic circulation) of many hydrophobic drugs therefore is determined by the rate at which the drug dissolves in the aqueous medium. Since dissolution profiles of the drug can be manipulated through physicochemical or formulation changes, the industrial pharmaceutical scientist has the flexibility to formulate the drug for maximum bioavailability.[4]

FACTORS AFFECTING DRUG DISSOLUTION

The rate and extent of the dissolution of a drug are governed by the *physicochemical properties* of the active ingredient, the *formulation factors*, and the *physiologic factors* of the human being. It is very important to analyze the physicochemical and the formulation variables to achieve reproducibility in results from batch to batch.

Physicochemical Properties of the Drug
Ionized Versus Unionized Forms

Since the salt (ionized) form of weak electrolytes is more water-soluble than the unionized form, drug dissolution typically occurs much faster with salt forms. The ionization state of the drug depends on the pH of the environment where dissolution is expected to occur. After oral administration, for instance, the pH varies from 1.5 to 3.0 in the stomach and from about 5.0 to 7.0 in the upper intestine. Weak bases dissolve readily in the low pH environment of the stomach, as there is a higher fraction of the protonated, ionized form. For weak acids, in contrast, the fraction ionized will be higher in the small and large intestines. It is important to note that although the dissolution rate increases with ionization, absorption of drugs is more efficient when a drug is in the unionized state. With some weakly basic drugs, the dissolution may occur in the stomach, but they are absorbed more effectively from the upper intestine, where they become unionized and are exposed to a greater absorptive surface area.

Particle Size

As noted in the Noyes-Whitney relationship, the rate of dissolution of particles (dM/dt) is directly proportional to the surface area (S). The *effective surface area* of the particles in the dissolution medium is inversely proportional to the particle size. *Decreasing*

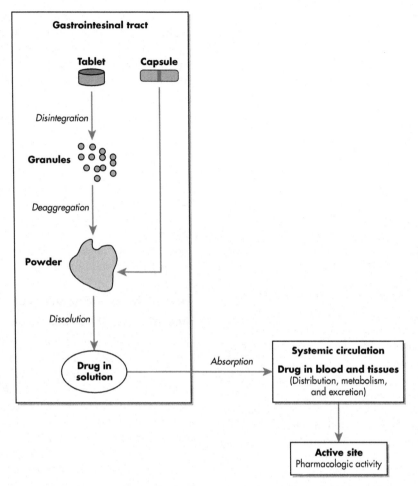

FIGURE 5-12 The relationships between dissolution, absorption, and availability of drug in the body from solid dosage forms.

the particle size therefore will *increase* the surface area. For poorly soluble drugs, decreasing the particle size can significantly improve the dissolution rate and bioavailability after oral administration. One method of decreasing particle size is *micronization*, which increases the effective surface area available for dissolution. A number of drugs are available in a micronized (<10 μm in diameter) form, including nitrofurantoin, griseofulvin, and chloramphenicol.

Crystalline State

Crystalline drugs tend to dissolve more slowly in water than do the amorphous forms because more energy is required to break the crystal habit to allow dissolution. Formulation scientists sometimes, therefore, prefer to have drugs in the amorphous form rather than the crystalline form. For example, lyophilized (freeze-dried) products are often amorphous, which enables quick dissolution upon reconstitution. Corticosteroids and many other drugs can undergo changes in the crystalline structure when in a suspension or a saturated solution. This phenomena is known as *polymorphism*. Polymorphic transition sometimes can lead to slower dissolving of a drug by changing from a relatively faster dissolving crystalline structure to a slower form. The β crystals of chloramphenicol, for instance, are more soluble in water and better absorbed than are those of the α form. Because of the relationship of crystalline form to dissolution rate, the bioavailability of drugs, like chloramphenicol, can be markedly affected by the specific crystalline form in which it exists. In some cases, it has also been observed that the anhydrous form of the drug dissolves faster than does the crystalline hydrate form. Ampicillin trihydrate, for instance, is significantly less absorbed after oral administration than is the anhydrous form, probably because of the faster dissolution of the anhydrous form.

Formulation Factors

Solid Dosage Forms

As was shown earlier, drugs in tablet and capsule formulations need to dissolve in the gastrointestinal fluid for absorption across the mucosal tissue and into the systemic circulation to occur. For tablets, the dissolution process, particularly for poorly soluble drugs, is dependent on disintegration and deaggregation to form fine particles. Tablet disintegration is affected by the addition of binding agents during wet granulation. In some instances, it has been shown that polymeric binders such as poly(N-vinylpyrrolidone) and sodium carboxymethylcellulose can increase the rate of dissolution of hydrophobic drug particles, probably through an enhanced wetting effect on the surface. The addition of some diluents, such as starch and lactose, also can increase the dissolution rate of hydrophobic drugs in tablet and capsule dosage forms. This probably is due to an increase in the effective surface area available for dissolution of the drug in the presence of diluents. Lubricating agents such as stearic acid and magnesium stearate, in contrast, decrease the dissolution rate of the drug by creating a hydrophobic layer on the particles that impedes their interfacial interaction with the aqueous medium. In a tablet formulation, the compression force is also an important variable that affects the disintegration as well as dissolution processes.[5]

For some solid dosage forms, formulation approaches are utilized to make something other than drug dissolution to be the rate-limiting step for drug dissolution. This is the case for many oral sustained- or controlled-release formulations, where drug release from the dosage form is the rate-limiting step, and for enteric-coated formulations, where a coating does not dissolve in low pH gastric fluid, but will dissolve in higher pH intestinal fluid, thereby delaying drug dissolution until the dosage form reaches the intestines. The USP test for enteric-coated tablets requires that the product not disintegrate in the gastric fluid (pH 1.2) for up to 1 h, and 75% of the drug should dissolve in 45 min in the intestinal fluid (pH 6.8).

Suspensions and Emulsions

Dissolution is an important quality control test for suspension formulations. For a suspension, it has been noted that dissolution of drugs is the rate-limiting step in the absorption process from the gastrointestinal tract. Several important factors influence the dissolution of drugs from suspensions, including settling, aggregation, and change in the crystalline structure upon aging of suspensions. An increase in the effective particle size upon settling and aggregation will decrease the dissolution rate. Aging of suspensions occurs when they have been on the shelf for a long time. In suspensions, there is an equilibrium between the solid drug and the saturated solution. As time progresses, the drug in a saturated solution can precipitate in a different crystalline form or as larger crystals, with a potential consequence of altered drug dissolution. It has been observed that drug particles in aged suspensions tend to dissolve more slowly. The viscosity of suspensions can also affect the dissolution of the drug, as solute diffusion decreases with increasing viscosity. For example, some suspensions intended for intramuscular and subcutaneous administration are designed to have higher viscosity after injection to slow drug release and dissolution.

Semisolid Dosage Forms

Semisolid dosage forms include creams, gels, and ointments. The basic components of these preparations include varying amounts of active ingredient in a hydrophobic (e.g., oleaginous) or hydrophilic (e.g., water-miscible) base. The dissolution characteristics of a drug in semisolid dosage forms are dependent on the type of base used. *In vitro* dissolution studies performed on semisolid dosage forms typically involve examination of the release rate of active ingredient from the system as a function of time, using *in vitro* skin permeation models such as the Franz cell. The data are analyzed according to the square-root-of-time relationship proposed by Higuchi:

$$Q = [2AC_s Dt]^{1/2}$$

where Q is the amount of drug released per unit surface area A (in mg/cm^2), C_s is the saturated solubility of the drug in the dissolution medium, D is the diffusion coefficient, and t is time. A plot of Q versus the square root of time ($t^{1/2}$) is constructed, and the slope of the line is used to determine the release rate.

Physiologic Factors

Gastric Emptying and Intestinal Transit Times

The time taken for nondigestable materials to pass through the stomach varies from about 15 min to more than 2 h, depending on the type of material and the presence or absence of food in the stomach. Gastric emptying time can have a marked effect on how much of the drug will dissolve before it moves farther along the GI tract. The emptying time is much shorter during the fasted state than it is during the fed state. Since most drugs are absorbed from the upper intestine (duodenum and jejunum), the transit times of solid dosage forms in these areas of the gastrointestinal tract also have a marked effect on dissolution, absorption, and bioavailability. Typical intestinal transit time varies from 1 to 4 h in the upper intestine. The overall effect of food on bioavailability is complex, as it can affect many relevant factors including tablet disintegration, drug dissolution and drug transit.

Variability of pH

Another important variable that can affect drug dissolution *in vivo* is the pH in the various regions of the gastrointestinal tract. The pH of a fasted stomach, for instance, can be as low as 1.2. During the fed state, the pH increases, depending on the amount and the type of food, to a value of 3.5 or above. In the intestinal tract, pH varies from about 5.4 in the small intestine to about 8.4 in the colon.

Weakly acidic drugs tend to dissolve less well in the acidic pH of the stomach, as less of the drug will be ionized that higher in the intestines. Weakly basic drugs tend to dissolve better in the stomach, where more of the drug will be ionized. The particular pH-dissolution relationship will be a function of the pKa of the drug.

PARTITIONING

General Partitioning Concepts

The partition law states that a solute will distribute itself between two immiscible solvents so that the ratio of its concentration in each solvent is equal to the ratio of its solubility in each one. It is therefore a constant for a given solute in the immiscible system at a given temperature.

$$K_d = \frac{C_0}{C_w}$$

where C_0 = molar concentration in organic layer

C_w = molar concentration in aqueous layer

K_d = partition coefficient or distribution constant

The partition law as given applies only to the concentration of solvents common to both phases (same chemical structure in both phases). The aqueous layer is generally a buffered aqueous solution. The organic layer is an immiscible organic solvent such as octanol or an oil layer.

For Strong Electrolytes as Solutes

Strong electrolytes are completely dissociated in aqueous solution and exist as cations and anions in the aqueous layer and are water-soluble. Partitioning into the organic layer does not occur to any great extent. The solubility of strong electrolytes in organic, nonpolar solvents is very low. The ions may form an ion pair in the organic phase, but this occurs only for large anions or cations. Without ion pairing, they do not partition into the organic layer. Therefore:

$$K_d \to 0$$

For Nonelectrolyte Solutes

The partition law for nonelectrolytes, which remain as single molecules in solution and do not form dimers or trimers, is the general partition law:

$$K_d = \frac{C_0}{C_w}$$

For Weak Electrolytes as Solutes

The partition law for weak electrolytes depends on pH. For weak electrolytes in solution with pH different from pK_a by at least 2 units on the uncharged molecule side (pH < pK_a for a weak acid; pH > pK_a for a weak base), K_d is written as follows:

$$K_d = \frac{C_0}{C_w} \text{ for a weak organic acid } K_d = \frac{(HA)_0}{(HA)_w}$$

and $K_d = \frac{C_0}{C_w}$ for a weak organic base $K_d = \frac{(B)_0}{(B)_w}$

For organic weak acid electrolytes when pH ≥ pK_a ± 2, dissociation is important. The aqueous phase is buffered at a pH at which the weak acid exists as associated and dissociated in the aqueous and as associated in the organic phase. The concentrations vary with pH just as the solubilities vary with pH (as discussed in the section on "Solubility"):

$$C_0 = (HA)_0$$
$$C_w = (HA)_w + (A^-)_w$$
$$K_d^1 = \frac{(HA)_0}{(HA)_w + (A^-)_w} = \frac{C_0}{C_w}$$

where K_d^1 is the *apparent partition coefficient* and its value depends on pH, compared to K_d, which is pH-independent, since it is a ratio of concentrations of only unionized species. Using the K_a expression, an equation can be developed that gives the hydrogen ion dependency.

For the organic weak acid:

$$HA \Leftrightarrow H^+ + A^-$$

where $K_a = \frac{[H^+][A^-]_w}{[HA]_w}$

Rearrangement, substitution for the equation for K_d^1, and factoring gives:

$$K_d^1 = \frac{[HA]_0 [H^+]}{[HA]_w (K_a + [H^+])}$$

Substitution of K_d for $\frac{(HA)_0}{(HA)_w}$ gives:

$$K_d^1 = K_d \frac{[H^+]}{K_a + [H^+]}$$

which is an equation in terms of hydrogen ion and the constants, K_d and K_a, that can be used to calculate the apparent partition coefficient for an organic weak acid in an aqueous solution buffered at pH = pK_a or higher.

For an organic weak base electrolyte buffered at pH equal to or less than pK_a, a similar expression can be developed to show the dependency of the apparent partition coefficient on the hydrogen ion concentration.

If the aqueous phase is buffered at a pH near pK_a, the weak base, which is represented as B:, exists both as associated and as dissociated in the aqueous phase and is dissociated only in the organic phase. Therefore:

$$C_0 = [B:]_0$$
$$C_w = [B:]_w + [BH^+]_w$$

and for partitioning:

$$K_d^1 = \frac{[B:]_0}{[B:]_w + [BH^+]_w} = \frac{C_0}{C_w}$$

where K_d^1 is the apparent partition coefficient and its value depends on pH.

For $BH^+ \rightarrow B: + H^+$

$$K_a = \frac{[B:]_w [H^+]}{[BH^+]_w}$$

Rearrangement, substitution for the equation for K_d^1, and factoring gives:

$$K_d^1 = \frac{[B:]_0}{[B:]_w \left(1 + \frac{[H^+]}{K_a}\right)}$$

Placing under a common denominator and simplifying gives:

$$\frac{K_a + [H^+]}{K_a}$$

and substitution of $K_d = \frac{[B:]_0}{[B:]_w}$ gives:

$$K_d^1 = K_d \frac{K_a}{K_a + [H^+]}$$

This is the equation to calculate the apparent partition coefficient for an organic weak base in an aqueous solution buffered at pH = pK_a or lower, given the [H$^+$] of the aqueous layer.

Applications of Partitioning (Distribution) Concepts

Drug Permeation and the pH-Partition Hypothesis

As described in Chapter 6, drug permeation across the lipid bilayer of biological membranes occurs most often by passive diffusion. Two of the important factors that will influence a drug's permeation through the bilayer are its solubility and its lipid/water partition coefficient (K_d). Solubility is important, because the drug must be dissolved (i.e., in its molecular form) to cross the membrane. The lipid/water partition coefficient is important, because the drug must partition through the lipid bilayer to get to the other side.

Also discussed in Chapter 6 was the pH-partition hypothesis, which explains that for ionizable drugs, the more lipid-soluble unionized (uncharged) form is the form that most readily crosses a lipid bilayer. Therefore, as the pH determines the extent of a drug's ionization, the pH at absorption or permeation sites plays a large role in determining the passive diffusion of a drug across a lipid bilayer. As described in Chapters 4 and 6, the Henderson-Hasselbalch equations can be used to calculate the percent of a drug's ionization, based on its pK_a and the pH of its environment.

For weak acids (WA):

$$pH = pK_a + \log \frac{I}{U}$$

where I = dissociated weak acid and U = associated weak acid.

And:

$$\% \text{ ionization} = \frac{100}{I + \text{antilog}(pK_a - pH)}$$

For weak bases (WB):

$$pH = pK_a + \log \frac{U}{I}$$

and:

$$\% \text{ ionization} = \frac{100}{I + \text{antilog}(pH - pK_a)}$$

Also described in Chapter 6, the pH-partition hypothesis helps to explain the influence of gastrointestinal pH and drug pK_a on the extent of drug absorption at different gastrointestinal sites, excluding other factors such as absorptive surface area. Other interesting applications of the pH-partition hypothesis to drug diffusion across biological membranes include drug excretion in urine, sweat, and breast milk.

Excretion of Drugs in Urine (pH 5 to 7): Weak acids with a pK_a in the range of pH 3.0 to 7.5, which have ionization change with a change in urine pH, are excreted at a high clearance in highly alkaline urine. Urine pH may be increased by intravenous (IV) administration of sodium bicarbonate. Barbiturate poisoning may be treated by increasing the pH of plasma and urine, which clears barbiturate ion from the body. Weak bases with pK_a in the range of 7.5 to 10.5, having their ionization increased by acidifying the urine, are excreted at higher clearance in an acidic urine. Urine pH may be decreased by IV administration of ammonium chloride or ascorbic acid.

Excretion of Drugs in Sweat (pH 5 to 7): Weak acids with higher pK_a will partition as the unionized form across sweat gland epithelial cells and have higher sweat to plasma concentration (s/p) ratios. The lower the pK_a of the weak acid ($pK_a < 7.4$), the lower the ratio, because the stronger acid will be more ionized, less able to partition, and more soluble in plasma. For example, consider two weak base sulfonamides: sulfanilamide (pK_a 10.4) and sulfathiazole (pK_a 7.1). The calculated s/p ratios are 0.7 and 0.13, respectively, at pH = 7 for sweat. Weak bases have higher s/p equilibrium values as they become more ionized at lower pH. Weak bases with a higher pK_a will partition as unionized, solubilize in sweat, and have higher s/p ratios. The lower the pK_a of the weak base ($pK_a < 7.4$), the lower the ratio. Stronger bases are more ionized in sweat.

Excretion of Drugs in Human Milk (pH 6.6): Weak acids have human milk to plasma concentration (m/p) ratios that depend on pK_a and have lower concentrations in human milk than in plasma at equilibrium than do weak bases. The lower the pK_a of the weak acid ($pK_a < 7.4$), the lower the ratio. Weak bases appear in human milk in concentrations greater than they do in plasma. The higher the pK_a, the higher the m/p ratio.

Preservation of Emulsions

Flavored aqueous portions of emulsions have to be preserved against yeast, mold, and bacteria. The total preservative added is the concentration added to the aqueous phase before the aqueous solution is equilibrated with the oil. For the simplest application, one can assume equal volumes of oil and water and concentrations with no dimerization of molecules. For organic weak acid preservatives such as benzoic acid and sorbic acid, only the associated molecule is active as a preservative. A minimum effective concentration as a preservative for benzoic acid is 25 mg/100 mL; for sorbic acid, it is 50 mg/100 mL.

The total preservative added = C, where $C = C_0 + C_w$ and:

$$C = [HA]_0 + [HA]_w + [A^-]_w$$

$$K_d = \frac{[HA]_0}{[HA]_w}$$

Rearrange and substitute to get:

$$[HA]_w = \frac{C}{K_d + 1 + \frac{K_a}{[H^+]}}$$

The concentration of preservative to use in the emulsion (C) can be calculated to give the preservative effect of $[HA]_w$ required for effective antimicrobial concentration in the water phase. For example, if sorbic acid is distributed between equal volumes of a vegetable oil and water, what must be the original concentration in the aqueous phase so that 0.50 mg/mL of associated sorbic acid remains in the aqueous phase buffered at pH 4.5? $K_d = 10.8$, $K_a = 1.7 \times 10^{-5}$.

Sorbic acid will partition between the aqueous and oil phases only as the associated molecule, HA. Use:

$$C = [HA]_w \left(K_d + 1 + \frac{K_a}{[H^+]} \right)$$

Convert pH into $[H^+]$:

$$C = 0.50 \text{ mg/mL} \left(10.8 + 1 + \frac{1.7 \times 10^{-5}}{3.16 \times 10^{-5}} \right)$$

$$= 6.17 \text{ mg/mL}$$

For this emulsion with equal volumes of oil and water, 6.17 mg/mL of sorbic acid would be the concentration added to the aqueous phase, and the emulsion would be preserved, having a concentration of 0.5 mg/mL of sorbic acid in the aqueous phase for the completely prepared emulsion.

SOLUBILITY PROBLEMS

Calculations for slightly soluble strong electrolytes using K_{sp}:

1. What is the molar solubility of silver chloride in water at 20°C if the $K_{sp} = 1.25 \times 10^{-10}$?

$$AgCl_{solid} \Leftrightarrow [Ag^+]_{aq} + [Cl^-]_{aq}$$

2. If the solubility product of silver chromate is 2×10^{-12} at 25°C, what is the solubility in moles per liter of silver chromate?

$$Ag_2CrO_{4\,solid} \Leftrightarrow 2[Ag^+]_{aq} + [CrO_4]_{aq}$$

3. Calculate the solubility of calcium fluoride in solution if the K_{sp} is 3.95×10^{-11}.

Calculations for weak electrolytes using pH_p:

1. The molar solubility of sulfathiazole in water is about 0.001 M.
   a. What is the lowest pH allowable for complete solubility of a 1% solution of sodium sulfathiazole?
   b. A 5% solution?
2. a. Calculate pH_p for a 1.0% solution of atropine sulfate monohydrate. The molar solubility of the base is 5.50×10^{-3}.
   b. Calculate pH_p for a 0.5% solution.
3. At what pH will free methicillin precipitate from a solution containing 1 g sodium methicillin monohydrate/2 mL solution if the solubility of methicillin is estimated to be 1×10^{-3} M?
4. a. Calculate the theoretical solubility S of benzoic acid ($S_0 = 0.028$ M) over the pH range 3 to 6.
   b. How does S change with pH?
5. a. What is the total solubility of promethazine HCl at pH 5.5? At pH 6.5?
   b. What is the change in S? ($S_0 = 1 \times 10^{-3}$ M)
6. a. What is the total solubility of sodium pentobarbital at pH 3?
   b. At pH 4?
   c. What is the change in S? ($S_0 = 1 \times 10^{-4}$ M)
7. Calculate the pH_p of benzoic acid if 1 g of sodium benzoate is added to purified water to make 60 mL. The molar solubility of the associated benzoic acid is 2.79×10^{-2} M.
8. What are two ways to increase the solubility of a weak acid drug in aqueous solution?
9. Calculate the pH_{ppt} of the drug in the following prescription. The molar solubility of nafcillin is estimated to be 4.8×10^{-3} M.

| Sodium nafcillin | 0.5 g |
| Purified water qs ad | 60 mL |

Calculations using mixed solvents:

1. What is the pH_p for a 2% solution of sodium phenobarbital in a hydroalcoholic solution containing 15% by volume of alcohol? The solubility of phenobarbital in 15% alcohol is 0.22%. The pK_a of phenobarbital in the mixed solvent system is 7.6.
2. A prescription calls for 0.45 g of phenobarbital in 60 mL of hydroalcoholic solution containing glycerin. How much alcohol must be added if the percent glycerin is 20%?

Solubility of Phenobarbital in 20% Glycerin, % w/v	% Alcohol
0.2	0
0.3	10
0.5	20
2.8	40

3. What is the minimum pH required for the complete solubility of phenobarbital in the following prescription?

Phenobarbital sodium (MW 254)	1%
Alcohol	10%
Purified water qs ad	120 mL

pK_a	$S_0 \times 10^3\,M$	Alcohol, % w/w
7.4	5	0
7.5	7.7	10
7.6	12.9	20
7.8	27.6	30

ANSWERS

For slightly soluble strong electrolytes using K_{sp}:

1. $K_{sp} = [Ag^+][Cl^-]$
   $S = [Ag] = [Cl^-]$
   $K_{sp} = S^2$
   $S = 1.12 \times 10^{-5}\ M$
2. $K_{sp} = [Ag]^2[CrO_4^=]$
   $S = [CrO_4^=]$
   $K_{sp} = (2S)^2(S) = 4S^3$
   $S = 7.9 \times 10^{-5}\ M$
3. $K_{sp} = [Ca^{2+}][F^-]^2$
   $S = [Ca^{2+}]$
   $K_{sp} = (S \times 2S)^2 = 4S^3$
   $S = 2 \times 10^{-4}\ M$

For weak electrolytes using pH_p:

1. **a.** $NaS \rightarrow S^- + Na^+$
   $S \Leftrightarrow SH + OH^-$
   $[SH]_{sat} = S_0 = 0.001\ M$
   $S = 1\% = \dfrac{1\ g}{100\ mL} = \dfrac{10\ g}{1000\ mL} = \dfrac{mole}{277\ g}$
   $= 0.036\ M$

   $pH_p = pK_a + \log\dfrac{S - S_0}{S_0}$
   $= 7.12 + \log\dfrac{0.036 - 0.001}{0.001}$
   $= 8.7$

   b. $S = 5\%\ \dfrac{5\ g}{100\ mL} = \dfrac{50\ g}{1000\ mL} + \dfrac{mole}{277\ g}$
   $= 0.18\ M$
   $pH_p = 7.12 + \log\dfrac{.18 - 0.001}{0.001}$
   $= 9.4$

2. **a.** $(AtH)_2 SO_4 \rightarrow 2\ AtH^+ + SO_4$
   $= AtH^+ \Leftrightarrow At: + H^+$
   $[At]_{sat} = S_0 = 5.5 \times 10^{-3}\ M$

   There are two molecules of atropines for each sulfate, and so the concentration of atropine in the salt is two times the concentration of the salt:

   $S = 1\% = \dfrac{1\ g}{100\ mL} = \dfrac{10\ g}{1000\ mL} \times \dfrac{mole}{694.8\ g}$
   $= 0.014 \times 2$
   $= 0.028\ M$

   $pH_{ppt} = pK + \log\dfrac{S_0}{S - S_0}$
   $= 9.65 + \log\dfrac{0.0055}{0.028 - 0.0055}$
   $= 9.04$

   b. $S = 0.5\% = \dfrac{0.5\ g}{100\ mL}$
   $= \dfrac{5\ g}{1000\ mL} \times \dfrac{mole}{694.8\ g}$
   $= 0.007 \times 2 = 0.0144\ M$
   $pH_{ppt} = 9.65 + \log 0.0055/0.144 - 0.0055$
   $= 9.44$

3. $NaM \rightarrow M^- + Na^+$
   $M^- \Leftrightarrow MH + OH^-$
   $[MH]_{sat} = S_0 = 1 \times 10^{-3}\ M$
   $S = \dfrac{1\ g}{2\ mL} = \dfrac{50\ g}{100\ mL} = \dfrac{500\ g}{1000\ mL} = \dfrac{mole}{420.4\ g}$
   $= 1.189\ M$

   $pH_{ppt} = pK_a + \log\dfrac{S - S_0}{S_0}$
   $= 3.01 + \log\dfrac{1.189 - 0.001}{0.001}$
   $= 6.1$

4. **a.** $S = 0.028\ M$
   $K_a^0 = 6.3 \times 10^{-5}$
   $S = S_0\left(1 + \dfrac{K_a}{[H^+]}\right)$

at pH 3 $[H^+] = 1 \times 10^{-3}$

$S = 0.028\left(1 + \dfrac{6.3 \times 10^{-5}}{1 \times 10^{-3}}\right)$

$S = 2.97 \times 10^{-2}$ M at pH 3

at pH 4 $[H^+] = 1 \times 10^{-4}$

$S = 0.028\left(1 + \dfrac{6.3 \times 10^{-5}}{1 \times 10^{-4}}\right)$

$S = 4.56 \times 10^{-2}$ M at pH 4

at pH 5 $[H^+] = 1 \times 10^{-5}$

$S = 0.028\left(1 + \dfrac{6.3 \times 10^{-5}}{1 \times 10^{-5}}\right)$

$S = 0.204$ M at pH 5

at pH 6 $[H^+] = 1 \times 10^{-6}$

$S = 1.79$ M at pH 6

4. b. $S \uparrow$ as pH $\uparrow$

5. a. At pH 5.5 $[H^+] = 3.16 \times 10^{-6}$

$S = S_0\left(1 + \dfrac{H^+}{K_a}\right)$, $K_a = 8.3 \times 10^{-10}$

$S_0 = 1 \times 10^{-3}$

$S = 3.8$ M

At pH 6.5, $[H^+] = 3.16 \times 10^{-7}$

$S = 0.38$ M

b. Solubility decreases by 10; $S \downarrow$ as pH $\uparrow$

6. a. $S = S_0\left(1 + \dfrac{K_a}{H^+}\right)$, $K_a = 7.94 \times 10^{-9}$

$S_0 = 1 \times 10^{-4}$ M

b. $S = 1 \times 10^{-4}$ M

c. No change

7. $K_a = 6.3 \times 10^{-5}$

$S_0 = 2.79 \times 10^{-2}$

$S = \dfrac{1\,g}{60} = \dfrac{x}{1000} = \dfrac{16\,\text{mole}}{144.11} = 0.1156\ M$

Sodium benzoate 1g $pH_p = 4.2$

$+ \log \dfrac{0.1156 - 0.279}{0.0279}$

Purified water qs ad 60 mL $= 4.2 + 4.9$

$pH_p = 4.7$

8. Increase pH; add alcohol

9. $S = \dfrac{0.5\,g}{60} \times \dfrac{x}{1000}$ $x = 8.33\,g \times \dfrac{\text{mole}}{436\,g}$

$= 1.91 \times 10^{-2}$ M

$pH_{ppt} = pK_a + \log \dfrac{S - S_0}{S_0}$

$= 2.65 + \log \dfrac{1.91 \times 10^{-2} - 4.8 \times 10^{-3}}{4.8 \times 10^{-3}}$

$= 3.12$

For mixed solvents:

1. $S = 2\% = \dfrac{2\,g}{100\,mL} = \dfrac{209}{1000\,mL} \times \dfrac{\text{mole}}{254\,g}$

$= 0.0787$ M

$S_0 = \dfrac{0.22\,g}{100\,mL} = \dfrac{2.2\,g}{1000\,mL} \times \dfrac{\text{mole}}{232\,g}$

$= 0.00948$ M

$pH_p = pK_a 15\% + \log \dfrac{S - S_0}{S_0}$

$= 7.6 + \log \dfrac{0.0787 - 0.00948}{0.00948}$

$= 8.46$

2. Estimate the answer to be between 20% and 40%. Graph the data.

0.45 g in 60 mL is 0.75 g in 100 mL (0.75%) from the graph; 0.75% requires ~25 mL for 60 mL; 15 mL is needed.

3. $S = 1\% = \dfrac{1\,g}{100\,mL} = \dfrac{10\,g}{1000\,mL} \times \dfrac{\text{mole}}{254\,g} = 0.04$

$S_0 \times 10^3\,M = 12.9$ at 20% alcohol
$S_0 = 12.9 \times 10^{-3}$
$pK_a = 7.6$ at 20% alcohol

$pH_p = pK_a + \log \dfrac{0.04 - 0.0129}{0.0129}$

$= 7.9$

DISSOLUTION PROBLEMS

1. According to the United States Pharmacopeia, a dissolution study is an extremely important quality control test for solid dosage forms. Why is it necessary to study the dissolution profile of solid dosage forms?
2. **a.** How does the particle size of a drug influence its dissolution properties? Explain your answer.
   b. How does the difference between the crystalline form and the amorphous form of a drug affect the dissolution properties of the drug after oral administration?
   c. Can the dissolution rate be a useful predictor of the absorption of a water-insoluble drug from the gastrointestinal tract? If so, how?
3. **a.** The Noyes-Whitney equation for the rate of dissolution of drugs is:

$$\frac{dM}{dt} = (DS/h)(C_s - C_b)$$

   Define each term in the equation and give the appropriate units.
   b. Using the Noyes-Whitney relationship, calculate the rate of dissolution (in mg/min) of theophylline particles with the following parameters: $C_s = 7.5$ μg/mL, $S = 3.75 \times 10^3$ cm^2, $D = 1.75 \times 10^{-8}$ cm^2/s, $h = 25$ μm, and $C_b = 0.13$ μg/mL.

ANSWERS

1. Dissolution studies are important to help predict *in vivo* absorption of drugs from solid dosage forms, to ensure batch-to-batch reproducibility, and they are very useful overall in quality control. Dissolution studies are, therefore, extremely important in formulation design.
2. **a.** According to the Noyes-Whitney equation (see question 3), the rate of dissolution of particles will increase with reduction in particle size (as a relatively greater amount of drug will reside at the interface between the solid and the dissolution medium).
   b. Amorphous materials will dissolve faster than crystalline materials.
   c. *In vitro* dissolution rate can be an excellent predictor of *in vivo* absorption of water-insoluble drug if dissolution is the *rate-limiting step* in the absorption process.
3. **a.** The Noyes-Whitney equation:
   $$dM/dt = (DS/h)(C_s - C_b),$$
   where dM/dt is the dissolution rate (units = amount/time), D is the diffusion coefficient (units = cm^2/s), S is the surface area (units = cm^2), h is the diffusion layer thickness (units = cm), C_s is the saturated solubility of the drug (units = amount/mL), and C_b is the bulk concentration of the drug (units = amount/mL).
   b. Using the Noyes-Whitney equation, the dissolution rate of theophylline particles will be: First, convert $h = 25$ μm into cm. Since 1 μm = 10^{-4} cm, $h = 2.5 \times 10^{-3}$ cm:
   $$dM/dt = [(1.75 \times 10^{-8})(3.75 \times 10^3)/(2.5 \times 10^{-3})](7.5 - 0.13)$$
   $$= 0.193 \text{ μg/s}$$
   $$= 0.016 \text{ mg/min}$$

PARTITIONING PROBLEMS

1. When boric acid was distributed between water and amyl alcohol at 25°C, the concentration in water was found to be 0.0510 mole/L, and in amyl alcohol it was found to be 0.0155 mole/L. What is the partition coefficient?
2. If 6 g of phenol (weak acid, pKa 9.9) is distributed between 100 mL of an aqueous buffered solution at pH 5 and 150 mL of mineral oil, how much phenol is left in the mineral oil layer? The partition coefficient for phenol in this system is 2.3.
3. What percent of benzoic acid (pKa 4.2) will remain undissociated in the aqueous phase of a 50% oil/water (o/w) emulsion if the initial concentration of benzoic acid added to the aqueous phase is 0.5%? The aqueous phase is buffered at pH 5, and the o/w partition coefficient is 5.33. Assume that benzoic acid remains as a monomer in the oil phase.
4. The drug phenylpropanolamine (weak base, pKa 9.4) is distributed between equal volumes of peanut oil and aqueous buffer at pH 7.4. The true partition coefficient is 10.3. What is the apparent partition coefficient?
5. Among the drugs diazepam (weak base, pKa 3.3), oxacillin (weak base, pKa 2.8), phenol (weak acid, pKa 9.9), and morphine (amphoteric, amine pKa 8.0, phenol pKa 9.6), which is the most highly ionized at a pH of 7.4?
6. A physician asks a pharmacist if rectal absorption of p-aminosalicylic acid (pK_a 3.3) could be anticipated from an aqueous enema buffered at rectal pH (pH 7.8). Is rectal absorption anticipated?
7. Among the drugs acetylsalicylic acid (weak acid, pKa 3.3), ephedrine (amphoteric, phenol pKa 8.7, amine pKa 9.9), chlordiazepoxide (weak base, pKa 4.8), and ethanol, which is the one expected to have the highest human milk to plasma concentration ratio?

ANSWERS

1. $K_d = 0.3$
2. 4.65 g
3. 0.04 g/100 mL
4. $K'_d = 0.1$
5. Oxacillin
6. No
7. Ephedrine

KEY POINTS

- Solubility, which can be defined as the concentration of solute in a saturated solution at equilibrium, is a fundamental drug property that can be expressed in different ways including as descriptive terms such as "freely soluble" and "practically insoluble."
- Among the many factors that can affect a drug's solubility are temperature, drug and solvent polarity, hydrogen bonding, particle size (for very small particles only), and state of ionization.
- Most drugs are weak acids or weak bases and the more polar ionized form of the drug is the more water-soluble form. Therefore, it is important to have a good understanding of how the drug's pKa(s) can influence its water solubility, depending on the pH of its medium.

- Equations are available to enable the pharmacist to readily predict the pH at which a drug is likely to precipitate based on its solubility, pKa, and its concentration.
- Dissolution is well modeled by the Noyes-Whitney relationship, which describes the relationship between the rate of dissolution and such properties as solute particle surface area (a function of particle size), the thickness of a stagnant solvent layer surrounding the drug particle, the concentration gradient between that in the stagnant layer and that in the bulk solvent, and the solvent viscosity.
- The USP specifies dissolution testing requirements for drugs, as determined by the use of various approved apparatuses, each of which agitate the dosage form over time, coupled with assays for dissolved drug.
- Among the most important physicochemical properties that affect the dissolution rate of a drug are its state of ionization, the crystalline form, and the drug particle size. Faster dissolution in aqueous media is more likely for the more polar ionized form compared to the unionized form, for the more weakly bound amorphous or metastable form compared to the more stable crystalline form, and for the smaller particles, with greater exposed surface area compared to larger particles.
- The dissolution of drug from solid dosage forms such as tablets is accelerated by their disintegration and disaggregation, which exposes the individual drug particles to the dissolution medium. Drug dissolution can be the rate-limiting step for drug absorption into the systemic circulation.
- A drug's partition coefficient is a measure of its concentration in a nonpolar organic phase to that in a polar aqueous phase.
- The pH-partition hypothesis explains that for ionizable drugs, the more lipid soluble unionized (uncharged) form is the form that most readily crosses a lipid bilayer. Therefore, as the pH determines the extent of a drug's ionization, the pH at absorption or permeation sites plays a large role in determining the passive diffusion of a drug across a lipid bilayer and can be used, along with knowledge of a drug's pKa, to predict drug absorption at different administration sites and the transport of drugs into different body fluids, such as urine, sweat, and breast milk. It can also be used to calculate the distribution of a solute between immiscible liquids such as oil and water.

CLINICAL QUESTIONS

1. Insulin glargine is a recombinant long-acting insulin analogue, which has arginine amino acids added to shift the isoelectric point (pH of zero electrical charge) from a pH of 5.4 to 6.7, so that at its formulated pH of 4, it is a clear solution. Explain how the low formulation pH enables glargine dissolution in the vial, and conjecture what physicochemical changes occur to the drug when it meets the physiological pH upon subcutaneous injection, which ultimately results in sustained and low plasma insulin levels.
2. In a study comparing one brand of the oral antidiabetic drug glyburide to another, it was discovered that the 3-mg tablet of one brand led to slightly higher and faster plasma levels compared to a 5-mg tablet of the other brand. The main difference between the two brands was that the 3-mg tablet contained smaller (micronized) glyburide crystals than the 5-mg tablet. Explain how the smaller particle size enabled the product with the lower dose to achieve slightly faster and higher levels than the higher-dose tablet, and discuss if it would be wise to keep the same dose when switching from one brand to another.

6 Mass Transport

Mansoor M. Amiji

Learning Objectives

After completing this chapter, the reader should be able to:

- Appreciate the role of thermodynamics and kinetics in the transport of molecules across a region.
- Understand the different forms of transport processes in biological and pharmaceutical systems.
- Distinguish between passive, facilitated, and active transport of solute molecules.
- Understand solvent transport and the process of osmosis.
- Describe the different membrane systems that are relevant to pharmacy.
- Identify the different terms and units of Fick's law of diffusion.
- Understand the differences between zero- and first-order diffusional release in pharmaceutical systems.
- Appreciate the role of the lag-time effect and burst effect on the diffusion of drugs from a delivery device.
- Appreciate the significance of diffusion in biological and pharmaceutical processes.
- Appreciate the role of osmosis in biological and pharmaceutical processes.

INTRODUCTION

Mass transport refers to the movement of molecules from one region to another as a result of a specific driving force. Two fundamental physical concepts are involved in the transport of substances across a region: the principles of thermodynamics and kinetics.[1]

Thermodynamics involves *energy* change in a process that determines directionality. According to thermodynamic principles, systems tend to go toward a decrease in free energy. This means that a particular reaction will occur in the direction stated only if the free energy of the product is lower than the free energy of the reactants. When a patient takes a tablet or capsule formulation orally, the drug dissolves in the stomach and the majority of the absorption of the solution occurs in the upper intestine (duodenum, jejunum). From a thermodynamic point of view, drug absorption from the gastrointestinal tract into the bloodstream occurs because the free energy of the molecules in the gastrointestinal tract is much higher than the free energy in the bloodstream. Free energy in this case is related to the concentration of the molecules at the two sites. The concentration of the drug in the gastrointestinal tract is much higher than that in the bloodstream. According to thermodynamic principles, movement of drug from the gastrointestinal lumen to the bloodstream decreases this imbalance of free energy in the two compartments of the body. If the drug is transported into the body, the free energy of the molecules decreases (concentration decreases) and the free energy of the molecules in the bloodstream increases (concentration increases). There is a point where the two free energies are equal and the change in free energy is zero. This is the point of *equilibrium* that all reversible processes strive to achieve. At the point of equilibrium, the movement of drug molecules from the gastrointestinal tract into the bloodstream will stop. In all transport systems, the free energy change has to be favorable for the process to occur.

Kinetics is based on the *effect of time* on a process or reaction (see Chapter 11 for further details). As was mentioned, in thermodynamics we want to know if the reaction will occur in the direction that is stated. From a kinetic point of view, we are interested in knowing how much time it will take for the reaction to proceed. In the example of drug absorption from the gastrointestinal tract, thermodynamics indicates that the absorption occurs to minimize free energy change (until it reaches zero or equilibrium). From a kinetic point of view, we want to know how fast the absorption will occur and whether in that time frame the drug molecules will remain in the upper intestine as a result of peristaltic contraction and the gastrointestinal transit time. Some reactions tend to be very fast, and some are very slow. Thermodynamics does not discriminate between fast and slow reactions. Kinetics, however, indicates whether the reaction will occur within a realistic time frame and be of any significance. For instance, according to thermodynamics, *all* matter flows. However, the flow behavior of solids such as metals is so slow that it does not have any practical utility. When metals are heated and there is an increase in the kinetic energy of the molecules, they melt and begin to flow instantaneously.

From a pharmaceutical perspective, this chapter focuses primarily on the transport of drug molecules from a delivery system, absorption into the bloodstream, distribution in the body, and in some instances uptake by the cells. In all these transport systems, it is important to keep in mind that drug movement from one region to another is governed by both thermodynamic and kinetic principles.

TRANSPORT SYSTEMS

Pharmaceutical transport systems can be divided into two major categories: solute transport and solvent transport. *Solute transport* refers to the movement of molecules across artificial or natural boundaries. *Solvent transport* is predominantly the movement of water across semipermeable membranes by the process of osmosis.[2]

Solute Transport

Passive Transport

The random Brownian movement of molecules based on the differences in the concentration (i.e., concentration gradient) across the two regions is called *diffusion*. When molecules are transported by the movement of a liquid or a gaseous vehicle, the process is called *convection*. Both diffusion and convection are passive transport systems in that they do not require any form of energy to move solute molecules from one region to another.

According to thermodynamic principles, the driving force for diffusion is the concentration gradient (also known as chemical potential). Although the movement process is random as dictated by Brownian motion, the system tries to conform to the lowest free energy state; therefore, the net effect is that solute molecules move from the region of higher concentration to that of lower concentration. Kinetic energy plays an important role in the movement of solute molecules as a function of temperature. When temperature is increased, the kinetic energy increases and the molecules tend to move faster. The point of equilibrium will occur when there is no concentration differential. Diffusion occurring with nonionic solute molecules is based purely on the concentration gradient. However, for charged molecules, the electrical gradient across the region also contributes to the movement.

There are many examples of diffusional transport in pharmaceutical systems. For instance, drug release from a transdermal patch occurs by diffusion. Absorption of drug from the skin, gastrointestinal tract, and other administration sites occurs predominantly by diffusion as well. Distribution of drugs in the body, especially by permeation through biological tissues such as the blood-brain barrier, occurs by diffusion. Additionally, the reabsorption of solutes with water in the nephron occurs by diffusion.

Convection is the other passive transport system, in which a liquid or gaseous carrier transports the solute. In this case, the transport of solute depends on the flow properties of the carrier fluid. The best example of convection in the biological system is the transport of oxygen, nutrients, drugs, and other materials by the blood. Transport of small water-soluble molecules across the cell occurs through the *paracellular* mechanism involving fluid-filled channels by convection. The transport of molecules through a porous matrix also occurs primarily through convection as the fluid in the pores or channels facilitates the movement of the molecules.

Facilitated Transport

There are several cases in which the transport of drug molecules, occurring by diffusion, can be facilitated when the molecule is attached to a carrier. There is greater permeability of the drug-carrier complex than of the drug alone, thus enhancing the transport. Facilitated transport is primarily a biological phenomenon, and the majority of examples involve the transport of molecules across biological membranes. After the drug-carrier complex crosses the membrane, it dissociates and the carrier can return for additional complexation or new carrier molecules can be synthesized and eliminated to keep the process continuous. The absorption of vitamin B_{12} from the gastrointestinal tract is an example of facilitated transport in which a carrier is required to shuttle the molecule from the lumen into the bloodstream. Facilitated transport can occur with uncharged and charged molecules. For charged drug molecules, ionic interactions with a carrier molecule can lead to facilitated transport of a neutral complex, which releases the active moiety by the ion-exchange mechanism. This process of transporting charged molecules also is known as *ion-pair formation*. For instance, the basic drug propranolol forms ion pairs with oleic acid, and the complex permeates faster in the gastrointestinal tract than does the drug alone.

Active Transport

Active transport, characterized by the movement of molecules against the concentration gradient—that is, from a region of low concentration to a region of high concentration—requires some form of chemical energy to accomplish its goals. Many examples of active transport processes are biological. The primary energy system is the conversion of adenosine triphosphate (ATP) to adenosine diphosphate (ADP). The rate of active transport is characterized by an asymptotic value or a *saturable process* as the concentration of the substrate increases. The transport rate as a function of substrate concentration follows the Michaelis-Menten kinetic profile (see "Enzyme Kinetics" in Chapter 11), as shown in Figure 6-1.

Cellular uptake of biochemically important electrolytes and nutritive molecules occurs through active transport systems. Various pumps exist in the cell to transport materials into the cells (influx) and excrete the metabolites and synthesized proteins outside the cells (outflux). One of the important cellular pump systems is the transport of sodium (Na^+) and potassium (K^+) ions from the extracellular to the

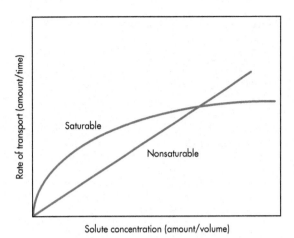

FIGURE 6-1 Comparison of rate profiles for nonsaturable and saturable (active) transport systems.

intracellular compartments. This is an example of active transport mediated by an ATP-dependent pump. The extracellular concentrations of K^+ and Na^+ are maintained at 3.5 to 4.5 mEq/L and 135 to 145 mEq/L, respectively. K^+ ions are actively transported into the intracellular compartment against the concentration gradient. An increase in the extracellular K^+ concentration from the normal range can cause serious toxicity in the body, resulting in cardiac arrhythmia. The proton (H^+) pump in the stomach is another example of active transport where the hydronium ions are pumped actively into the stomach lumen. Proton-pump inhibitors such as omeprazole (Prilosec) exert their effect by inhibiting the proton-pump action and decreasing the secretion of hydronium ions in the lumen of the stomach.

Active transport also occurs in the absorption and elimination of drugs from the body. Many transport proteins in the gastrointestinal tract actively pump the nutrients and drugs in or out of the mucosal layer. Each system appears to be highly specific to a particular chemical compound. For example, the L-amino acids and monosaccharides are actively absorbed from the intestinal region, but by different carrier systems. Bile transporter also has been studied in detail for its ability to recycle bile salts and as a means to enhance the permeability of certain drugs. In addition to structural specificity, there appears to be site specificity in the presence of these carriers along the gastrointestinal tract. The

TABLE 6-1 Some Actively Transported Drugs in the Gastrointestinal Tract

Transport System	Drug
Amino acid transporter	Methyldopa Gabapentin Baclofen Cycloserine
Oligopeptide transporter	Cephalexin Cefadroxil Cefixime Cephadrine Lisinopril Captopril
Phosphate transporter	Foscarnet Fostomycin
P-glycoprotein efflux system	Cyclosporin Etoposide Vinblastine

Source: Adapted from **Shargel L, Yu ABC:** *Applied Biopharmaceutics and Pharmacokinetics*, 4th ed. Stamford, CT: Appleton & Lange, 1999.

antitumor agent 5-fluorouracil is actively transported across the intestinal mucosa by the pyrimidine transport system. Other drugs that are actively absorbed from the intestine include riboflavin and several cephalosporin antibiotics (Table 6-1).[3]

P-glycoprotein (Pgp) is a transmembrane protein that has been identified in the intestine and many tissues in the body (e.g., tumors). Pgp actively effluxes drugs from the epithelial cells to reduce the apparent permeability of lipophilic and cytotoxic drugs (Table 6-1). The efflux transport protein has been implicated as one of the major contributors to multiple drug resistance in tumor cells.

Renal secretion is an example of an active transport process in that the movement of molecules occurs against the concentration gradient and the process requires energy. Penicillins and similar antibiotics are actively secreted by the *weak acid* transport system. Probenecid competes with penicillin for the secretion and therefore is used to increase the duration of the therapeutic activity of penicillins in the body.

Cellular Transport

Pinocytosis, endocytosis, and phagocytosis are transport systems that allow molecules to be taken up into a cell. *Pinocytosis* refers to the uptake of liquid matter into the cell, and *endocytosis* refers to the uptake of colloidal particles and macromolecules up to a diameter of 100 to 200 nm. Almost all cells are capable of pinocytosis and endocytosis. The cells of the immune system, such as macrophages, also are capable of *phagocytosis*, or ingesting larger particles in the range of 2 to 5 μm. During endocytosis and phagocytosis, the cell membrane captures the material and surrounds it to create a vesicle. The vesicle then is internalized into the cell and moves through the cytoplasm as an *endosome*. Fusion of the endosome with the lysosomal vesicle exposes the endosomal content to enzyme and an acidic pH medium that digests and detoxifies the foreign materials, such as microbial organisms. *Exocytosis* refers to the efflux of synthesized proteins and metabolic waste products from the cell. Insulin secretion from the β islet cells of the pancreas is an example of an exocytotic process.

Solvent Transport

The most important solvent for biological and pharmaceutical purposes is water. The transport of water molecules across membranes that are *selectively permeable* (or semipermeable) as a result of the osmotic pressure gradient is known as *osmosis*. A semipermeable membrane is permeable to water but not to the solute molecules.

Osmotic pressure is one of the four colligative properties discussed in Chapter 3. (The others are vapor pressure lowering, boiling point elevation, and freezing point depression.) Like other colligative properties, osmotic pressure depends on the concentration of dissolved solute. To equalize the pressure gradient, solvent (e.g., water) molecules move from the solvent chamber to the solution chamber, as shown in Figure 6-2. At the point of equilibrium, the osmotic pressure difference between the two compartments is nonexistent and there is no driving force for the movement of solvent.[4]

Osmotic pressure (π) is measured by using the equation:

$$\pi = CRT$$

where C is the concentration of solute in solution in moles per liter, R is the universal gas constant

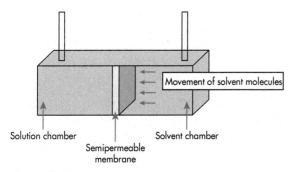

FIGURE 6-2   Diagram depicting the process of osmosis.

(0.0821 l.atm/mol.K), and T is absolute temperature (i.e., 273.15 + °C). Based on this equation, the units of osmotic pressure will be in *atmospheres*.

First, if we convert the units of osmotic pressure from atmospheres to dynes/cm^2 using the factor of 1.013×10^6 [(dynes/cm^2)/atm], we can replace the equation of pressure with:

$$\pi \text{ (in dynes/cm}^2\text{)} = h\rho g$$

where h is the height of the rise in solution volume in centimeters, ρ is the density of solution in g/cm^3, and g is acceleration resulting from gravity (i.e., 980 cm/s^2). Combining the two equations for measurements of osmotic pressure:

$$h\rho g = (1.013 \times 10^6)CRT$$

and

$$h = [(1.013 \times 10^6)CRT/\rho g]$$

In dilute aqueous solutions, the density is approximately equal to 1.00 g/cm^3, and using the value of g as 980 cm/s^2:

$$h = (1.034 \times 10^3)CRT$$

where 1.034×10^3 is $[1.013 \times 10^6/(1.00)(980)]$.

EXAMPLE 1 ▶▶▶

Calculate the height of rise of solvent (in centimeters) resulting from the osmotic pressure exerted by 0.1% (w/v) poly(ethylene glycol) (PEG, molecular weight 1000 d) aqueous solution at 25°C (77°F).

Solution

First, determine the concentration (C) of PEG in moles per liter:

 0.1% (w/v) PEG solution = 0.1 g of PEG per 100 mL of solution

Therefore, in 1 L of solution, there will be 1.0 g of PEG.

 Concentration (C) = number of moles per liter of solution

Therefore, $C = [(1 \text{ g}/1000)/1 \text{ L}]$
       $= 1 \times 10^{-3}$ moles/L

Using the equation $h = (1.034 \times 10^3)CRT$:

$h = (1.034 \times 10^3)(1 \times 10^{-3})(0.0821)(298)$
  $= 25.3$ cm

Osmosis is important in the development of pharmaceutical products that come in contact with cells and mucosal surfaces. When the osmotic pressure of the drug solution is approximately equal to that of biological fluid (~310 mOsmol/L), the solution is termed *isotonic*. In contrast, if the osmotic pressure of the solution is greater or less than that of the biological fluid, the solution is *hypertonic* or *hypotonic*, respectively. Parenteral, ophthalmic, and nasal solutions should be isotonic relative to the osmotic pressure of blood. Sodium chloride equivalent and other methods of adjusting the tonicity of pharmaceutical products are discussed in Chapter 3.

DIFFUSION THROUGH A MEMBRANE

Natural and Synthetic Membranes

A *membrane* is defined as a physical barrier that separates two or more regions. For the purposes of this chapter, membranes can be natural (biological) or synthetic (polymeric). We are all familiar from biochemistry with the Sanger model of a cell membrane that is composed of phospholipids and proteins. Absorption of drugs from the gastrointestinal tract occurs through the mucosal membranes. The blood-brain barrier is another example of a biological membrane that is significant for drug transport into the central nervous system. Skin is also an example of a membrane for the entry of drugs by transdermal absorption. Many biological membranes, such as the blood-brain barrier and the skin, also serve as formidable barriers to the entry of matter from the outside into the brain and the bloodstream, respectively. Their protective function is dependent on the integrity of the membrane.

Synthetic membranes used in pharmaceutical applications are made from polymeric material and are used mainly for controlled or sustained drug delivery. For instance, silicone rubber is used to prepare membranes for transdermal patches that regulate the rate of drug release from these devices. Synthetic membranes can be fabricated to be completely nonporous or highly porous or to be a network of intertwined polymeric chains. Transport of solute molecules through a nonporous membrane occurs exclusively by diffusion. In the case of porous membranes, transport can occur by both diffusion and convection. For membranes made by the intertwining of polymeric chains, transport usually occurs predominantly by convection through the tortuous channels.

Fick's Law of Diffusion

Diffusion through a membrane can be modeled by Fick's law of diffusion. According to this law, the amount of material M (units = grams or moles) crossing a unit area S (units = cm^2) in time t (units = seconds) is known as the *flux* (J). That is:

$$J = \frac{dM}{Sdt}$$

The flux, in turn, is proportional to the concentration gradient:

$$J \alpha \frac{dC}{dx}$$

where dC/dx is the change in concentration over infinitely small distances. To change from the proportionality sign to an equal sign, a constant is added:

$$J = -D \, dC/dx$$

where D is known as the *diffusion coefficient* or diffusivity (units = cm^2/s). The diffusion coefficient is a measure of the ease of permeability of molecules across the region. The negative sign shows that the concentration decreases as a function of distance. However, the flux will always be a positive number. Table 6-2 shows the values of the diffusion coefficients of some drugs in model systems.

Therefore, combining the previous two equations:

$$\frac{dM}{dt} = -DS(dC/dx)$$

TABLE 6-2 Diffusion Coefficients of Drugs in Model Systems

Drug	Diffusion Coefficient, cm^2/s	Temperature, °C	Conditions
Water	2.80×10^{-10}	37	Transport into the human skin layer
Butyl *p*-aminobenzoate	2.70×10^{-6}	37	Release from silicone rubber membrane
Ethynodiol diacetate	3.94×10^{-7}	25	Release from silicone rubber matrix
Fluocinolone acetonide	1.11×10^{-8}	25	Release from 30% propylene glycol aqueous solution through a polyethylene membrane
Medroxyprogesterone acetate	3.70×10^{-7}	25	Release from silicone rubber matrix
Salicylates	1.69×10^{-6}	37	Transport across cellulose membrane

Source: Adapted from **Martin A, Bustamante P:** *Physical Pharmacy: Physical Chemical Principles in the Pharmaceutical Science,* 4th ed. Philadelphia: Lea & Febiger, 1993.

The concentration gradient across the membrane (dC/dx) from the donor side to the receptor side can be simplified as:

$$\frac{dC}{dx} = [(C_1 - C_2)/h]$$

where C_1 is the concentration of the material in the membrane at the donor side and C_2 is the concentration of the material in the membrane at the receptor side, as shown in Figure 6-3.

Therefore, the rate of transport (dM/dt) will be given as:

$$\frac{dM}{dt} = (DS/h)(C_1 - C_2)$$

The concentrations C_1 and C_2 cannot be measured since these are values inside the membrane. The concentrations C_1 and C_2 in the membrane are related to the donor (C_d) and receptor (C_r) concentrations by the *partition coefficient K*:

$$K = C_1/C_d \quad \text{or} \quad C_2/C_r$$

Therefore:

$$C_1 = KC_d \quad \text{and} \quad C_2 = KC_r$$

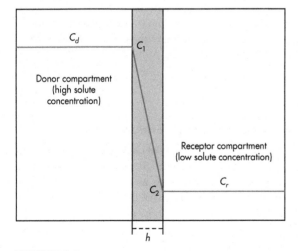

FIGURE 6-3 The concentration gradient of solute across the membrane. C_1 and C_2 are the concentrations of the drug in the membrane at the donor and receptor sides, respectively. C_d and C_r are the concentrations of the drug in the donor and receptor compartments, respectively, and h is the thickness of the membrane.

For efficient transport of solute molecules from the donor to the receptor compartment, the value of the partition coefficient (K) must be equal to or greater than 1.0. That is, the concentration in the membrane at the donor side at any given time must be greater than the concentration in the donor compartment. Typically, for hydrophobic drugs and hydrophobic polymeric membranes such as silicone rubber, the partition coefficient values tend to be between 10 and 100.

Substituting into the previous equation, the rate of transport in a diffusion system is:

$$\frac{dM}{dt} = (DSK/h)(C_d - C_r)$$

Notice that the rate of drug transport in a diffusional system is dependent on the magnitude of the concentration gradient (i.e., $C_d - C_r$). For any given system, the other parameters (i.e., DSK/h) are constant.

Sink Condition Approximation

In most pharmaceutical systems, the concentration of drug in the receptor side (C_r) is significantly lower than that at the donor side. For instance, when a transdermal patch is applied to the skin, the concentration of drug inside the skin at any time will be very low compared with the concentration in the patch. This is the case because the drug in the skin is absorbed continuously into the bloodstream, transported to the active site, and eliminated from the body. When the concentration C_r is approximately zero, this state is defined as the *sink condition*. Sink conditions occur when the rate of exit of drug from a compartment is much greater than the rate of entry. Thus, there is no accumulation of the drug in the compartment. These conditions are analogous to a sink, where the rate of water exit through the drain is much greater than the rate of entry from the tap and thus there is no accumulation of water in the sink.

Under sink conditions, it can be assumed that $C_r = 0$ and:

$$\frac{dM}{dt} = (DSK/h)C_d$$

or:

$$\frac{dM}{dt} = PSC_d$$

where $P (= DK/h)$ is the *permeability coefficient* (cm/s).

This equation can be written as:

$$dM = PSC_d\, dt$$

Integrating the equation from zero to infinity, it is possible to find the equation for the amount of drug transported through a membrane as a function of time:

$$M = PSC_d t$$

This equation suggests that the amount of drug transported is constant over time. A graph of the amount transported as a function of time, as shown in Figure 6-4, will be a straight line with the slope equal to PSC_d. This type of transport process is known as a *zero-order process*. A zero-order process, as defined in Chapter 11, occurs when the rate is constant. It can occur only if C_d does not change with time. To maintain C_d as a constant, most zero-order systems are designed with excess drug in the donor compartment. The transport of drugs from diffusional delivery systems such as transdermal patches occurs by a zero-order process. Transdermal patches have a suspension of the drug in the reservoir so that the concentration of dissolved drug does not change significantly with time.

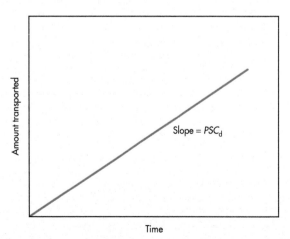

FIGURE 6-4 The amount of drug transported as a function of time for a zero-order process.

EXAMPLE 2 ▶▶▶

To study the oral absorption of paclitaxel from an oil-in-water emulsion formulation, an inverted closed-loop intestinal model was used. The drug was instilled in the intestine, and the system was maintained at 37°C (98.6°F) in an oxygen-rich buffer medium. The surface area available for diffusion was 28.4 cm², and the concentration of paclitaxel in the intestine was 1.50 mg/mL. Calculate the amount of paclitaxel that will permeate the intestine in 6 h of study if the permeability coefficient was 4.25×10^{-6} cm/s. Assume zero-order transport under sink conditions.

Solution

Using the equation $M = PSC_d t$, the amount of paclitaxel permeated in 6 h (21,600 s) will be:

$$M = (4.25 \times 10^{-6})(28.4)(1.50)(21,600)$$
$$= 3.91\,mg$$

If the donor concentration changes with time, the permeability will follow *first-order transport*, as described by:

$$\ln(C_d)t = \ln(C_d)_0 - (PS/V_d)\,t$$

where $(C_d)_t$ is the donor concentration at any time t, $(C_d)_0$ is the initial donor concentration, and V_d is the volume of the donor compartment in mL.

Lag-Time and Burst Effects

Release of drug from most controlled-release systems (e.g., transdermal patches) does not occur exclusively by zero-order processes as shown. The release is slightly different because of the time required to saturate the membrane (lag-time effect) or the initial rapid release of the drug (burst effect) from presaturated membranes.

The lag-time effect is observed typically in systems when they first contact the skin or mucosal tissue. It is the time required by the drug to saturate the tissue before it is absorbed into the bloodstream.

The lag-time effect (t_L) is dependent on the thickness of the membrane and the diffusion coefficient of the drug:

$$t_L = \frac{h^2}{6D}$$

where h is the membrane thickness in centimeters and D is the diffusion coefficient in cm²/s. Therefore, correcting for the lag time into the equation of the amount released as a function of time:

$$M = PSC_d(t - t_L)$$

The burst effect is observed in systems that have been stored for a long time and the rate-controlling membrane is presaturated with the drug. The burst effect (t_B) is also dependent on the thickness of the membrane and the diffusion coefficient and is expressed as:

$$t_B = h^2/3D$$

Correcting for the burst effect, the equation for amount released is written as:

$$M = PSC_d(t + t_B)$$

Figure 6-5 shows the amount of drug released as a function of time from zero-order devices that display lag-time and burst effects.

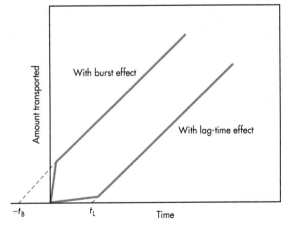

FIGURE 6-5 Zero-order release in the presence of lag-time and burst effects.

EXAMPLE 3 ▶▶▶

The lag time of methadone, a drug used in the treatment of heroin addiction, at 25°C (77°F) through a silicone membrane transdermal patch was calculated to be 4.65 min. The surface area and thickness of the membrane were 12.53 cm² and 100 μm, respectively.

a. Calculate the permeability coefficient of the drug at 25°C (77°F) (K = 10.5).
b. Calculate the total amount in milligrams of methadone released from the patch in 12 h if the concentration inside the patch was 6.25 mg/mL.

Solution
a. The permeability coefficient (P) is given by DK/h. Therefore, the first thing to determine is the diffusion coefficient D. Using the lag-time equation ($t_L = h^2/6D$):

$$D = h^2/6t_L$$
$$= (1.00 \times 10^{-2})^2/(6)(279)$$
$$= 5.97 \times 10^{-8} \text{ cm}^2/\text{s}$$

Using the permeability coefficient equation ($P = DK/h$):

$$P = [(5.97 \times 10^{-8})(10.5)]/(1.00 \times 10^{-2})$$
$$= 6.27 \times 10^{-5} \text{ cm/s}$$

b. The amount released after 12 h (43,200 s) is given by the equation

$$M = PSC_d(t - t_L)$$
$$= [(6.27 \times 10^{-5})(12.53)(6.25)][(43,200) - (279)]$$
$$= 210.8 \text{ mg}$$

SIGNIFICANCE OF DIFFUSION

Drug Release by Diffusion

Diffusion is a main driving force for the release of drugs from many different types of sustained-release and controlled-release products. Diffusional release occurs when the amount of drug released over time is dictated by the transport properties of a rate-controlling membrane. There are two major types of diffusional release systems: *reservoir systems* and *matrix systems*.[5]

Reservoir Systems

As the name implies, these controlled-release devices are manufactured with the drug placed centrally in a core that is surrounded by a hydrophobic rate-controlling polymer membrane. The release from a reservoir system is governed by Fick's law and is predominantly zero-order, because within the system is a saturate solution reservoir that enables constant drug concentration, thus providing a constant concentration gradient. Figure 6-6 shows the mechanism of drug release from a reservoir system. Reservoir systems offer a number of benefits, including predictable release by zero-order processes and variability of release properties by changing the membrane parameters. However, these systems are designed with a large excess of drug that can lead to dangerous toxicologic complications when the integrity of the device is compromised. Instantaneous release of a large excess of the drug from a reservoir device is known as *dose dumping*. When dispensing these medications, it is important to mention to the patients that diffusional reservoir systems should not be cut, broken, or chewed. In such instances, all of the drug will be released and will lead to serious toxicity.

There are many reservoir delivery systems currently marketed in the United States (Table 6-3). They are available for many routes of administration, including oral, transdermal, parenteral, intravaginal, and intrauterine. As examples of how they are made, oral formulations are often manufactured by coating the drug reservoir particles with a hydrophobic polymer membrane, which are then made into a tablet or capsule. Typically, ethylcellulose is used for the rate-limiting membrane. Transdermal systems are manufactured as adhesive patches with the drug reservoir, backing membrane, rate-controlling membrane, and adhesive layer. Drug diffusion therefore occurs through the rate-controlling membrane and the adhesive layer and subsequently through the skin for systemic delivery.

Matrix Systems

In matrix systems, the drug is evenly dispersed throughout the matrix, as shown in Figure 6-7. The diffusional release matrix is made from either hydrophobic polymers or wax materials. Drug diffusion in a matrix system occurs more slowly than it does in a reservoir system because initially the drug molecules in the periphery of the matrix will be released. Drug molecules from the center of the matrix need to diffuse to the periphery before they can be released. The release profile therefore follows a square-root-of-time relationship, as described by the *Higuchi equation*.

The equation for amount released (M) as a function of time (t) from a matrix system is:

$$M = [C_s D_m (2C_0 - C_s)t]^{1/2}$$

where C_s is the saturated concentration of the drug in the matrix, C_0 is the total amount of drug in the matrix, and D_m is the diffusion coefficient of the drug in the matrix. For porous or granular matrix systems, the equation is modified to include *porosity* (σ) and *tortuosity* (ε) terms:

$$M = [D_s C_a (\sigma/\varepsilon)(2C_0 - \sigma C_a)t]^{1/2}$$

where D_s is the diffusion coefficient of the drug in the release medium and C_a is the solubility of the drug in the release medium.

For data analysis, both of these equations can be simplified:

$$M = kt^{1/2}$$

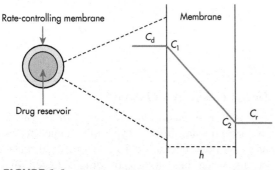

FIGURE 6-6 Representation of drug release from a diffusional reservoir system.

TABLE 6-3 Examples of Diffusional Reservoir Delivery Systems

Route of Administration	Product	Active Ingredient	Therapeutic Indication(s)
Oral	Naprelan	Naproxen sodium	Arthritis
	Micro-K, Micro-K LS, K-Dur	Potassium chloride	Potassium replacement
	Toprol XL	Metoprolol succinate	Hypertension, stable angina, heart failure
Transdermal	Catapres-TTS	Clonidine	Hypertension
	Androderm	Testosterone	Hypogonadism
	Estraderm	Estradiol	Postmenopausal symptoms and osteoporosis
	Nicoderm CQ	Nicotine	Smoking cessation
	Transderm-Scop	Scopolamine	Motion sickness
	Transderm-Nitro	Nitroglycerin	Angina
Interuterine	Mirena	Levonorgestrel	Contraception
Intravaginal	Nuvaring	Etonogestrel and ethinyl estradiol	Contraception
	Estring	Estradiol	Symptoms related to postmenopausal vaginal atrophy
Implant	Suprellen LA	Histrelin acetate	Central precocious puberty
	Vantas	Histrelin acetate	Advanced prostate cancer
	Implanon	Etonorgestrel	Birth control

where k is a constant that takes into account all the parameters in the two equations.

As shown in Table 6-4, several currently available oral, transdermal, and parenteral products are manufactured as matrix diffusional systems. Matrix formulations are less expensive to manufacture than reservoir systems, and there is no problem with dose dumping. Oral formulations can be as prepared as a matrix of drug in a water-insoluble system or in a biodegradable matrix system. These systems are made into tablets or capsules to provide uniform doses of drug.

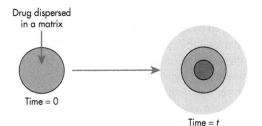

FIGURE 6-7 Representation of drug release from a diffusional matrix system.

Drug Absorption and Distribution

The absorption of the majority of drugs from the gastrointestinal tract and other sites of administration occur by diffusion.[6] The drug molecule is transported passively from a region of high concentration to a region of low concentration. In the gastrointestinal tract, the oral formulations may need to disintegrate and dissolve. Although there are some cases of particulate absorption (e.g., oral polio vaccine), the majority of drugs are absorbed in the intestine in a solution form. Absorption of drug molecules in the gastrointestinal tract is governed by several factors, including their concentration, size of the molecule, partition coefficient, ionization state, surface area of the site, and blood flow to the region. The concentration of drug in the gastrointestinal tract can be mainly a function of dose for highly water-soluble compounds. However, drugs with low aqueous solubility have a tendency to dissolve slowly in gastrointestinal fluids, and this slow speed of dissolution can limit the amount of drug absorbed (i.e., only dissolved drug is absorbed). Thus, for these drugs, the dissolution step becomes the rate-limiting step in the

TABLE 6-4 Examples of Diffusional Matrix Delivery Systems

Route of Administration	Product	Active Ingredient	Therapeutic Indication(s)
Oral	Slow K	Potassium chloride	Potassium replacement
	Slow-Fe	Ferrous sulfate	Iron supplement
	Paxil CR	Paroxetine HCl	Major depressive disorder
	Dilacor XR	Diltiazem HCl	Hypertension, stable angina
Transdermal	Nitrodur	Nitroglycerin	Angina
	Duragesic	Fentanyl	Pain
	Lidoderm	Lidocaine HCl	Postherpetic neuralgia
	Ortho Evra	Norelgestromin/ ethinyl estradiol	Contraception
	Climara	Estradiol	Vasomotor symptoms associated with the menopause
Implant	Zoladex	Goserelin acetate	Prostate cancer
Injectable microspheres	Lupron Depot	Leuprolide acetate	Prostate cancer

absorption process. Controlled- and sustained-release formulations are designed specifically to release the drug at a slow rate so that the absorption process is governed by the availability of drug at the site. In this way, the release rate of the drug becomes the rate-limiting step. The molecular weight (or size) of the drug also affects its absorption characteristics. For instance, high-molecular-weight compounds, such as polymeric materials, are not absorbed when administered orally. The partition coefficient, discussed in Chapter 5, is a measure of the hydrophobicity of drug molecules, and is also an important predictor of diffusion through the lipid bilayer of the cell membrane. Ideally, drugs with partition coefficient values of 10 to 100 are considered optimal for oral absorption. Absorption occurs primarily in the duodenum and jejunum of the small intestine. The intestine offers a large surface area for absorption through the microvilli. Transport of drugs from the lumen into the mucosal epithelial cells can occur by either *paracellular* or *transcellular* routes. In the paracellular route, the dissolved drug moves through the aqueous channels between cells, primarily by convection. In the transcellular route, the drug is transported through the cell by diffusion. Blood flow to the region is also important for the transport of drugs. Once the drug reaches the circulatory system, diffusion is no longer the primary transport mode. The drug is then transported by convection. After absorption, drug molecules are transported to the liver by the hepatic portal vein, a process known as the *first-pass effect*. (See Chapter 1.) During the first pass, some drugs, such as corticosteroids, can be extracted by as much as 50% to 60% of the dose absorbed.

The factors influencing drug absorption from other sites of administration, such as lung, mucosal tissues, and skin, are similar to those for oral administration. In the case of the lung, for instance, delivery into the alveolar region is essential for optimal systemic absorption. Current research is focused on the administration of dry powders of protein drugs into the deep area of the lung as a noninvasive drug delivery technique for chronic therapy. Absorption of even protein drugs is very efficacious because of the large surface area of the alveolar region and the very efficient blood flow. Ocular, nasal, buccal, vaginal, and other mucosal sites are also important sites for the administration of drugs. An advantage of mucosal delivery of drugs for systemic therapy is that there is no hepatic first-pass effect after absorption. Skin is an attractive site for the systemic delivery of therapeutic agents. The transdermal absorption of drugs is influenced by transport across the *stratum corneum*, the hydrophobic layer on the skin surface, which is composed predominantly of dead cells. Drugs available in transdermal patches (nitroglycerin, nicotine, etc.) can permeate the stratum corneum and diffuse into the epidermal and dermal layers of the skin. Absorption through the skin is affected by the thickness of the skin. Patients therefore are often recommended to place the patch either in the abdominal area or on the upper arm, where there is uniform thickness. There is a great deal of scientific interest in increasing the absorption of hydrophilic and high-molecular-weight drugs (e.g., proteins, heparin) through the skin by using chemical permeation enhancers (a zone, dimethylsulfoxide, etc.), low-voltage electric current, and ultrasound.

Since the majority of drugs are weak electrolytes, an important factor governing absorption from the gastrointestinal tract and other sites of administration is the relationship between the pH of the environment and the dissociation constant (K_a or K_b) of the drug. According to the *pH partition hypothesis*, it has been suggested that since biological membranes are mainly hydrophobic, drugs penetrate in the *unionized* (or undissociated) form. Knowing the pH of the environment and either the pK_a or the pK_b of the drug, one can predict whether the drug will be absorbed efficiently.

From the Henderson-Hasselbalch equation for a weak acid (acetylsalicylic acid, penicillins, etc.; (See Chapter 4 and 5),

$$pH = pK_a + \log\{[\text{ionized}]/[\text{unionized}]\}$$

or   % ionized = $100/[1 + \text{antilog }(pK_a - pH)]$

and   % unionized = $100 - $ (% ionized)

For a weak base (codeine, pilocarpine, etc.), the Henderson-Hasselbalch equation is written as:

$$pH = pK_a + \log\{[\text{unionized}]/[\text{ionized}]\}$$

where $pK_a = pK_w - pK_b$ (or $14 - pK_b$)

or   % ionized = $100/[1 + \text{antilog }(pH - pK_a)]$

and   % unionized = $100 - $ (% ionized)

EXAMPLE 4 ▶▶▶

Calculate the percent ionized and unionized of morphine, a weak base ($K_b = 7.4 \times 10^7$), in the plasma where the pH is 7.4 at 37°C (98.6°F).

Solution

First, $pK_b = -\log(K_b)$ is calculated to be:

$$pK_b = -\log(7.4 \times 10^7)$$
$$= 6.13$$

and $pK_a = 14 - pK_b$ is:

$$pK_a = 14 - 6.13$$
$$= 7.87$$

Using the equation % ionized = $100/[1 + \text{antilog }(pH - pK_a)]$ for a weak base:

% ionized = $100/[1 + \text{antilog }(7.4 - 7.87)]$
% ionized = 74.7%

And % unionized = $100 - $ (% ionized), so:

% unionized = 25.3%

Based on the percent unionized for a weak acid or base at any given pH, one can predict the absorption profile, especially in the gastrointestinal tract, where there are pH differences from the stomach to the colon. The pH of gastric fluid in a fasted stomach, for instance, is about 1.2 to 2.0. During the fed state, the pH may increase to about 3.0 to 4.0, depending on the composition of the meal. In the duodenum and jejunum, the pH of the intestinal fluid is around 6.5. For a weak acid such as acetylsalicylic acid, it is clear from Table 6-5 that the majority of molecules are unionized in acidic pH (<2.0). However, as the pH increases above 4.0, the percent ionization increases proportionally.

Based on the data in Table 6-5, one can predict that acetylsalicylic acid probably will be absorbed

TABLE 6-5 Percent Ionized and Unionized Acetylsalicylic Acid ($pK_a = 3.5$) as a Function of pH

pH	% Ionized	% Unionized
1.2	0.50	99.5
2.0	3.10	96.9
3.5	50.0	50.0
5.0	96.9	3.10
6.5	99.9	0.10
7.4	99.98	0.02

from the stomach. The surface area available for absorption in the stomach is significantly less than that in the upper intestine. In addition, the presence of food can interfere in the absorption process. Therefore, even though the majority of drug is ionized in the intestinal pH, the predominant absorption of acetylsalicylic acid still occurs in this area. However, once the drug is absorbed, the percent ionization in the plasma at pH 7.4 is close to 100%.

Absorption of drugs from mucosal surfaces and skin also is influenced by the pH partition hypothesis. For instance, when pilocarpine, a weak base (pK_a = 6.8), is administered for the treatment of glaucoma, the majority of the drug is in the ionized form at the pH of the tear fluid (~7.3). The permeability of ionized pilocarpine through the corneal tissue is very low. In addition, since the ionized drug is highly water-soluble, it is excreted rapidly by the efficient tear drainage system. Several methods of increasing the permeability of pilocarpine have been utilized, including the design of hydrophobic prodrugs.

After absorption, the drug enters the systemic circulation and transport to the various tissues occurs by convection. The amount of drug reaching the site of action is dependent on the concentration and the rate of blood flow to the tissue. Highly vascularized tissues such as the lung, liver, and kidneys receive a large fraction of the dose compared with poorly vascularized tissues such as the skin. In addition, there are other factors, such as plasma protein binding and the ionization state, that affect the distribution of the drug in the body. Plasma protein binding plays a very important role in the transport of drugs to the site of action. Only the unbound form of the drug can interact with the receptor to induce the pharmacologic effect. The equilibrium between protein-bound and protein-unbound drug is dependent on the concentrations of each and the association constant. When the unbound drug is eliminated from the body, more is released from the protein-bound reservoir. The administration of a second drug or an endogenous compound that has a higher affinity for the binding site can displace the bound drug and increase the concentration of unbound drug in the plasma. In some cases, such as warfarin, which is extensively protein-bound (>99%), serious toxicity results from displacement by another drug.

The ionization state of the molecule is also important for distribution in the body and ultimately for reaching the site of action. One can exploit the pH/pK_a relationship to increase the permeability of drugs across biological membranes. Parkinson's disease and related disorders are due to a lack of dopamine secretion in the brain. Dopamine is a weak base with a pK_a of around 10.6 for the primary amine group. Thus, in the plasma, dopamine is significantly ionized and does not penetrate the blood-brain barrier. To increase the permeability across the blood-brain barrier, levodopa, a precursor to dopamine, is administered. Once levodopa enters the brain tissue, it is converted to dopamine by a decarboxylase enzyme. The product Sinemet contains a combination of levodopa and carbidopa, a decarboxylase inhibitor that prevents conversion to dopamine in the systemic circulation before it reaches the brain.

From the pH/pK_a relationship, it is clear that to increase the permeability of drugs across biological membranes, the drugs should be in their unionized form. To have a higher fraction unionized for a weak acid, the pK_a should be higher than the pH of the environment. In contrast, to have a higher fraction of unionized weak base, the pK_a (of the conjugate acid) should be lower than the pH. Slight modifications in the chemical structure of the drug can provide the necessary pK_a for optimal absorption. Alternatively, the pharmaceutical scientist can design a prodrug that has the optimal pK_a for absorption and is converted into the active species in the systemic circulation or, preferably, at the site of action.

Weak acids tend to be highly ionized at pH 7.4 in plasma. The ionized drug tends to be less permeable in the biological membrane and more soluble in water. Phenobarbital poisoning, for instance, is treated by intravenous administration of sodium bicarbonate to increase the pH of plasma. When the pH is increased, the percent of ionized phenobarbital increases, and it becomes less permeable through the blood-brain barrier and thus less able to induce toxicity in the central nervous system. The ionized form is also more soluble in water and therefore is more readily excreted in the urine. Alternatively, for a weak base such as cocaine, the plasma pH can be lowered to increase the percent ionization. This method of alleviating the toxicity of drugs is known as *ion trapping*.

SIGNIFICANCE OF OSMOSIS

Drug Release by Osmosis

Based on the principles of solvent transport through a semipermeable membrane as a result of the osmotic pressure gradient, an elegant drug delivery system, originally designed by the Alza Corporation and marketed as OROS (Oral Osmotic System), is the Gastrointestinal Therapeutic System (GITS). The product is designed as a tablet for zero-order release of the drug in the gastrointestinal tract. As shown in Figure 6-8, the tablet contains a central core of active ingredient and an osmotic agent [e.g., sodium chloride or poly(ethylene oxide)] surrounded by a semipermeable membrane. At the top of the tablet, there is a small hole drilled by a laser beam.

When the tablet is ingested orally, water from the gastrointestinal tract permeates the semipermeable membrane into the core because of the osmotic pressure gradient. Once water is inside the matrix, the drug dissolves and is released gradually as a solution through the laser-drilled hole on the top of the tablet. The amount of drug released (M) over time (t) from a GITS system is expressed by:

$$M = [(S/h)(k'\pi C_s)t]$$

where S is the surface area of the semipermeable membrane (in cm²), h is the thickness of the membrane (in centimeters), k' is the permeability of water through the membrane (in cm²/atm.h), π is the osmotic pressure (in atmospheres), and C_s is the saturated solubility of drug in water (in mg/mL). Based on the preceding equation, it is clear that the release of drug from the osmotic system is zero-order when the parameters $(S/h)(k'\pi C_s)$ are maintained constant.

EXAMPLE 5 ▶▶▶

Calculate the amount of drug released from a GITS preparation in 12 h if the surface area (S) is 2.5 cm², the thickness of the membrane (h) is 0.03 cm, the permeability coefficient of water is 3.2×10^{-6} cm²/atm.h, and the osmotic pressure is 200 atm. The concentration of saturated solution of the drug (C_s) is 275 mg/mL.

Solution
Using the equation $M = [(S/h)(k'\pi C_s)t]$, the amount released in 12 h will be:

$$M = [(2.5/0.03)(3.2 \times 10^{-6})(200)(275)(12)]$$
$$= 176 \text{ mg}$$

The GITS formulation offers several advantages in terms of controlled delivery of the drug. The release is governed primarily by the osmotic pressure gradient and the solubility of the drug in water. The osmotic pressure gradient can be varied easily by appropriate selection of the osmotic compound. The dosage form can be used for the delivery of molecules of any size, and the release rate is independent of the pH or ionic strength of the environment. Additionally, the product can be tailored for complete release to occur in a few hours or more, depending on the drug properties. The GITS dosage form, however, has a number of disadvantages. The laser-drilled hole can get plugged, and initial evaluation by the pharmacist is strongly encouraged. In addition, dispensing a GITS product offers an excellent opportunity for counseling the patients. These dosage forms should not be broken or chewed. Any damage to the integrity of the semipermeable membrane will result in dose dumping. Furthermore,

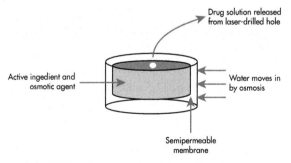

FIGURE 6-8 Representation of drug release from a Gastrointestinal Therapeutic System (GITS).

TABLE 6-6 Examples of Oral Osmotic Delivery Systems

Product	Active Ingredient	Therapeutic Indication(s)
Concerta	Methylphenidate	Attention deficit–hyperactivity disorder
Ditropan XL	Oxybutynin chloride	Overactive bladder
Glucotrol XL	Glipizide	Insulin-independent diabetes mellitus
Procardia XL	Nifedipine	Angina and hypertension
Volmax	Albuterol	Bronchospasm

patients may see the empty shell (ghost) in their stool and think that the tablet did not function properly. It is important to mention that for effective release and absorption, GITS products do not need to disintegrate as conventional tablets do. Table 6-6 gives examples of pharmaceutical products that are currently marketed in GITS formulations.

Several variations of GITS have been designed, including an implantable minipump system (Alzet) that can be filled with drug solution and a microcapsule formulation whose membrane ruptures after water uptake by osmosis, resulting in release of the encapsulated drug. However, these formulations are not currently marketed.

Isotonicity

It is important in the formulation of drugs that contact the mucous membrane for nasal and ocular dosage forms, as well as those that are administered into the systemic circulation, to have the same osmotic pressure as the blood. Using 0.9% (w/v) or 0.15 M sodium chloride in water as a standard, drug solutions tend to be generally hypotonic. Their tonicity needs to be adjusted with sodium chloride or other compounds, as discussed in Chapter 3.

If one carries out a simple experiment with erythrocytes (red blood cells) suspended in different solutions, one will notice that solutions that are hypotonic to the intracellular fluid will cause the erythrocytes to rupture (lysis) and release the hemoglobin. Alternatively, solutions that are hypertonic will cause the cells to shrink by pulling water from the inside to the outside. Solutions, such as 0.9% (w/v) sodium chloride in water, that are isotonic will not cause any damage to the erythrocytes. The administration of a hypotonic drug solution in the bloodstream will cause the release of hemoglobin, which is nephrotoxic. Therefore, it is important that solutions for parenteral administration be adjusted for isotonicity.

Intravenously administered mannitol solution, used as an osmotic diuretic for the treatment of cerebral edema and glaucoma, is an exception. The solutions with concentrations ranging from 5% (w/v) to 25% (w/v) are obviously hypertonic. However, the mechanism of diuresis of mannitol is based on an elevation of plasma osmotic pressure so that the excess fluids from the body can be excreted, resulting in a reduction in intracranial and intraocular pressures.

PROBLEMS

1. The lag time for the diffusion of medroxyprogesterone acetate at 25°C (77°F) through a patch of rat skin was calculated to be 45 s. The surface area and thickness of the skin were 18.5 cm² and 15 μm, respectively. Calculate the permeability coefficient (P) of the drug through the skin if the partition coefficient was 2.05.

2. A nicotine transdermal patch (Habitrol) is designed with a cross-sectional area of 10.5 cm² and has a membrane thickness of 0.075 cm. The lag time of diffusion of nicotine was calculated to be 12.5 min. The concentration of nicotine inside the patch is 3.0×10^{-5} g/cm³.
   a. Calculate the diffusion coefficient of nicotine from this patch.
   b. If the permeability of nicotine is 3.47×10^{-5} cm/s, calculate the amount of nicotine in micrograms released from the patch after 12 h.

3. Fluocinolone acetonide is a topical corticosteroid drug with anti-inflammatory properties. The release of fluocinolone acetonide from a patch with a cross-sectional area of 12.5 cm² and a membrane thickness of 75 μm was investigated. The lag time of diffusion of fluocinolone acetonide was calculated to be 8.45 min. The concentration of fluocinolone acetonide inside the patch is 4.5×10^{-5} g/cm³.
   a. Calculate the permeability coefficient (in cm/s) of fluocinolone acetonide from this patch ($K = 2.43$).
   b. Calculate the total amount of fluocinolone acetonide in micrograms released from the patch in 24 h.
4. To study the oral absorption of cyclosporin from an oil-in-water emulsion formulation, an inverted closed-loop intestinal model was used. The drug was instilled in the intestine, and the system was maintained at 37°C (98.6°F) in an oxygen-rich buffer medium. The lag time for the permeation of cyclosporin through the intestinal wall of 75-μm thickness was 7.85 min. The surface area available for diffusion was 28.4 cm², and the concentration of cyclosporin in the intestine was 1.35×10^{-3} g/cm³.
   a. Calculate the partition coefficient of cyclosporin in this system if the permeability coefficient was 4.25×10^{-6} cm/s.
   b. Determine the cumulative amount of cyclosporin in milligrams that permeated through the intestinal wall in 6 h of the study.
   c. Explain what is meant by zero-order release from a diffusional delivery system.

ANSWERS

1. Using the lag-time equation (i.e., $t_L = h^2/6D$):

   $D = h^2/6t_L$
   $D = (1.5 \times 10^{-3})^2/(6)(45)$
   $D = 8.33 \times 10^{-9}$ cm²/sec

   The permeability coefficient (i.e., $P = DK/h$) will be:

   $P = [(8.33 \times 10^{-9})(2.05)]/(1.5 \times 10^{-3})$
   $P = 1.14 \times 10^{-5}$ cm/sec

2. a. Using the lag-time equation (i.e., $t_L = h^2/6D$):
   $D = h^2/6t_L$
   $D = (0.075)^2/(6)(750)$
   $D = 1.25 \times 10^{-6}$ cm²/sec
   b. Amount released after 12 h [i.e., $M = PSC_d (t - t_L)$]
   $M = [(3.47 \times 10^{-5})(10.5)(3 \times 10^{-5})(43,200 - 750)]$
   $M = 4.64 \times 10^{-4}$ g
   $M = 464$ μg

3. a. Using the lag-time equation (i.e., $t_L = h^2/6D$):
   $D = h^2/6t_L$
   $D = (7.5 \times 10^{-3})^2/(6)(507)$
   $D = 1.85 \times 10^{-6}$ cm²/sec

   The permeability coefficient, P, is given by
   $P = (DK/h)$
   $P = [(1.85 \times 10^{-6})(2.43)/(7.5 \times 10^{-3})]$
   $P = 5.99 \times 10^{-6}$ cm/sec
   b. Amount released after 24 h [i.e., $M = PSC_d (t - t_L)$]
   $M = [(5.99 \times 10^{-6})(12.5)(4.5 \times 10^{-5})(86,400 - 507)]$
   $M = 2.90 \times 10^{-4}$ g
   $M = 0.290$ mg

4. a. Using the lag-time equation (i.e., $t_L = h^2/6D$):
   $D = h^2/6t_L$
   $D = (7.5 \times 10^{-3})^2/(6)(471)$
   $D = 1.99 \times 10^{-8}$ cm²/sec

   The permeability coefficient, P, is given by
   $P = (DK/h)$. Therefore, $K = (Ph/D)$
   $K = (4.25 \times 10^{-6})(7.5 \times 10^{-3})/(1.99 \times 10^{-8})$
   $K = 1.60$
   b. Amount permeated in 6 h [i.e., $M = PSC_d (t - t_L)$]
   $M = [(4.25 \times 10^{-6})(28.4)(1.35 \times 10^{-3})(21,600 - 471)]$
   $M = 3.44 \times 10^{-3}$ g
   $M = 3.44$ mg

c. Zero-order diffusional release means that the amount of drug being released from the device over a finite time range is constant. In contrast, in first-order, the amount released will depend on the concentration of drug remaining at the donor side.

KEY POINTS

▶ Mass transport refers to the movement of molecules of solutes or solvents from one region to another; for example, the movement of drug solute or water solvent across a biological membrane.

▶ Mechanisms of mass transport include active, passive, and facilitated processes.

▶ Active transport through membranes is important biologically for the transport of many endogenous molecules (e.g., sodium and potassium). It is also important for the transport of many drugs across biological membranes. Facilitated transport (e.g., of ion pairs) also has biological importance.

▶ Passive transport includes the passive diffusion of a solute down its concentration gradient, and convection—the carrying (conveying) of a solute by the movement of its solvent from one region to another.

▶ Passive diffusion is the most common way drugs cross biological and synthetic membranes. It is well modeled by Fick's law of diffusion, which contains important parameters for estimating the speed of movement (the *flux*) of a substance down its concentration gradient. These parameters include membrane thickness, the surface area of the membrane, and a diffusion coefficient.

▶ For passive diffusion of weak acid drugs or weak base drugs across lipid bilayer membranes (e.g., of gastrointestinal epithelial cells), it is the unionized (more lipophilic) form that diffuses through more readily. The Henderson-Hasselbalch equation can be used to calculate the fraction of unionized form as a function of pH, and can therefore be used to predict optimal absorption sites based on the pH of the relevant biological fluid. However, following Fick's law the surface area of the absorption site (e.g., of the small intestine) can be more important for drug transport than the local pH.

CLINICAL QUESTIONS

1. Nicotine gum contains nicotine bound to a resin that releases the nicotine when it is chewed. Among the excipients are buffers that adjust the salivary pH in a manner that increases its bioavailability through the buccal membrane. Given that nicotine is a weak base, explain the direction that the buffers adjust the salivary pH and how that adjustment can increase the absorption of nicotine.

2. Metformin is an oral antidiabetic drug that is eliminated mainly in the kidneys by glomerular filtration and tubular secretion. The drug transporter, organic cation transporter-2 (OCT2), which resides in the basolateral membrane (the blood side) of renal tubular epithelial (RTE) cells plays an important role in this process by actively transporting metformin prior to its secretion into the urine. With this information, describe what role OCT2 plays in the RTE, uptake or efflux. Also, conjecture how the drug cimetidine, through its effect on OCT2, causes a decrease in the renal clearance of metformin.

7 Complexation and Protein Binding

Mansoor M. Amiji

Learning Objectives

After completing this chapter, the reader should be able to:

- Understand the significance of complexation in pharmaceutical products.
- Appreciate the fundamental forces that are related to the formation of drug complexes.
- Differentiate between coordination and molecular complexation.
- Understand the mechanism of coordinate bond formation leading to the formation of coordinate complexes.
- Appreciate the biological and pharmaceutical roles of coordinate complexes.
- Describe the mechanism of inclusion complex formation, with special emphasis on drug-cyclodextrin complexes.
- Relate the formation of drug-cyclodextrin complexes with improvements in the physicochemical properties and bioavailability of drugs.
- Determine the values of the association constant and the stoichiometry of association.
- Understand the importance of the ion-exchange mechanism

INTRODUCTION

Complexation, a term with a broad definition, is used in the context of this chapter to characterize the *covalent* or *noncovalent* interactions between two or more compounds that are capable of independent existence.[1] The *ligand* is a molecule that interacts with another molecule, the *substrate*, to form a complex. Drug molecules can form complexes with other small molecules or with macromolecules such as proteins. Once complexation occurs, the physical and chemical properties of the complexing species are altered.[2] These properties include solubility, stability, partitioning, energy absorption and emission, and conductance of the drug. Drug complexation, therefore, can lead to beneficial properties such as enhanced aqueous solubility (e.g., theophylline complexation with ethylenediamine to form aminophylline) and stability (e.g., inclusion complexes of labile drugs with cyclodextrins). Complexation also can aid in the optimization of delivery systems (e.g., ion-exchange resins) and affect the distribution in the body after systemic administration as a result of protein binding. The topic of drug-protein binding is covered in depth in the later part of the chapter. In some instances, complexation also can lead to poor solubility or decreased absorption of drugs in the body. For example, the aqueous solubility of tetracycline decreases substantially when it complexes with calcium ions, and coadministration of some drugs with antacids decreases absorption from the gastrointestinal tract. For some drugs, complexation with certain hydrophilic compounds can enhance excretion. Finally, complexes can alter the pharmacologic activity of the agent by inhibiting interactions with receptors.

In general, complexation of a ligand with a substrate molecule can occur as a result of coordinate covalent bonding or one or more of the following noncovalent interactions: (1) van der Waals forces, (2) dipolar forces, (3) electrostatic forces, (4) hydrogen bonding, (5) charge transfer, and (6) hydrophobic interaction. The details of some of these forces are described in Chapter 2.

- and its role in drug delivery and therapy.
- ▶ Appreciate the significance of protein-ligand interactions.
- ▶ Understand the significance of plasma protein binding for the distributive properties of drugs in the body.
- ▶ Identify the important properties of plasma proteins and the mechanism of their interactions with drugs.
- ▶ Appreciate equilibrium dialysis and other techniques for *in vitro* analysis of drug-protein binding.
- ▶ Analyze protein-binding data by the double-reciprocal method and determine the values of the association constant and the number of binding sites.
- ▶ Analyze protein-binding data by the Scatchard method and determine the values of the association constant and the number of binding sites.
- ▶ Appreciate the advantages of the Scatchard method over the double-reciprocal method of analysis with respect to multiple binding affinities.

TYPES OF COMPLEXES

Depending on the type of interaction involved in complexations, ligand-substrate interactions are classified as *coordination complexes* or *molecular complexes*.

Coordination Complexes

A coordination complex (or compound) consists of a complex ion, typically a transition metal ion, with one or more attached ligands, or counterions, which are either anions or cations that will result in an electrically neutral complex. The interaction between the metal ion and the ligand often is classified as a Lewis acid-base reaction, in which the ligand (a base) donates a pair of electrons (:) to the metal ion (an acid) to form the coordinate covalent bond.[3] For example:

$$Ag^+ + 2(:NH_3) \rightarrow [Ag(NH_3)_2]^+$$

where silver ion (Ag^+) is the central metal ion interacting with ammonia (NH_3) to form the silver-ammonia $[Ag(NH_3)_2]^+$ coordinate complex. The complex will be neutralized with Cl^- to form $[Ag(NH_3)_2]Cl$.

The *coordination number* is defined as the maximum number of atoms or groups that can combine in the coordination sphere with the central metal atom. For instance, in $K_3[Fe(CN)_6]$, the coordination number is 6 since six cyano groups are complexed with the central iron atom.

EXAMPLE 1 ▶▶▶

What are the central metal ion, the ligand, and the coordination number in the $[Cu(H_2O)_6]^{2+}$ complex ion?

Solution
The central metal ion is Cu^{2+}. The ligands are water molecules (H_2O), and since there are six water molecules bound to Cu^{2+}, the coordination number is 6.

Ammonia, which has a single pair of electrons (basic group) for bonding with metal ion, is called a *unidentate* ligand. Other ligands, such as ethylenediamine ($H_2N-CH_2-CH_2-NH_2$), with two basic groups, are known as *bidentate*. Ethylenediaminetetraacetic acid (EDTA), as shown in Figure 7-1, has a total of six points (4 :O and 2 :N) for the attachment of metal ions and is referred to as *hexadentate*. Additionally, ligands with multiple binding sites (e.g., polymers) are *multidentate* or *polydentate*. If the same metal ion binds with two or more sites on a multidentate ligand, the

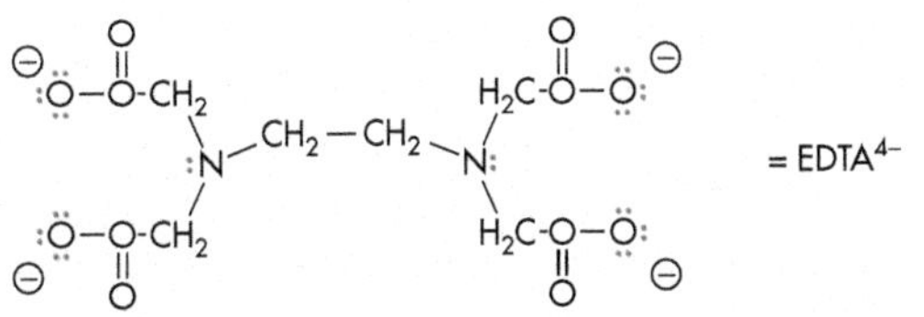

FIGURE 7-1 The structure of ethylenediaminetetraacetic acid (EDTA), a hexadentate molecule.

complex is called a *chelate* (from the Greek word *kelos*, "claw").

Three main theories have been propounded to describe coordinate complexes: *crystal field theory*, *molecular orbital theory*, and *valence bond theory*. Crystal field theory focuses on the electrostatic interactions between ligands and transition metal ions. In an isolated atom or ion, all five *d*-orbitals of the transition metal have the same energy regardless of their orientations. However, when the atom or ion is surrounded by ligands, this is no longer the case. The perturbations exerted by the ligands affect the energy of the *d*-orbitals. Crystal field theory provides a clear explanation of the spectral and magnetic properties of the complexes. Crystal field theory does not take into account the covalent bond formed by metal ions and ligands in the formation of coordinate complexes. To take this into account, molecular orbital theory was proposed; it shows how electrons are oriented to form bonds in coordinate complexes. Valence bond theory explains the nature of hybridization and the geometry of the molecule.

Molecular Complexes

Molecular complexes are formed as a result of noncovalent interactions between ligand and substrate. The interactions can occur through oppositely charged ions (electrostatic forces), van der Waals forces, charge transfer, hydrogen bonding, or hydrophobic effects. These complexes can occur between a small molecule and a small molecule (e.g., ethylenediamine and theophylline to form aminophylline), between a small molecule and a large molecule (e.g., protein binding of drugs and the combination of iodine and Povidone—polyvinylpyrrolidone—to form Povidone-iodine), ion pairs (e.g., ion-exchange resins), self-association to form aggregates (e.g., surfactant micelles), and inclusion complexes (e.g., cyclodextrin complexes).

METAL-ION COORDINATE COMPLEXES

Iron Complexes

The ability of metal ions to coordinate with and then release ligands in some processes and to oxidize and reduce them in other processes makes them ideal for use in biological systems. The most common metal used in the body is iron, and it plays a central role in almost all living cells. For example, heme proteins of *myoglobin* and *hemoglobin* are iron complexes that are essential for the transport of oxygen in the blood and tissues. *Cytochrome c* is another heme protein that is involved in both photosynthetic and respiratory systems.

Myoglobin is a monomeric heme protein that is found mainly in muscle tissue, where it serves as an intracellular storage site for oxygen. During periods of oxygen deprivation, *oxymyoglobin* releases its bound oxygen, which is used for metabolic purposes. Each myoglobin molecule contains one heme group inserted into a hydrophobic cleft in the protein. Each heme residue contains one central coordinately bound iron atom that is normally in the ferrous oxidation state (Fe^{2+}). The oxygen carried by heme proteins is bound directly to the ferrous iron atom of the heme group. Oxidation of the iron to the ferric oxidation state (Fe^{3+}) renders the molecule incapable of normal oxygen binding. Hydrophobic interactions between the tetrapyrrole ring and the hydrophobic amino acid in the interior of the cleft in the protein strongly stabilize the heme protein conjugate. In addition, a nitrogen atom from a histidine group above the plane of the heme ring is coordinated with the iron atom, further stabilizing the interaction between the heme and the protein (Fig. 7-2).

Hemoglobin is a tetrameric heme protein [$\alpha(2):\beta(2)$] that is found in erythrocytes (red blood cells), where it is responsible for binding oxygen in the lungs and transporting the bound oxygen throughout the body, where it is used in aerobic metabolic pathways. Each subunit of a hemoglobin tetramer has a heme prosthetic group that is identical to that described for myoglobin. When oxygen binds to an iron atom of deoxyhemoglobin, it pulls the iron atom into the plane of the heme. Since the iron also is bound to histidine, this residue also is pulled toward

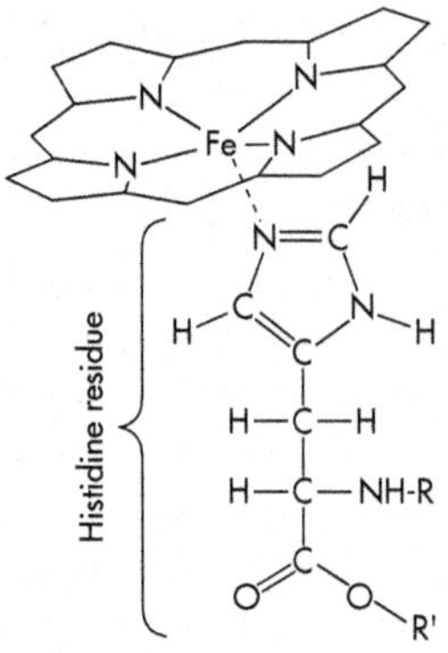

FIGURE 7-2 The structure of heme illustrating the coordinated complex between central iron atom and the nitrogens of histidine residues.

the plane of the heme ring. The conformational change at histidine is transmitted throughout the peptide backbone, resulting in a significant change in the tertiary structure of the entire subunit. Conformational changes at the subunit surface lead to a new set of binding interactions between adjacent subunits. These changes include the disruption of salt bridges and the formation of new hydrogen bonds and new hydrophobic interactions, all of which contribute to the new quaternary structure. In both oxymyoglobin and oxyhemoglobin, the remaining bonding site on the iron atom (the sixth coordinate position) is occupied by the oxygen, whose binding is stabilized by a second histidine residue. Carbon monoxide (CO) also binds coordinately to heme iron atoms in a manner similar to that of oxygen, but the binding of carbon monoxide to heme is 200 times stronger than that of oxygen. The preferential binding of carbon monoxide to heme iron is largely responsible for the asphyxiation that results from carbon monoxide poisoning.

Acute iron overdose is treated with chelating agents such as *deferoxamine* (Fig. 7-3). Chelation of ferric iron (Fe^{3+}) ion with deferoxamine results in a strong coordinated water-soluble ferrioxamine complex that can be excreted through the kidneys. Deferoxamine has very low affinity for divalent ions, including Fe^{2+} and Ca^{2+}, but binds very tightly to trivalent cations such as Fe^{3+}. Theoretically, a 100-mg dose of deferoxamine can bind with 8.5 mg of Fe^{3+}.

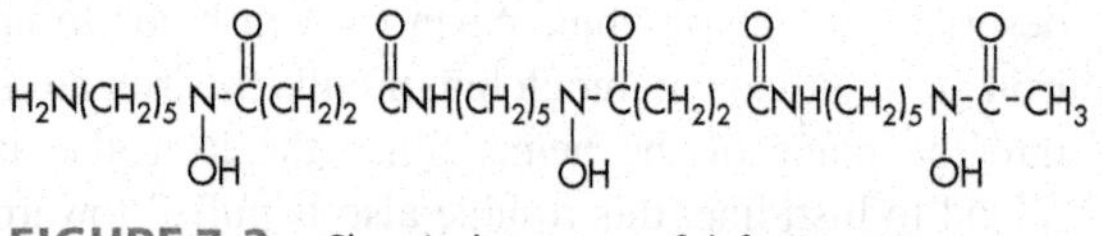

Platinum Complexes

Cisplatin and *carboplatin* are platinum (II) complexes that have proved to be useful agents in the treatment of cancer. Cisplatin and carboplatin have been shown to form identical types of adducts with DNA and have similar activities against ovarian and lung tumors; however, carboplatin is less toxic to the peripheral nervous system and the kidneys. The lower toxicity of carboplatin compared with cisplatin is believed to be due to the structure of carboplatin. The presence of the bidentatedicarboxylate ligand (Fig. 7-4) in carboplatin slows the degradation of carboplatin into potentially damaging derivatives. Indeed, in physiologic conditions, the retention half-life of carboplatin in blood plasma is 30 h, whereas that of cisplatin is only 1.5 to 3.6 h. In addition to the lower toxicity of carboplatin, it has been shown to work in some cases when cisplatin has failed. The decreased toxicity of carboplatin and the activity of carboplatin against cisplatin-resistant tumors have led to greater use of carboplatin.

A variety of six-coordinate platinum (IV) complexes were tested in cell culture, using six human ovarian cancer cell lines. Researchers found that as the ring size of the ligand increased, the complexes became increasingly more effective than cisplatin at killing cancer cells. Moreover, like carboplatin, these complexes were toxic to certain cancer cell lines that were resistant to cisplatin. Researchers believe that

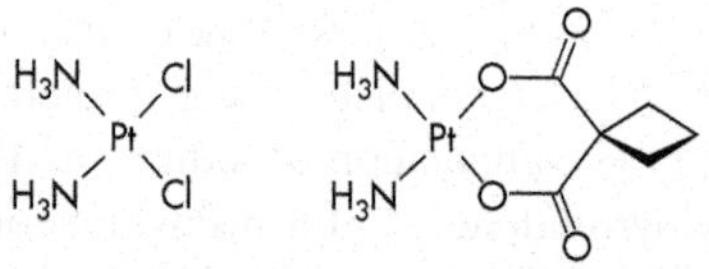

FIGURE 7-3 Chemical structure of deferoxamine.

FIGURE 7-4 The chemical structures of cisplatin and carboplatin.

these drugs are more effective than cisplatin because they are taken up into the cell in greater concentrations than is cisplatin.

Copper and Cobalt Complexes

Copper ion is present in a variety of important proteins and enzymes, including *hemocyanin, superoxide dismutase*, and *cytochrome oxidase*. Copper is typically in the Cu(I) state in the body, forming colorless tetrahedral complexes. Hemocyanin, an oxygen-binding protein, is a colorless Cu(I) complex that turns blue when bound to oxygen, indicating a transition to the Cu(II) state.

Like copper, cobalt exists in two oxidation states [Co(II) and Co(III)] in solution. The biological role of cobalt is largely confined to *vitamin B_{12} (cyanacobalamin)*. In vitamin B_{12}, the cobalt ion is situated in the center of the conjugated corrin ring structure (Fig. 7-5).

Zinc Complexes

Zinc is an important metal ion that is present in many proteins and confers structure and stability. Zinc is the only metal ion found in crystalline *insulin*. The insulin hexamer (six associated insulin molecules) can bind up to nine atoms of zinc. Recent studies also have shown that certain zinc-binding proteins interact

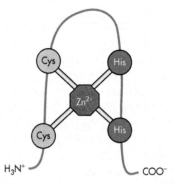

FIGURE 7-6 Illustration of a zinc ion tetrahydrally bound to histidine and cysteine residues of protein to form the zinc finger.

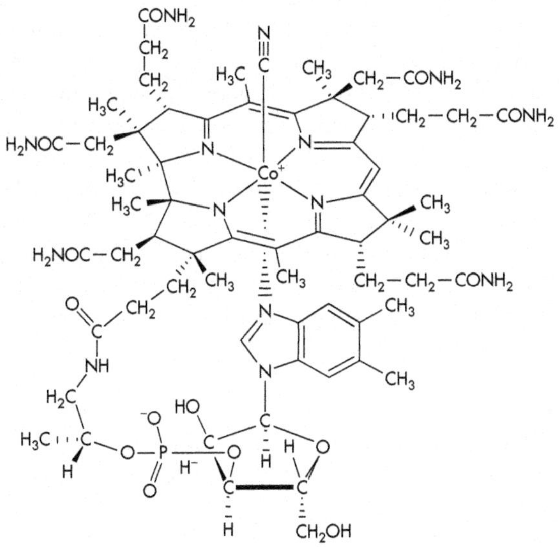

with DNA at specific sequences. The divalent zinc ion tetrahydrally bound to the two histine and two cysteine residues of the protein to form a loop (*zinc finger*), as shown in Figure 7-6, which can fit into the major groove of DNA. Zinc is also present in the enzymes *carboxypeptidase* and *carbonic anhydrase*.

Toxic Heavy Metal Complexes

Because of the presence of lead in older paints and water (from lead pipes and solder) and that of mercury in thermometers, poisoning incidences involving these heavy metal ions are very common, especially in the pediatric population. The incidence of lead and mercury poisoning is most significant in many developing countries that have lax regulation of the sources of these heavy metals. Lead and mercury toxicity commonly is treated by the administration of chelating agents such as *dicalcium salt of EDTA* and *2,3-dimercaptopropanol*, also known as BAL (British Anti-Lewisite) (Fig. 7-7).

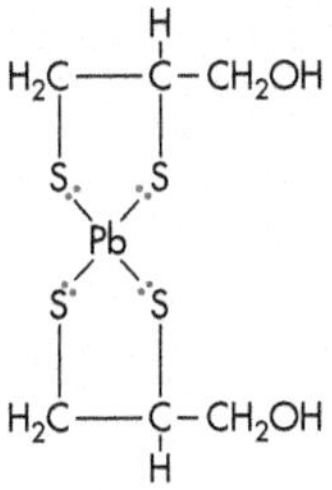

FIGURE 7-5 The chemical structure of vitamin B_{12} (cyanocobalamin).

FIGURE 7-7 The chemical structure of 2,3-dimercaptopropanol (BAL)–lead complex.

CYCLODEXTRIN COMPLEXES

One of the most important molecular complexations is the interaction between "guest" molecules and cyclodextrins to form reversible inclusion complexes. Cyclodextrins were isolated in 1891 by Villiers as degradation products of starch and were characterized as cyclic oligosaccharides in 1904 by Schardinger.[4] As shown in Figure 7-8, cyclodextrins are *donut-shaped molecules* of D-glucopyranose. There are three types of natural cyclodextrins—*alpha* (α), *beta* (β), and *gamma* (γ)—with six, seven, and eight residues of D-glucose, respectively.[5] The cyclodextrin cavity dimension, which is important for guest-host interaction, ranges from 5.0 Å for alpha cyclodextrins to 8.0 Å for gamma cyclodextrins.

Cyclodextrins are metabolized in the body and some have been found to be relatively nontoxic.[6] Six months of oral chronic toxicity of β-cyclodextrin, for instance, was studied in rats at a dose of up to 1.6 g/kg body weight. No sign of toxicity was observed with respect to weight gain, food consumption, or any of the clinical and biochemical values. One of the major limiting features of cyclodextrins is the low aqueous solubility of β-cyclodextrins (1.9 g/100 mL), which can limit their usefulness. Additional properties of cyclodextrins are summarized in Table 7-1. To overcome the solubility constraints, many derivatives of β-cyclodextrins, including methyl-, dimethyl-, and 2-hydroxypropyl-substituted forms, are synthesized as well. The reaction of β-cyclodextrin with propylene oxide forms 2-hydroxypropyl-β-cyclodextrin, with an aqueous solubility in excess of 60 g/100 mL.

Assuming that the guest molecule or its segment is of sufficient size to fit in the cyclodextrin cavity, complexation is mediated primarily by van der Waals attraction and hydrophobic interaction. The nonpolar nature of the cavity allows for the inclusion of a hydrophobic segment or a whole molecule in the cavity. The surface of cyclodextrins is highly hydrophilic because of the multiple hydroxyl (—OH) functional groups that are involved in hydrogen bonding with water. Cyclodextrin complexation, therefore, can serve as an effective means to increase the aqueous soubility of hydrophobic compounds.

Cyclodextrins are used in a number of different applications, ranging from food, cosmetics, and agriculture to pharmaceuticals.[7] In food preparations, cyclodextrins are used to stabilize flavors and eliminate unpleasant tastes and odors. Cyclodextrins have been used in cosmetics for the preparation of long-acting deodorants and emulsion bases and in dentifrices. Complexes of cyclodextrins with pesticides allow for a decrease in volatility and decomposition, allow the conversion of liquids into microcrystalline powders for easier application, increase water solubility with a significant decrease in toxicity and

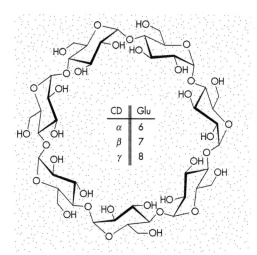

FIGURE 7-8 The chemical structure of cyclodextrins.

TABLE 7-1 Summary of Various Properties of Cyclodextrins

Cyclodextrin Type	No. Glucose Residues	Molecular Weight	Internal Cavity Dimension, Å	Melting Point, °C	Aqueous Solubility, g/100 mL
Alpha (α)	6	973	5	275	15
Beta (β)	7	1135	6	280	1.9
Gamma (γ)	8	1297	8	275	23

TABLE 7-2 Summary of Improvements in Properties of Selected Drug Compounds by Complexation with Cyclodextrins

Property	Drug Examples
Enhanced aqueous solubility	Prostaglandins; ketoprofen and other nonsteroidal anti-inflammatory drugs (NSAIDs); digoxin and digitoxin; progesterone, testosterone, and other steroid hormones; barbiturates; taxanes; chloramphenicol; sulfonamides; phenytoin; benzodiazepines; coumarin anticoagulants; some diuretics
Improved stability	Hydrolysis (aspirin, atropine, procaine, digoxin, and prostaglandins), oxidation (chlorpromazine and epinephrine), photodecomposition (phenothiazines, some antibiotics, and vitamins), dehydration (prostaglandin E)
Enhanced absorption and bioavailability	Aspirin, phenytoin, digoxin, ketoprofen and other NSAIDS, barbiturates, sulfonamides, and some diuretics
Change from liquid to solid	Oil-soluble vitamins (A, D, K), phenols, clofibrate, nitroglycerine, methyl salicylate, and essential oils
Decreased volatility	Iodine, camphor, menthol, salicylic acid, and chlorobutanol
Improved taste and odor	Chloral hydrate, prostaglandins, NSAIDS, thymol, and chloramphenicol
Decreased stomach irritation	Aspirin, indomethacin, and other NSAIDS
Inhibit red blood cell lysis	Phenothiazines and other basic drugs, fluphenamic acid and other acidic drugs, antibiotics, and menandione
Prevention of incompatibilities	Vitamins

negative environmental consequences, and allow safer handling during transportation and application. Cyclodextrins also have been used in analytic chemistry for enhancement of signal during spot analysis of thin-layer chromatography plates. Additionally, signal enhancement of cyclodextrin with fluorophores can be very useful in fluorescence spectroscopy.

In pharmaceutical applications, cyclodextrin complexation with drugs can yield significant benefits, such as the enhancement of the aqueous solubility, absorption, and bioavailability of hydrophobic drugs; stabilization of compounds that otherwise could degrade by heat, moisture, oxidation, or light; conversion of liquid products into powders; and the masking of unpleasant taste and odors (Table 7-2). In some instances, such as aspirin and indomethacin, cyclodextrin complexation can decrease the toxicity of the drug in the gastrointestinal tract. Figure 7-9 shows the interaction between cyclodextrin and some drugs by a 1:1 (drug:cyclodextrin) stoichiometry. Drugs also can complex with cyclodextrin in a 1:2 (sandwich form) and 1:3 complex. Calculations involving the complexation of cyclodextrins and their solubilizing effects are found in the Appendix of this chapter.

ION-EXCHANGE RESINS

Ion exchange is a method of complexation that is based on *electrostatic interactions* between the surface-bound ions on a solid particle (or resin) and the oppositely charged ions in aqueous solution. Ion-exchange resins are organic (polymer-based) or inorganic (minerals) solid particles with a positive or negative surface charge. Organic resins frequently are made of polystyrene that is cross-linked with divinylbenzene. The naturally occurring minerals, based on aluminum silicate chemistry, are referred to as *zeolites*. Ion-exchange resins are classified as *cation exchangers* or *anion exchangers* depending on the types of ions they can replace. Cation exchangers replace surface-bound positively charged ions with similarly charged ions in solution. In contrast, anion exchangers replace negatively charged ions on the surface with similarly charged ions in solution. Ion-exchange resins are used mainly for *purification* and *drug delivery* purposes.

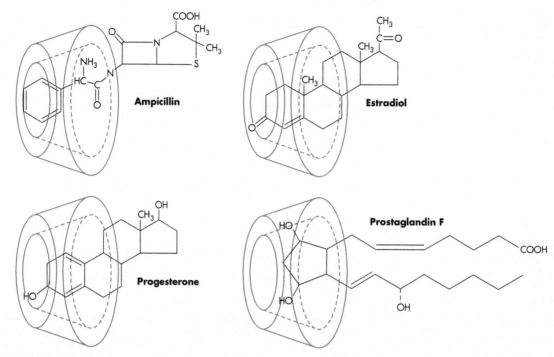

FIGURE 7-9 The structures of drug-cyclodextrin complexes.

In ion-exchange chromatography (i.e., separation), chemicals can be removed selectively from a solution by passing them through a solid resin bed (stationary phase). Certain ions in the solution replace ions or groups of ions in the resin, from which they then can be eluted, or washed out. By controlling the pH, ionic strength, and composition of the solution and the nature of the resin, ions in solution are more or less selectively exchanged for the ions that are on the resin. Hardness in water, which is caused by calcium and magnesium ions, which form insoluble compounds, is removed by ion exchange. The water is filtered through an artificial zeolite such as Permutite, and the sodium in the zeolite replaces the undesirable ions in the water. When the zeolite is saturated with these metallic ions, it is washed with salt solution, which restores the sodium. A wide variety of important drugs, such as antibiotics and vitamins, are isolated and purified from fermentation by using Amberlite ion exchange and XAD adsorbent resins in the critical purification step.

For drug delivery purposes, the drug-resin complex is prepared by continuous exposure of the resin to the drug to saturate all available ion-exchange sites on the resin. The complex is washed, dried, and packaged into the appropriate dosage form (powder, tablet, capsule, suspension, etc.). When the product is administered orally, as shown in Figure 7-10, the drug-resin complex will release the free drug by exchanging with similarly charged ions in the gastrointestinal tract. Drug release kinetics from the ion-exchange resins are dependent on the pH and ionic strength of the fluid in the gastrointestinal tract.

Most drug formulations that employ ion-exchange principles are made with cross-linked poly(styrene sulfonate) resins complexed with cationic drugs, such as hydrocodone and chloropherinamine in *Tussionex*. The Pennkinetic system developed by Pennwalt further encapsulates the drug-resin complex in a water-insoluble polymer coating to create a sustained-release oral liquid formulation. The long-acting cough suppressant *Delsym* is an example of the Pennkinetic system in which the dextromethorphan-resin complex is coated with ethylcellulose to provide 12 h of relief per dose.

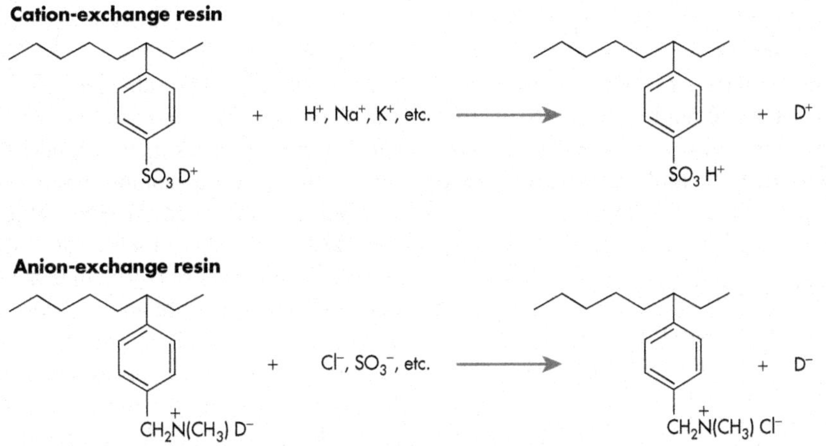

FIGURE 7-10 Principle of ion exchange in drug delivery systems. Cation exchangers such as poly(styrene sulfonate) will complex with positively charged drug (D⁺) and release by exchanging with a cation in the gastrointestinal fluid. Anion exchangers complex with negatively charged (D⁻) drugs and release by exchanging with anions.

Ion-exchange resins also can perform additional functions in the formulation, such as taste masking and enhanced disintegration and dissolution of tablets.

Several drugs employ ion-exchange principles for their therapeutic action. Sodium polystyrene sulfonate, a cation exchanger, in 70% sorbitol suspension (*Kayexalate*) is used for the treatment of elevated potassium levels (hyperkalemia). *Renagel* (cross-linked allylamine hydrochloride), an anion exchanger, binds with phosphate ions in the gastrointestinal tract and is used for the treatment of hyperphosphatemia in renal-impaired patients. *Cholestyramine* and *colestipol* resins, which are made with cross-linked polystyrene with a positive surface charge, are used for binding with bile acids in the gastrointestinal tract and to decrease serum cholesterol levels.

PROTEIN-LIGAND INTERACTION

The interaction between small molecules such as drugs and proteins is a widely studied biochemical phenomenon. Protein-ligand interaction is important in drug binding to its receptor for pharmacologic activity, enzyme-substrate interaction in catalysis, antibody-antigen recognition, and the interaction between drugs and proteins in plasma that affects the distribution profile of a drug in the body. Many of the interactions between protein and low-molecular-weight compounds occur in a *reversible manner* according to the following equilibrium:

$$[P] + [L] \rightleftharpoons [PL]$$

where $[P]$ is the molar concentration of the protein, $[L]$ is the molar concentration of the ligand (or drug), and $[PL]$ is the molar concentration of the protein-ligand complex.

The association constant (K_a), a measure of the *affinity* between the protein and the ligand, is described as follows:

$$K_a = \frac{[PL]}{[P][L]}$$

The association constant has units of $1/M$ (M^{-1} or liters per mole). The association constant is the reciprocal of the dissociation constant (K_d) (i.e., $K_a = 1/K_d$). The other important parameter to consider is the capacity of binding, which is defined by the number of ligand-binding sites on the protein (v).

Ligand-binding sites on proteins are usually depressions on the surface of the protein, with the size and shape dependent on the nature of the ligand. Very small ligands such as metal ions frequently are carried inside the protein, whereas larger ligands are bound to larger depressions. The basic concept behind ligand binding is that of complementarity between the ligand and the binding site. This is manifested as *steric complementarity* (the shape of the ligand is mirrored in the shape of the binding site) and *physicochemical complementarity*, allowing molecular interactions between the two.

Mechanisms of Interactions

Since proteins are molecules composed of different types of amino acids, the interactions between proteins and small molecules can occur through one or more types of noncovalent forces. The interactions between a protein and ligands are generally similar to those seen within the protein (i.e., hydrogen bonding, electrostatic interactions, van der Waals interactions, and hydrophobic interactions). Proteins can interact with small molecules as a result of *hydrogen bonding* between donor or acceptor functional groups in the amino acid sequence and donor or acceptor groups in the small molecule. In some instances, complex hydrogen bonding also can occur between the ligand and the protein involving water molecules as intermediates. *Electrostatic interactions* occur between charged amino acids (arginine, lysine, aspartic acid, and glutamic acid) with oppositely charged ligand molecules. Electrostatic interaction between the protein and the ligand, therefore, is affected by pH and the ionic strength of the solution. Dipole-dipole, dipole-induced dipole, and dispersion forces collectively are called *van der Waals forces*. Dipole-dipole interactions occur between uncharged polar molecules that carry a dipole moment. These attractive forces occur at small distances. *Hydrophobic interactions*, an interfacial phenomenon, occur as a result of attraction between nonpolar (hydrophobic) groups with subsequent release of water molecules in an entropically favorable manner. Hydrophobic interaction is the major stabilizing force in three-dimensional protein structure, where nonpolar amino acids (e.g., glycine, alanine, proline, phenylalanine, and valine) are in the interior of the molecule and have little or no contact with water. The polar amino acids (e.g., aspartic acid, glutamic acid, arginine, and lysine), by contrast, are on the surface of the protein. Physical unfolding (or denaturation) of proteins, as shown in Figure 7-11, results from perturbation in the environment of the molecule that can alter the hydrophobic interactions. Ligand binding with protein molecules also can occur through hydrophobic interactions, such as the association of bilirubin, a by-product of heme metabolism, with albumin.

Experimental Analysis of Protein-Ligand Interactions

Protein-ligand interactions are studied by using various experimental techniques, including *equilibrium dialysis*, *ultracentrifugation*, *gel filtration*, and *spectroscopic methods*. A brief description of each of the methods and how it is applied to protein-ligand binding studies follows.

Equilibrium dialysis, as shown in Figure 7-12, involves separation of compounds according to size difference (based on molecular weight), using membranes that have a specific molecular weight cutoff point.[8] Dialysis membranes made from cellulose, for instance, are commercially available that can allow compounds with a molecular weight lower than 6000 to 8000 to pass through. Proteins having significantly higher molecular weight than the cutoff point of the

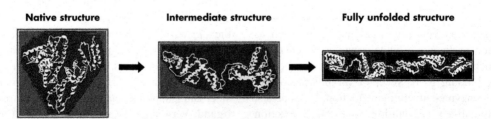

FIGURE 7-11 Illustration of protein unfolding from its native three-dimensional state to the unfolded state by physical denaturation.

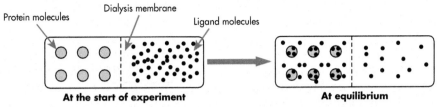

FIGURE 7-12 Experimental setup for equilibrium dialysis for analysis of protein-ligand interactions.

membrane will be retained in the dialysis chamber. In an equilibrium dialysis experiment, a known concentration of protein (P_t) is placed in the dialysis tubing, and the tubing is closed. When the protein-containing dialysis tubing is placed in the ligand solution (at constant temperature, pH, and ionic strength), the ligand will diffuse into the dialysis tubing and bind with the protein. At equilibrium, the concentrations of the free ligand inside and outside will be equal. Based on the concentration of the free ligand outside (L) and the concentration of the protein-ligand complex inside (PL), the association constant (K_a) and the number of binding sites (v) are determined as discussed later in this chapter (see "Analysis of Plasma Protein Binding" in the Appendix of this chapter.). A modified form of equilibrium dialysis is called *dynamic dialysis*, in which the known concentrations of protein and ligand are placed in the dialysis tubing. Periodically, the outside buffer solution is replaced with fresh solution to maintain *sink conditions*. Dynamic dialysis significantly accelerates the diffusion of free ligand into the buffer solution and thus can shorten the dialysis time.

Ultrafiltration is a method by which compounds with different molecular weights are separated with the aid of a high centrifugal force. Ultracentrifuges can spin protein-ligand mixtures at speeds in excess of 20,000 to 50,000 rpm, and the separation of bound ligand from free ligand occurs on the basis of the relative settling of compounds with different molecular weights in the centrifuge tube. Different types of weight gradients (e.g., cesium chloride or Ficoll) also can be used to facilitate the separation between the protein-ligand complex and free ligand in the tube. The main advantages of the ultracentrifuge over the equilibrium or dynamic dialysis method are the economy of time and the small amount of material needed. After separation, the concentrations of free ligand and protein-ligand complex are measured.

Gel filtration method was first adapted for measurement of protein-ligand interactions by Hummel and Dryer.[9] In the procedure, they equilibrated Sephadex (cross-linked dextran beads) in a solution of the ligand and packed it into a chromatographic column. A protein dissolved in small volume of the same ligand solution then is applied to the top of the column. The protein is eluted from the column, using the same ligand solution. Since Sephadex beads have a different elution profile that corresponds to the molecular weight of the compounds, the protein-ligand complex elutes first as it is excluded completely from the column. The area under the peak is related to the concentration of the protein-ligand complex. Other investigators have used radiolabeled ligands and found that the results from gel filtration are comparable to those obtained by dialysis or ultracentrifugation.

In recent years, a number of spectroscopic methods, including *ultraviolet* (UV) and *visible absorption spectroscopy*, *fluorescence spectroscopy*, *optical rotatory dispersion* (ORD), *circular dichroism* (CD), *nuclear magnetic resonance* (NMR), and *electron spin resonance* (EPR), have become very popular methods of analyzing protein-ligand interactions. In each of these methods, there is a change in the information provided by the spectroscopic method when the complex is formed relative to that of either the ligand or the protein alone. A very important advantage of spectroscopic methods is that they not only can measure the interaction between the protein and the ligand but also can provide useful information about the mechanism of interactions.[10] UV or visible absorption, for instance, is related to the polarity of the ligand-binding site of the protein. CD can provide very useful information about the changes in the three-dimensional conformation of the protein upon binding, while NMR indicates which functional group or part of the molecule is involved in the binding process. Fluorescence spectroscopy is also very useful, since the aromatic amino acids (tyrosine, tryptophan, and phenylalanine) are intrinsically fluorescent. The fluorescence of these amino acids is markedly altered upon binding with ligands. In addition, many ligands (especially drugs) are fluorescent as well. When the anticoagulant drug warfarin, for instance, binds with serum albumin, the efficiency of fluorescence, as measured by quantum yield, increases markedly and the maximum emission shifts to a shorter wavelength (higher energy). The increase in quantum yield and the shift in emission wavelength indicate that warfarin interacts with albumin at hydrophobic sites on albumin.

PLASMA PROTEIN BINDING

Upon systemic administration of therapeutic agents, those agents encounter a massive pool of plasma proteins, such as albumin and α_1-acid glycoproteins, that can interact with drug molecules and affect the biological properties of these molecules. In the present discussion, we will examine the properties of the important plasma proteins and the significance of drug binding.

Plasma Proteins

Human plasma is composed of about 200 known proteins. Proteins in plasma function in different capacities, including maintenance of osmotic pressure between intracellular and extracellular fluids (albumin), coagulation of blood (clotting factors, thrombin, fibrinogen, etc.), immune reactions (antibodies), and transport of endogenous and exogenous compounds; they also function as enzymes or hormones. For the purpose of this discussion, the proteins that are important in binding to drugs in the plasma include *albumin*, α_1-*acid glycoprotein*, and *lipoproteins* (Table 7-3).

As a result of its abundance in plasma, serum albumin is the most important protein in terms of drug binding. Almost all *in vitro* studies of protein binding have used albumin, and the results are generalized to the whole plasma. Albumin is an acidic protein (isoelectric point 4.8) with a molecular weight of 69,000.[11] It is very soluble in water and thus has a concentration of 30 to 50 mg/mL in

TABLE 7-3 Properties of Major Plasma Proteins That Bind with Drugs

Plasma Protein	Molecular Weight	Normal Concentration Range, mg/mL
Albumin	69,000	35–50
α_1-acid glycoprotein	44,000	0.4–1.0
Lipoproteins	200,000–3,400,000	Variable

normal human plasma. Albumin is also a very stable protein because of the 17 disulfide bonds in the tertiary structure. It has an elimination half-life of 17 to 18 days. As shown in Figure 7-13, albumin has an ellipsoidal structure in solution with an overall dimension of 40 × 140 Å. The primary function of albumin is to serve as an osmotic agent for the regulation of pressure differences between intracellular and extracellular fluids. It also serves as an important transport protein in the plasma for both endogenous and exogenous compounds.

The most salient aspect of albumin is its ability to bind to many different types of molecules. The majority of ligands bind with albumin by hydrophobic interactions because of the immense flexibility in the protein molecule, which allows for conformation change to accommodate the different shapes of the ligands. Long-chain fatty acids such as oleic, stearic, linoleic, and palmitic acids are the most tightly bound organic ligands to albumin. The interaction of bilirubin, a by-product of heme metabolism, with albumin has been found to occur at two different sites. One of the binding sites has a very high affinity ($K_a = 5.01 \times 10^7 \, M^{-1}$) for bilirubin, while the other binds with

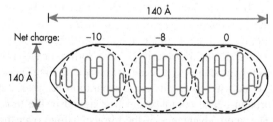

FIGURE 7-13 The structure of serum albumin based on physical properties (adapted from **Peters T:** Serum albumin. *Adv Protein Chem* 37:161, 1985).

slightly lower affinity ($K_a = 3.16 \times 10^5\ M^{-1}$). In some instances, the interactions with albumin can be very specific. For instance, the affinity of L-tryptophan is 100-fold higher than that of the D isomer. Albumin also can interact by electrostatic forces, and this is responsible for the binding of different cations, including metal ions. More than 50% of plasma calcium ion is bound to albumin with a relatively low binding affinity ($K_a = 900\ M^{-1}$). Other metal ions, including copper, nickel, manganese, cobalt, and cadmium, also are bound to albumin with varying degrees of affinity. Many weakly acidic drugs, such as salicylates and penicillins, interact preferentially with albumin.

α_1-acid glycoprotein is a globular protein with a molecular weight of 44,000 daltons. The plasma concentration of α_1-acid glycoprotein is very low, and it preferentially binds with basic drugs such as imipramine, propranolol, and lidocaine. Lipoproteins are high-molecular-weight complexes of proteins with lipids and are classified according to their densities and separation in an ultracentrifuge. Lipoproteins are classified as very low density (VLDL), low density (LDL), or high density (HDL). They are primarily responsible for the binding and transport of plasma lipids; however, they do interact with drugs, especially when the albumin-binding sites are saturated. Other serum proteins, such as immunoglobulins and transferrin, also are involved in drug binding to a minor extent.

Significance of Plasma Protein Binding

The interactions between drugs and plasma proteins after systemic administration can have profound implications for the therapeutic outcomes. The relationship between plasma protein binding, physiologic distribution, and elimination is shown in Figure 7-14. Highly protein-bound drugs tend to remain mainly in the systemic circulation, as opposed to binding with adipose tissue, and have a relatively lower volume of distribution. These drugs also tend to have a longer biological half-life in the body.[12] The two most important parameters of protein binding are *affinity* and *capacity*. Affinity of binding, or the association constant (K_a), is a measure of the strength of the interaction between the protein and the drug molecule. Since the binding is reversible, a drug with higher affinity (or a larger K_a value) can displace another drug from the same binding site. Displacement of bound drug can lead to significant toxicity as the free drug interacts with the active site (receptor or enzyme) to produce the pharmacologic activity. Some drugs (e.g., warfarin) are extensively bound to plasma proteins, but with relatively weak affinity. Therefore, many nonsteroidal anti-inflammatory agents can displace warfarin from its binding site and cause serious hemorrhagic complications. Also important in the consideration of protein binding are the changes in concentrations of available proteins caused by disease, age, malnutrition, trauma, and the like. If the protein concentration decreases, the relative increase in free drug concentration will cause significant toxicity at the same dose.

Mathematical analysis of protein-drug interactions is performed to evaluate the binding affinity or association constant (K_a) and the capacity, as measured by the number of binding sites (v). Details of this analysis, including two common methods—*double-reciprocal* and *Scatchard* methods—are described in the Appendix of this chapter.

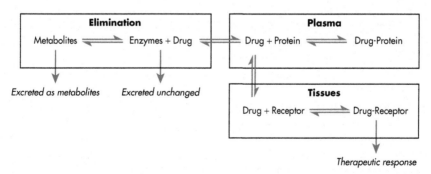

FIGURE 7-14 Effect of protein binding on drug distribution into the tissues and elimination from the body.

PROBLEMS

1. Identify the central metal ion, the ligands, and the coordination numbers of the following coordinate complexes:
   a. $[Co(NH_3)_6]^{3+}$
   b. $[Cr(H_2O)_6]^{3+}$
   c. $[Ag(NH_3)_2]^{+}$
   d. $[Fe(CN)_6]^{4-}$

2. The following data show the solubility of aspirin in the presence of β-cyclodextrin. (Refer to "Calculations of Cyclodextrin Complexation" in the Appendix of this chapter.)

β-Cyclodextrin Conc., mmole/L	Aspirin Conc., mmole/L
1.50	3.91
3.00	5.07
5.00	6.61
10.00	10.47
15.00	14.33
20.00	18.19
30.00	25.91
50.00	41.35

   a. Plot the data according to the Higuchi-Connors phase-solubility relationship and, using regression analysis, determine the slope and the y intercept.
   b. Determine the intrinsic solubility of aspirin (D_0) and the association constant (K_a), assuming that aspirin forms a 1:1 complex with β-cyclodextrin.

3. a. The binding of drugs to plasma proteins can have a profound effect on therapy. Discuss the important consequences of drug-protein interactions in the body for therapeutic outcomes.
   b. *In vitro* study of plasma protein binding is done by performing an equilibrium dialysis experiment. With the aid of a diagram, explain the process of equilibrium dialysis.
   c. Although there are more than 200 identified proteins in the plasma, serum albumin is considered the most important for drug binding. Explain why albumin is the most important protein in plasma for studying drug binding.

4. The following data were obtained in the *in vitro* binding study of a new sulfonamide antibiotic with human serum albumin at 37°C. (Refer to "Analysis of Plasma Protein Binding" in the Appendix of this chapter.)

r	[D], ηmole/L
0.60	1.00
1.10	1.43
1.70	2.00
1.90	2.69
2.50	4.29

   a. Plot the data on a linear graph paper according to the double-reciprocal model.
   b. From linear regression analysis, calculate the association constant (K_a) and the number of binding sites (v).

5. The following data were obtained in the *in vitro* binding study of a new antipsychotic medication with human serum albumin at 37°C. Refer to "Analysis of Plasma Protein Binding" in the Appendix of this chapter.

r	[D], μmole/L
0.512	4.125
0.836	8.137
1.489	19.240
2.245	53.192

   a. Plot the data on a linear graph paper according to the Scatchard model.
   b. From linear regression analysis, calculate the association constant (K_a) and the number of binding sites (v).
   c. If the association constant of aspirin with albumin is $1.67 \times 10^6 \, M^{-1}$, discuss the possible complications of aspirin coadministration, if any, in a patient who is already on this drug.

ANSWERS

1. **a.** $[Co(NH_3)_6]^{3+}$
   The central metal ion is cobalt (Co).
   The ligands are ammonia (NH_3).
   The coordination number is 6.
   b. $[Cr(H_2O)_6]^{3+}$
   The central metal ion is chromium (Cr).
   The ligands are water molecules (H_2O).
   The coordination number is 6.
   c. $[Ag(NH_3)_2]^+$
   The central metal ion is silver (Ag).
   The ligands are ammonia (NH_3).
   The coordination number is 2.
   d. $[Fe(CN)_6]^{4-}$
   The central metal ion is iron (Fe).
   The ligands are cyano (CN).
   The coordination number is 6.

2. **a.** Higuchi-Connors phase-solubility profile of aspirin in the presence of β-cyclodextrin: Based on the equation of the regression line (i.e., $y = 0.772x + 2.75$), the slope is 0.772 and the y-intercept is 2.75.
   b. The intrinsic solubility of aspirin is given by the y-intercept and is equal to 2.75×10^{-3} moles/L.
   The association constant (K_a) is calculated from the slope:

   $$K_a = \text{slope}/[D_0(1-\text{slope})]$$
   $$K_a = 0.772/[2.75 \times 10^{-3}(1 - 0.772)]$$
   $$= 1{,}231.3\ M^{-1}$$

3. **a.** Protein binding can affect the biological distribution, metabolism, and excretion of drugs from the body. Additionally, since protein binding is reversible, a drug with higher affinity will displace a weakly bound drug from the protein binding site, resulting in elevation of free drug concentration in the body. In many cases, the increase in free drug concentration will cause toxicity.
   b. Equilibrium dialysis involves passage of low-molecular-weight compound across a dialysis membrane with a specific molecular weight cutoff point. For a diagram and additional discussion of equilibrium dialysis, see the section titled "Experimental Analysis of Protein-Ligand Interactions."
   c. Albumin is the important protein for plasma protein binding because:
   a) It is the most abundant protein in plasma
   b) It specifically has sites for different types of endogenous ligand binding (e.g., bilirubin and calcium ions)
   c) Albumin is very stable"

4. **a.** According to the double-reciprocal method, the binding of sulfonamide to serum albumin at 37°C will be as follows:

r	[D] (nmole/L)	1/r	1/[D] (L/mole × 10^{-9})
0.60	1.00	1.6667	1.000
1.10	1.43	0.9091	0.699
1.70	2.00	0.5882	0.500
1.90	2.69	0.5263	0.372
2.50	4.29	0.4000	0.233

   The double-reciprocal plot will be:

   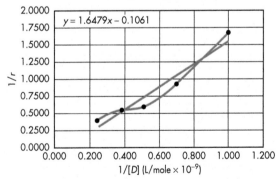

   b. From the equation of the line (i.e., $y = 1.6479x - 0.1061$), the values of K_a and v are calculated from the slope and y-intercept.

   Slope = $1/vK_a = 1.65 \times 10^{-9}$ (mole/L)
   y-intercept = $1/v = 0.1061$

   Therefore, the number of binding sites (v) = 1/0.1061 or 9.4 (~9.0). And $1/vK_a = 1.65 \times 10^{-9}$ (mole/L) and $v = 9.4$, therefore,

   $$K_a = 6.45 \times 10^7\ (L/mole)$$

5. a. According to the Scatchard equation, the data will be:

r	[D] (μmole/L)	r/[D] (L/mole × 10^{-6})
0.512	4.125	0.124
0.836	8.137	0.103
1.489	19.240	0.077
2.245	53.192	0.042

The Scatchard plot of the antipsychotic medication is given as follows:

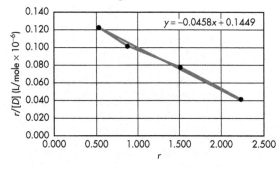

b. The slope = $-K_a$. Therefore, $-K_a = -0.046 \times 10^6$

$$K_a = 4.6 \times 10^4 \text{ L/mole}$$

The y-intercept = vK
Therefore, $vK = 1.45 \times 10^5$

$$v = (1.45 \times 10^5)/(4.6 \times 10^4)$$
$$v = 3.15 \ (\sim 3 \text{ binding sites})$$

c. Since aspirin has a higher dissociation constant than that of the anti-psychotic medication, it will displace the anti-psychotic drug molecules from albumin binding sites.

KEY POINTS

▶ Complexation involves the formation of a complex between a ligand and a substrate. Mechanisms of complexation include both covalent and noncovalent interactions.
▶ The two main types of complexes are coordination complexes and molecular complexes.
▶ Coordination complexes usually involve the covalent binding of a metal ion to an oppositely charged ligand. Examples of important coordination complexes in pharmacy include chelators such as EDTA (used to bind oxidation-initiating metals) and desferroxamine (used in iron overdoses).
▶ Molecular complexes result from the noncovalent interaction of a substrate with its ligand. They are widespread in pharmacy and include cyclodextrin complexation, ion-exchange resins, and protein binding.
▶ Among the most important examples of protein binding is the binding of drug ligand to plasma albumin. Drug bound to albumin cannot interact with its receptor.
▶ The ability of some drugs to displace other drugs from plasma albumin is a very important type of drug interaction that pharmacists must be aware of, because it can lead to an increased concentration of free drug to higher levels than expected.

CLINICAL QUESTIONS

The product Commit is a nicotine lozenge used for smoking cessation. It is a compressed formulation of cationic nicotine bound to polacrilex, a weak acid cation exchange resin, plus other excipients. It's meant to be slowly dissolved in the mouth to release nicotine over about a 30-minute time period. Conjecture what causes the nicotine to be displaced from the resin during the time that it remains in the mouth.

Albumin is a well-known binder of many drugs in plasma, including lipophilic drugs. That interaction has been put to an advantage in an injectable product that utilizes albumin to carry the poorly soluble lipophilic drug paclitaxel. It is indicated for the treatment of breast cancer. Conjecture what part of the protein, interior or exterior, this lipophilic drug is likely to interact with and what type of binding forces are likely involved.

APPENDIX

Calculations of Cyclodextrin Complexation

Measurement of the association constant (K_a, also known as the stability constant) is important, since this is the index of the physicochemical properties of the drug upon inclusion. One of the most useful empirical analyses of cyclodextrin inclusion complexes is the *phase-solubility profile* that initially was proposed by Higuchi and Connors[13] In this analysis, the effect of the solubilizer or ligand (i.e., cyclodextrin) in improving the solubility of the substrate (i.e., drug) is examined. Experimentally, the drug of interest is added in excess to different vials, and incremental concentrations of cyclodextrin aqueous solution are added to the vials. After mixing at constant temperature until equilibrium is reached, the excess solid drug is filtered and the molar concentration of the complexed drug is determined. A phase-solubility diagram is plotted with the total molar solubility of the drug (i.e., intrinsic solubility and solubility increase as a result of complexation) on the y axis and the molar cyclodextrin concentration on the x axis. These phase-solubility diagrams not only allow for qualitative assessments of the formed complexes but also can be used to determine the association constant between the drug and the cyclodextrin. Based on the equilibrium between drug $[D]$ and cyclodextrin $[C]$:

$$m[D] + n[C] \rightleftharpoons [D_m - C_n]$$

the association constant K_a:

$$K_a = [D_m - C_n]/([D]^m [C]^n)$$

Given that the instrinsic solubility of the drug (in the absence of cyclodextrin) is equal to D_0, the total solubility (D_t) of cyclodextrin in the presence of drug will be:

$$D_t = D_0 + m[D_m - C_n]$$

Given the intrinsic solubility of cyclodextrin to be C_0, the total solubility (C_t) in the presence of cyclodextrins will be:

$$C_t = C_0 + n[D_m - C_n]$$

From these equations, the values of $[D_m - C_n]$, $[D]$, and $[C]$ can be obtained as:

$$[D] = D_0$$
$$[D_m - C_n] = (D_t - D_0)/m$$
$$[C] = C_t - n[D_m - C_n]$$

EXAMPLE 2 ▶▶▶

The inclusion complexation of *p*-aminobenzoic acid with β-cyclodextrins was studied according to the solubility method. The concentration of *p*-amino-benzoic acid complex (in μmoles per liter) as a function of β-cyclodextrin concentration (mmoles per liter) was reported as follows:

β-Cyclodextrin concentration (mmoles/L): 0.50, 1.00, 1.50, 2.00, 2.50, 3.00, 4.00, 5.00

p-Aminobenzoic acid concentration (mmoles/L): 1.74, 2.23, 2.72, 3.21, 3.70, 4.19, 5.17, 6.16

a. Plot the phase-solubility profile of p-aminobenzoic acid concentration (y axis) as a function of β-cyclodextrin concentration (x axis) and, by regression analysis, determine the equation of the straight line.

b. Compute the intrinsic solubility (D_0) of p-aminobenzoic acid.

c. Calculate the association constant K_a of the p-aminobenzoic acid–β-cyclodextrin complex from the slope of the line according to the equation $K_a = \text{slope}/(D_0(1 - \text{slope}))$.

Solution

a. The phase-solubility profile is as follows:

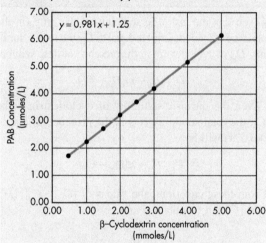

b. Based on the equation of the regression line (i.e., $y = 0.981x + 1.25$), $y = 0.981x + 1.21$, the intrinsic solubility of p-aminobenzoic acid (solubility in the absence of β-cyclodextrin) is given by the y intercept and is equal to 1.25×10^{-3} moles/L.

c. Using the equation, $K_a = \text{slope}/(D_0(1 - \text{slope}))$:

$$K_a = 0.981/(1.25 \times 10^{-3}(1 - 0.981))$$

$$= 41{,}305\ M^{-1}$$

Substituting these parameters into the association constant K_a equation, the total solubility of the drug (i.e., D_t) will be:

$$D_t = D_0 + \{(mK_aD_0^m)/(1 + K_aD_0^m)\}(C_t)$$

Notice that this equation is of the form $y = b - mx$ (straight line), and when D_t (total drug solubility) is plotted as a function of C_t (total cyclodextrin concentration), a straight-line relationship is obtained with the slope equal to $\{(mK_aD_0^m)/(1 + K_aD_0^m)\}$ and the y intercept equal to D_0. If m and n, the stoichiometry of cyclodextrin complexation, are known, the association constant K_a can be calculated easily. For a 1:1 (i.e., one molecule of drug complex with one molecule of cyclodextrin), the K_a is given by:

$$K_a = \text{slope}/(D_0(1 - \text{slope}))$$

Table 7-4 shows the association constants and the stoichiometry (drug:cyclodextrins) of some drugs in the presence of α, β, and γ cyclodextrins.[14]

Analysis of Plasma Protein Binding

As stated previously, mathematical analysis of protein-drug interactions is performed to evaluate the binding affinity or association constant (K_a) and the capacity, as measured by the number of binding sites (v). *In vitro* analysis of protein-drug interactions usually is studied with a single protein solution (e.g., albumin) in conditions that simulate the physiologic environment (aqueous buffer of pH 7.4, ionic strength 0.16, and temperature 37°C, or 98.6°F). *In vivo* analysis of protein binding is discussed in terms of the *fraction unbound* (f_u) in the plasma.

Binding Equilibria

The interactions between protein and drugs are described according to the law of mass action:

$$\text{Protein }[P] + \text{drug }[D] \underset{K_r}{\overset{K_f}{\rightleftharpoons}} \text{protein-drug complex }[PD]$$

where K_f is the forward rate constant and K_r is the reverse rate constant. At equilibrium, the rate of forward reaction is equal to the rate of the reverse reaction; therefore, the association constant (K_a) is described as:

$$K_a = [PD]/([P][D]) \text{ or } K_f/-K_r$$

where $[PD]$ is the molar concentration of the protein-drug complex, $[P]$ is the molar concentration of

TABLE 7-4 The Association Constant (K_a, M^{-1}) and Stoichiometries of Drug-Cyclodextrin Complexes in Aqueous Solution at 25°C

	α-Cyclodextrin		β-Cyclodextrin		γ-Cyclodextrin	
Drug	K_a (M^{-1})	Stoichiometry	K_a (M^{-1})	Stoichiometry	K_a (M^{-1})	Stoichiometry
Digoxin	180		11,200		12,200	1:4
Digitoxin	290		17,000		63,600	1:4
Diazepam	25		220		120	2:3
Fenbufen	30	1:2	440		180	1:1
Flubiprofen	20		5100	1:1	460	1:1
Ketoprofen	13		740	1:1	110	1:1
Ibuprofen	26		1030	1:1	80	1:1
Isosorbidedinitrate	30	1:2	120	1:1	20	2:1
Prostaglandin E$_1$	1430		1700		530	1:2
Hydrocortisone	60		1720	1:2	2240	2:3
Testosterone	130		7540	2:3	16,500	1:2
Progesterone	150		13,300	1:2	24,000	2:3
Prednisolone	300		3600	1:2	3240	2:3
Triamcinolone	120		2370	1:2	9920	2:3
Spironolactone	960		27,500	1:2	7600	2:3
Thiopental	260		1960		320	1:1
Tolbutamide	70		320	1:2		
Vitamin k	90	1:2	190		80	1:1

Source: Adapted from **Uekama K, Otagiri M:** Cyclodextrins in drug carrier systems. *Crit Rev Ther Drug Carrier Syst* 3:1, 1987.

protein with free binding sites, and [D] is the molar concentration of the free drug. The magnitude of the association constant (K_a) represents the affinity of drug-protein interactions.

From empirical observations of protein binding analysis (e.g., by equilibrium dialysis), the ratio r, which relates to the concentration of bound drug (i.e., [PD]) to the total protein concentration [P_t], is determined:

$$r = \frac{[PD]}{[P_t]}$$

By mass balance, the total number of moles of protein (i.e., [P_t]) is equal to the moles of bound drug (i.e., [PD]) plus moles of protein with free binding sites (i.e., [P]). The equation for r can be written as:

$$r = \frac{[PD]}{[P]} + [PD]$$

Using the equilibrium equation

(i.e., $K_a = [PD]/([P][D])$),
$[PD] = (K_a[P][D])$

Substituting the previous equation for [PD] in the r equation:

$$r = (K_a[P][D])/([P] + (K_a[P][D]))$$
or
$$r = K_a[D]/(1 + (K_a[D]))$$

This equation describes the simplest situation, where 1 mole of protein binds with 1 mole of drug to form a 1:1 complex. However, if the number of binding sites (v) is greater than one, the equation is modified as:

$$r = \frac{vK_a[D]}{1 + (K_a[D])}$$

where v represents the number of drug-binding sites on the protein that have the same affinity. The

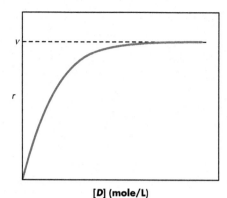

FIGURE 7-15 Langmuir isotherm of the ratio of the molar concentration of bound drug per total moles of protein (i.e., $r = [PD]/[P_t]$) as a function of the molar concentration of free drug (i.e., $[D]$).

equation is similar to the Langmuir adsorption isotherm (see Chapter 9), and the plot of r versus $[D]$, as shown in Figure 7-15, will yield a hyperbolic curve.

Analysis of Binding by the Double-Reciprocal Method

To determine the magnitude of the association constant (K_a) and the number of binding sites (v), the Langmuir equation is converted to a straight-line form. Using the double-reciprocal method:

$$1/r = \frac{(1 + (K_a[D]))}{v(K_a[D])}$$

or

$$1/r = \frac{1}{(vK_a[D])} + 1/v$$

From this equation, it is clear that the graph of $1/r$ versus $1/[D]$ will yield a straight line (Fig. 7-16). The slope of the line will be equal to $1/(vK_a)$, and the y intercept will be equal to $1/v$.

EXAMPLE 3 ▶▶▶

The following data were obtained in the *in vitro* binding study of naproxen, a nonsteroidal anti-inflammatory agent, with human serum albumin at 37°C.

[D] (μmole/L): 0.02, 0.04, 0.06, 0.08, 0.10, 0.12

r: 1.89, 1.94, 1.96, 1.97, 1.97, 1.98

a. Plot the data according to the double-reciprocal model of the protein-binding equations as follows:

$$1/r = (1/vK_a)(1/[D]) + 1/v$$

where r is the ratio of the concentration of bound drug to the total protein concentration ($[PD]/[P_t]$), K_a is the association constant, $[D]$ is the molar concentration of free drug, and v is the number of binding sites per protein molecule.

b. After linear regression analysis, from the slope and the y intercept, calculate the binding constant (K_a) and the number of binding sites (v) of naproxen on albumin.

Solution

a. For the plot of $1/r$ versus $1/[D]$ according to the double-reciprocal model, the values of $1/r$ and $1/[D]$ are

[D] (μmole/L)	0.02, 0.04, 0.06, 0.08, 0.10, 0.12
$1/[D]$ (L/mole $\times 10^{-6}$)	50.0, 25.0, 16.7, 12.5, 10.0, 8.33
r	1.89, 1.94, 1.96, 1.97, 1.97, 1.98
$1/r$	0.53, 0.52, 0.51, 0.50, 0.50, 0.50

The plot of $1/r$ (y axis) vesus $1/[D]$ therefore will be:

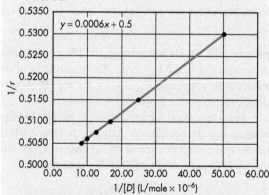

b. From the equation of the line ($y = 0.0006x + 0.5$), the values of K_a and v are calculated from the slope and the y intercept.

Slope $= 1/vK_a = 6.0 \times 10^{-10}$ (mole/L)

y intercept $= 1/v = 0.5$

Therefore, the number of binding sites (v) = $1/0.5$, or 2.0. And $1/vK_a = 6.0 \times 10^{-10}$ (mole/L) and $v = 2.0$:

$$K_a = 8.33 \times 10^8 \text{ L/mole}$$

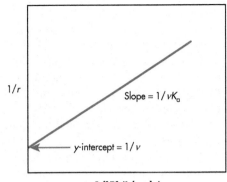

FIGURE 7-16 Plot of 1/r (y axis) versus 1/[D] (x axis) to describe the double-reciprocal relationship in drug-protein binding analysis.

Analysis of Binding by the Scatchard Method

According to the Scatchard method, the equation:

$$r = \frac{vK_a[D]}{1+(K_a[D])}$$

is converted into a straight-line form by cross-multipying:

$$vK_a[D] = r(1+(K_a[D]))$$

or $\quad vK_a[D] = (r + rK_a[D])$

Bringing r to the left side and dividing both sides by $[D]$:

$$r/[D] = (vK_a[D])/[D] - (rK_a[D])/[D]$$

or $\quad r/[D] = vK_a - rK_a$

This is a straight-line equation when $r/[D]$ (y axis) is plotted as a function of r (x axis). The slope is equal to $-K_a$, and the y intercept is vK_a (Fig. 7-17).

EXAMPLE 4 ▶▶▶

The interactions of a new anti-HIV drug used for the treatment of AIDS with plasma proteins have important therapeutic implications. The following data were obtained for the binding of the anti-HIV drug to human plasma proteins at 37°C.

[D] (nmole/L)	3.00, 6.00, 17.2, 50.8
r	0.40, 0.72, 1.35, 2.00

a. Plot the data according to the Scatchard equation as follows:

$$r/[D] = vK_a - rK_a$$

where r is the ratio of the concentration of bound drug to the total protein concentration ($[PD]/[P_t]$), K_a is the association constant, $[D]$ is the concentration of free drug, and v is the number of binding sites per protein molecule.

b. From the slope and the y intercept of the straight line, calculate the association constant (K_a) and the number of binding sites (v).

c. Using the value of K_a obtained in b, explain the significance of the displacement of the anti-HIV drug from the binding site for the therapeutic outcome in AIDS.

Solution

a. For the Scatchard plot, it is necessary to create a third row in the data of $r/[D]$ values as follows:

[D] (nmole/L)	3.00, 6.00, 17.2, 50.8
r	0.40, 0.72, 1.35, 2.00
$r/[D]$ (L/mole × 10⁻⁹)	0.13, 0.12, 0.08, 0.04

The Scatchard plot of anti-HIV drug binding to plasma proteins will be:

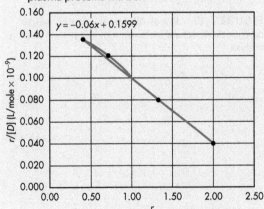

b. The slope = $-K_a$ (the association constant). Therefore, $-K_a = -0.06 \times 10^9$

$$K_a = 6.00 \times 10^7 \text{ L/mole}$$

The y intercept = vK

Therefore, $vK = 1.60 \times 10^8$

$$v = (1.60 \times 10^8)/(6.00 \times 10^7)$$

$$v = 2.67 \; (\sim 3 \text{ binding sites per molecule})$$

c. The binding association constant of 6.00 × 10^7 M^{-1} means that any drug with a K_a value greater than 6.00×10^7 M^{-1} will be able to displace the anti-HIV drug from the binding site. The increase in free anti-HIV drug concentration in the plasma can result in severe toxicity. In an AIDS patient, it is also important to note that the plasma protein concentrations, including that of albumin, are significantly lower than they are in healthy individuals.

In the protein-drug interaction studies discussed, we have assumed that even if there is more than one binding site for the drug, the binding affinity (i.e., K_a) is the same for all the drug molecules. However, in some cases, it is clear that drugs that have multiple binding sites also could have more than one binding association constant (or affinity). Two moles of bilirubin, as discussed, bind to 1 mole of albumin with one high-affinity site and one relatively low-affinity site. When the binding affinities are different, the plot of the Scatchard analysis shows a nonlinear profile, as depicted in Figure 7-18.

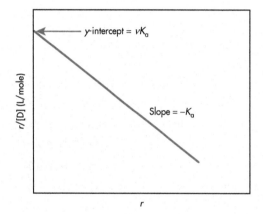

FIGURE 7-17 Plot of $r/[D]$ (y axis) versus r (x axis) to describe the Scatchard relationship in drug-protein binding analysis.

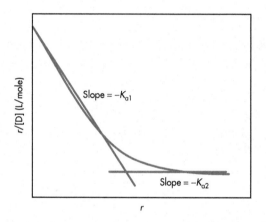

FIGURE 7-18 Nonlinear plot according to Scatchard analysis that represents multiple binding affinities (K_{a1} and K_{a2}).

8 Dispersed Systems

W. Cary Mobley

Learning Objectives

After completing this chapter, the reader should be able to:

▶ Describe the difference between homogenous and heterogeneous dispersions.

▶ Describe classifications of dispersed systems, based on the phases of the components and based on the size of the dispersed particles.

▶ Describe the properties and list examples of colloidal dispersions and coarse dispersions.

▶ Describe the natural forces involved in the movement and interaction of dispersed particles.

▶ Describe the clinical importance and mechanisms of destabilization of pharmaceutical dispersions.

▶ Apply knowledge of the parameters involved in dispersion destabilization to recognize, and prevent or resolve, the formulation, preparation, or patient use challenges associated with dispersed dosage forms.

Pharmaceutical dispersions are systems where one substance is dispersed within another substance. In this sense, these would include *homogenous molecular* dispersions—the true solutions discussed in Chapter 3. However, the dispersions discussed in this chapter differ from true solutions in that they are *heterogeneous* systems—as exemplified by suspensions and emulsions—where the dispersed phase, typically a particle of some type, is physically distinguishable from the medium in which it is dispersed. In pharmacy, dispersions are found in a wide variety of dosage forms and in nearly all routes of drug administration. Examples range from solutions of very large molecules (macromolecules) such as albumin and polysaccharides, to liquid suspensions of "nano"-sized crystals (nanocrystals) and of "micro"-sized droplets (microemulsions), to *coarse* (larger particle) emulsions and suspensions. Having physically distinct phases endows dispersions with properties that are different from true solutions, and it's these properties, such as particle settling and aggregation, that are the main concepts discussed in this chapter.

Terms used to describe dispersion components include *internal* or noncontinuous phase to describe the *dispersed phase* component (the particles) and the *external* or *continuous* phase to describe the *dispersion medium*. As for the nature of the dispersion mixture, the different phases can be the same state of matter, such as with an oil droplet-in-water (o/w) emulsion, or different states of matter, such as a suspension of solid crystals in water. How dispersions are classified can also vary, with the two main approaches to classification being (1) the nature of the dispersion's internal and external phases (e.g., solid, liquid, or gas) and (2) the size range of its dispersed particles (*colloidal* versus *coarse*).

DISPERSED SYSTEMS CLASSIFIED BY THEIR PHASES

Table 8-1 lists several examples of dispersions classified by the states of matter that comprise the dispersed phase and the dispersion medium. In addition to the macromolecular dispersions,

TABLE 8-1 Examples of Dispersions Based on the States of Matter of the Dispersed and Continuous Phases

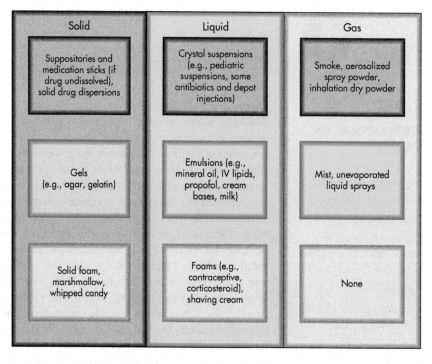

liquid-in-liquid emulsions, and solid-in-liquid suspensions mentioned, they include solid-in-solid suspensions (e.g., some suppositories), semisolid emulsions (e.g., creams), and gas-in-liquid foams. Adding to the spectrum of available dispersions are those that have more than two phases. Common examples of these include the typical over-the-counter hydrocortisone cream, in which the solid hydrocortisone crystals are suspended in an oil-water emulsion cream base, and ice cream, which contains a mixture of solid (ice crystals), liquid (fat globules), and gas (air bubbles). Contributing to the potential complexity of dispersed dosage forms is that the phases can be different during manufacture compared to the resulting product that sits on the shelf. For example, certain semisolids such as many creams, and some solids such as suppositories, are prepared while the medium is in the melted (liquid) form. They are then allowed to cool to the more solid forms used by the patients. This is important because liquid dispersions in particular can have problems with dispersed particle aggregation and settling. While concerns about these problems resolve to different degrees after melted forms of semisolid and solids solidify, the concerns persist for the entire shelf life of many liquid dispersions, because of continuing particle mobility.

Fortunately, issues of settling, aggregation, and other problems with dispersions can be explained by scientific principles that can be used to guide pharmaceutical scientists in product formulation and manufacture, and to guide pharmacists in knowing how to best utilize commercial products and to prepare quality compounded formulations. Thus, the science underlying dispersed systems will be an important part of pharmacists' basic science knowledge base and can prove critical for patient care, because poor formulations or destabilized good formulations can affect many properties important for product usage and drug effects. These properties range from esthetic acceptability (e.g., appearance and smoothness of creams), to drug bioavailability (e.g., based on crystal particle size and/or dissolution relationships), to accurate drug dosing (e.g., based on uniformity of the dispersion).

DISPERSED SYSTEMS CLASSIFIED BY THEIR PARTICLE SIZE

In addition to describing dispersed systems based on the nature of the dispersed phase and dispersion medium, another common and convenient approach is to classify the systems based on the size of the particles that make up the dispersed phase. The two main categories in this context are *colloidal dispersions*, which roughly encompass the particle diameter range of 0.1 to 1000 nm (1 µm), and *coarse dispersions*, which have larger particles (e.g., 1 to 200 µm). The reason that dispersions are often classified based on particle size is that dispersions of colloidal-sized particles can exhibit somewhat different behavior than particles in coarse dispersions. For example, coarse dispersions are more prone to settling problems because of their larger particle size and their relative lack of influence by *Brownian motion*. These behaviors will be discussed later, but it must be kept in mind that although there is a scientific basis for classifying dispersed systems by their dispersed phase particle sizes, there is not a clear line of demarcation separating colloidal from coarse dispersions. For example, a given dispersion typically has a *range* of particle sizes that will include representatives of the other class size (e.g., a coarse dispersion can have colloidal size particles in its distribution). Furthermore, some of the properties attributed more to one class, such as *Brownian motion* for colloidal systems, can still exist, but to a lesser degree in the other class.

Colloidal Dispersions

Colloidal dispersions are those in which very small particles—*at least one dimension* between approximately 1 and 1000 nm in length—are dispersed in a continuous phase of a different composition. The importance of colloidal dispersions is far-reaching, as they are found in a myriad of systems ranging from biological to industrial. Moreover, the nature of dispersed systems in general, such as how particles move and interact in their dispersion medium, is largely understood based on scientific inquiry into colloidal dispersions. Colloids can be classified in different ways, but there are three general classifications based primarily on how and to what degree they interact with their medium: (1) *lyophilic*, (2) *lyophobic*, and (3) *association colloids*.

Lyophilic and Lyophobic Colloids

Lyo*philic* colloids are those that have a strong affinity with their medium, whereas lyo*phobic* colloids have little or no affinity with their medium. The prefix *lyo* refers to the solvent or medium. Since for most pharmaceutical dispersions the medium is water, the terms *hydrophilic* and *hydrophobic* are used.

Hydrophilic colloids have polar regions (e.g., from hydroxyl groups or ionizable groups) that enable them to become hydrated in contact with aqueous environments. Common examples of these are certain *macromolecules*, such as proteins (e.g., albumin and gelatin) and polysaccharides (e.g., natural gums and semisynthetic cellulose derivatives). An important fact about macromolecular dispersions is that they are solutions, since they are solvated by their medium. However, their size (e.g., several nanometers in one direction) places them in the colloidal range. Accordingly, macromolecular dispersions are sometimes referred to as *colloid solutions* to distinguish them from *crystalloid solutions*, which are true solutions of smaller molecules such as electrolytes. One of the most common usages of the terms colloid and crystalloid solutions in medicine is in the field of plasma volume expansion, where the choice to increase a patient's plasma volume (e.g., in circulatory shock) includes intravenous crystalloid solutions (e.g., sodium chloride or lactated Ringer's) and intravenous colloid solutions (e.g., albumin or hydroxyethyl starch). Though there is controversy as to which type of solution is better for plasma volume expansion, the utility of colloid solutions for this purpose illustrates one of the important properties of macromolecular colloids: Their inability to cross membranes enables them to osmotically draw in water.

Polysaccharides and other hydrocolloids, such as gelatin, have the capacity to exist in the *sol* and *gel* state. In the sol (solution) state, these hydrocolloids move freely in their medium with no chemical or physical bonding between them. The gel state is created when the hydrocolloids interact in different ways, for example, by forming a polymeric network interspersed with the liquid medium. This capacity to form a network of interconnecting macromolecules allows them to increase a medium's viscosity and to form the gel topical dosage form.

Hydrophobic colloids do not have sufficient surface hydrophilicity to enable them to interact well with water. Examples of hydrophobic colloidal dispersions include milk, intravenous lipid emulsions, and nanocrystal suspensions. Intravenous lipid emulsions are used primarily to provide calories as part of an intravenous nutrition strategy. One of the problems of hydrophobic colloids is their tendency to aggregate in an aqueous environment. This is a significant concern for intravenous lipid emulsions, because it's important that the lipid particles do not aggregate to a size that can block small blood vessels. So, again, this illustrates that a basic understanding of the interparticulate forces that can lead to aggregation (discussed later) will be an important element of the pharmacist's knowledge base. Nanocrystals are becoming increasingly used as an approach to improve the solubility of poorly soluble drugs. For example, the nanocrystal suspension formulation of megestrol acetate (Megace® ES), a progesterone receptor agonist, showed a dramatic improvement in absorption compared to its larger crystal predecessor, because its nanocrystals dissolved more readily.

To summarize some important points, the affinity of hydrophilic and hydrophobic colloids with their medium is fundamental to explaining much of their behavior in the dispersion. For example, hydrophilic colloids will typically mix easily with water to form physically stable dispersions, whereas hydrophobic colloids, with their relative lack of affinity for the medium, may present difficulties when mixing them with water, and, in aqueous environments, will have a tendency to aggregate. These are problems that they share with the larger, coarse particle dispersions, such as emulsions and suspensions.

Association Colloids

Association colloids are those formed by the association of dissolved molecules of a substance to create particles of colloidal dimensions, most commonly termed micelles. They include the surfactant micelle (described in Chapter 9), as well as liposomes (bilayer spheres formed mainly from phospholipids), and microemulsions (Fig. 8-1). The surfactants of classic surfactant micelles and the phospholipids of liposomes are amphiphiles that spontaneously self-associate when their concentrations reach a certain level known as their *critical micelle concentrations*. They interact in such a way, so as to minimize contact between the lipophilic portion of the amphiphile and water, usually resulting in a spherical shape to the micelle. Surfactant micelles are typically 2 to 5 nm in diameter, whereas the commercially available liposomes (e.g., liposomal doxorubicin and daunorubicin) are about 50 to 100 nm. (Though liposomal drug products are typically colloidal, liposomes exceeding colloidal dimensions can be prepared.) Association colloids can also be more complex than these amphiphile-derived micelles. For example, the casein micelles in milk are colloidal-size complexes of casein proteins with calcium phosphate.

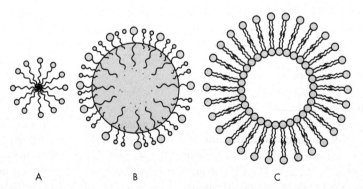

FIGURE 8-1 Examples of association colloids: (A) surfactant micelle, (B) microemulsion, and (C) liposome.

Another type of association colloid is the microemulsion. Like conventional coarse emulsions (described below), microemulsions are also dispersions of oil droplets in water (or water-in-oil). However, they differ from conventional emulsions in that the droplets are much smaller (e.g., 10 to 100 nm diameter), the dispersions are transparent, and they are considered thermodynamically stable. This achievement of these properties owes to the particular blends of oil, surfactant, and cosurfactant. The cosurfactant, often a four to seven carbon-chain alcohol, helps reduce surface tension and adds sufficient flexibility to the interfacial film to enable it to attain the high curvature needed to cover the tiny droplet. Another property of the microemulsion formation is that it allows for spontaneous emulsion formation when in contact with aqueous body fluids (i.e., they are self-emulsifying). Microemulsions join surfactant micelles, nanocrystals, liposomes, and other approaches in the formulator's arsenal, to help solubilize lipophilic drugs that might otherwise be too difficult to achieve the solubility needed to reach their therapeutic potential. A commercial example of the use of a microemulsion to help solubilize a drug is the product Neoral® (cyclosporine), an immunosuppressant.

Some Unique Properties of Colloids

The colloidal classification of dispersions exists because dispersions of colloidal-sized particles exhibit properties that the smaller molecular dispersions (true solutions) and larger coarse dispersions either do not exhibit or exhibit to a lesser degree. Among the unique properties of colloidal dispersions are light-scattering properties (*Tyndall effect*), osmotic properties (of colloid solutions), electrical properties, and dispersed particle movement (*Brownian motion*). Brownian motion and elements of their electrical properties will be discussed later in the section on physical stability.

Atoms or molecules in true solutions are too small to scatter ultraviolet or visible light. However, the particles in colloidal dispersions have dimensions that approximate ultraviolet and visible wavelength (390 to 750 nm), so that light passing through the dispersion is scattered by the particles, with the intensity of the scattered light being inversely proportional to the fourth power of the wavelength. So, for example, the light scattered by particles in the atmosphere yield a blue sky, because of its lower wavelength than yellows and reds. This light-scattering effect is termed the *Tyndall effect*, after the 19th-century scientist who discovered it. An important aspect of the Tyndall effect is that the light scattered by colloidal particles is larger than the particle itself, forming a visibly observable cone (the *Tyndall cone*). The presence of the Tyndall cone forms the basis for *ultramicroscopy*, which allows the individual colloidal particles to be "observed" as discrete bright spots corresponding to the Tyndall cone that they produce.

Coarse Dispersions

Coarse dispersions are those that have larger particles than colloidal dispersions, usually between 1 and about 200 μm in diameter. They include emulsions and the commonly used suspensions. Examples of pharmaceutical coarse dispersions include antibiotic suspensions, pediatric suspensions (e.g., ibuprofen and acetaminophen), antacid suspensions, and mineral oil and cod-liver oil emulsions.

Emulsions

Coarse emulsions, termed simply *emulsions*, are mixtures of immiscible liquids, in which one phase is dispersed as droplets within the other phase. The two main classifications of two-phase emulsions are oil-in-water (o/w), where oil droplets are suspended in an aqueous continuous phase and water-in-oil (w/o), which is the opposite configuration. Emulsions can also be more complex in that droplets of one phase can exist within droplets of another phase, which exist in a continuous phase; for example, a water-in-oil-in-water emulsion. These are called *multiple emulsions*. Oral emulsions are invariably the more palatable and digestible o/w type. More common than oral emulsions are topical emulsions, which include many lotions and most creams (o/w and w/o).

Suspensions

Coarse suspensions, or simply *suspensions*, are coarse dispersions where drug crystals are suspended in a medium in which they are insoluble. They are relatively common dosage forms that very often have specific particle-size requirements. For example, particle sizes on the lower end of the spectrum are desirable for pulmonary inhalation suspensions (about 3 to 5 µm for deep lung penetration), whereas particles approaching the upper end of the scale are usually acceptable for oral suspensions. In between will be recommended particle sizes of less than 100 µm for dermatologicals to minimize palpability, and less than 10 µm for ophthalmic suspensions to minimize their palpability and to facilitate their dissolution before being removed by the eye's tear drainage system. The inverse relationship between dissolution rate and particle size is a generally important parameter for formulators to consider when preparing suspension dosage forms, particularly when the drug's dissolution rate is slow compared to how long it has access to absorption sites before being removed.

The dispersion mediums for pharmaceutical suspensions are most often aqueous, which is the case for nearly all oral suspensions, many injectable suspensions (such as some longer-acting insulins), many lotions, and many o/w creams that have suspended crystals. Examples where the medium is nonaqueous include propellant-based inhaler suspensions, some injectable oil suspensions (e.g., fluphenazine decanoate), some topical oil suspensions (e.g., calcipotriene and betamethasone dipropionate topical suspension), and some creams and ointments. (Recall that suspensions can also include solid dosage forms such as suppository formulations that contain undissolved drug.) Aqueous dispersions, including aqueous suspensions, will be emphasized in the subsequent material of this chapter, which will focus on the physical stability of dispersed systems.

PHYSICAL STABILITY OF DISPERSED SYSTEMS

Among the important issues for dispersed systems is that they retain their physical integrity (i.e., they are physically stable) for their entire shelf life. For dispersions, this means that the particles retain their original size and that they remain uniformly distributed, or they easily regain their uniformity with simple agitation. To reemphasize, particle size can be critical to patient care for several reasons that depend on the dosage form and route of administration. For example, as described elsewhere in this textbook (e.g., Chapter 5), dissolution rate is particle-size-dependent, and therefore particle size affects a drug's bioavailability in many cases. Other examples of the importance of particle size include respiratory aerosols (both intranasal and pulmonary), where size dictates the particle's aerodynamics; and the relatively uncommon intravenous dispersions, where particles too large can be trapped in pulmonary vasculature, resulting in an embolism. Particle size is also important for determining how quickly particles will settle or rise in a dispersion and therefore how quickly the product loses its uniformity, which if not agitated back to uniformity, can lead to the wrong dose for the patient. The major physicochemical properties that affect the dispersed particle's size and movement over time are discussed next.

Dispersion Uniformity

As mentioned, dispersion uniformity (or the ease at which it can be agitated back to uniformity) must be maintained for the product's shelf life. Two of the more significant forces affecting the movement of the particles in a dispersion are Brownian motion and gravity.

Brownian Motion

Brownian motion is the random, irregular motion of particles caused by the momentum induced by being bombarded with molecules of the dispersion medium. This type of particle movement depends on many factors including temperature, medium viscosity, and dispersed particle size. Since the motion of the molecules of the medium results from thermal energy, a decrease in temperature will cause a decrease in Brownian motion. (Molecular motion, and therefore Brownian motion, stops at absolute zero.) Other factors that can decrease Brownian motion include increasing dispersed particle size and increasing viscosity of the medium. Although there is no definitive particle size above which Brownian motion becomes less important for dispersed systems, colloidal-sized particles tend to exhibit Brownian motion, which enables one of the distinctive features of most colloidal dispersions: the ability to resist gravitational forces and remain suspended for prolonged time periods. In other words, with very small particles, the mixing effect induced by Brownian motion can counteract the sedimentation forces induced by gravity, thereby keeping the particles suspended.

Gravitational Forces

In liquid dispersions with particles that are larger than colloidal dimensions (which include aggregates of colloidal particles), gravity plays a critical role in determining particle movement, either by sedimentation downward as with coarse suspensions or creaming upward as with coarse o/w emulsions. The velocity of particle movement under the force of gravity is described by *Stoke's law*. Although this equation has strict requirements for its mathematical accuracy, the parameters of the equation are useful for explaining the tendency for coarse dispersions to become nonuniformly distributed, thereby enabling the formulator and pharmacist to understand the potential effects of formula changes on a dispersion's uniformity. The equation is given as follows:

$$v = \frac{2r^2(\rho_1 - \rho_2)g}{9\eta_0}$$

where v is the velocity of sedimentation of spherical particles of radius r, g is the acceleration due to gravity, ρ_1 is the density of the particles, ρ_2 is the density of the medium, and η_o is the viscosity of the medium. This equation embodies the intuitive concepts that—all else being equal—larger particles (including aggregates of small particles) will settle faster than small particles, the speed a suspended particle rises or falls under the force of gravity depends on the relative densities of the dispersed particles and the dispersion medium, and increasing the viscosity of the dispersion medium will slow down particle movement.

Dispersion Particle Growth

Aggregation and Its Consequences

A near universal stability concern of dispersed systems is that the dispersed particles are driven to aggregate—reversibly in some cases and irreversibly in others. An irreversible particle growth can compromise formulations in many ways including those described in the introduction to this section on physical stability.

With the common dispersed systems of suspensions and emulsions, molecules at the interface of the dispersed phase and dispersion medium are exposed to a different environment than molecules far removed from the interface, which are bound to neighboring identical molecules. Thus, the interfacial molecules will possess a lesser overall binding energy, which corresponds to a positive surface energy. The magnitude of the positive surface energy depends on the strength of the cohesive forces within each phase compared to the strength of the adhesive forces between the opposing molecules at the interface of the two phases. With a fixed amount of dispersed phase, as the particles become smaller, the total exposed surface area and the number of molecules at the interface increases, and consequently, so does the system's total surface *free energy*. The relationship between change in surface free energy, ΔG, as a function of change in surface area, ΔA (i.e., change in particle size) can be expressed by the following equation:

$$\Delta G = \gamma \Delta A$$

where γ is the interfacial tension. The importance of the increase of the system's free energy associated with a reduction in particle size is that when particle size reduction occurs during the processes of emulsion and suspension production, the system becomes less thermodynamically stable. This, in turn, drives the particles to aggregate to reduce the interfacial area and surface free energy. In other words, particles in a dispersion are thermodynamically driven to aggregate to reduce the system's free energy.

For emulsions, the aggregation of droplets can lead to an irreversible uniting (*coalescence*) of droplets and as this process continues, ultimately the emulsion can undergo a *complete phase separation*, known as emulsion *breaking*, with the oil phase residing on top of the aqueous phase. The product is no longer usable at this point. For suspensions, the aggregation of crystals can lead to a permanent uniting of crystals. For example, the process of *caking* is somewhat analogous to the process of emulsion breaking. With caking, the suspended particles slowly form a sediment of irreversibly bound (*coagulated*) crystals at the bottom of the container (See "Applying the Basic Knowledge to Product Formulation"). As with emulsion breaking, suspension caking makes the product unusable. To clearly understand the potential for particle aggregation and its irreversible consequences, and what can be done to minimize them, requires a basic understanding of the nature of the forces of particle interaction.

Particles moving in a dispersion will collide with each other with varying frequency and force, as functions of temperature, particle size, particle concentration, dispersion medium viscosity, and other properties. Each collision can lead to particle cohesion with varying cohesive strength, depending on the attractive nature of the particle surfaces. If attractive forces between particles dominate over repulsive forces, particles may aggregate. But, if repulsive forces dominate, then aggregation can be prevented unless those forces are overcome.

DLVO Forces of Particle Interactions

The attractive and repulsive forces of interaction between particles in suspension have been partially modeled using DLVO theory, named after Derjaguin, Landau, Verwey, and Overbeek. Though the theory was developed to describe colloidal particles specifically, its principles apply to different degrees to other dispersions. The focus here will be to describe DLVO forces of interaction, but it's important to also realize that there are other "non-DLVO" forces that can be important for particle interactions in some systems. These will be briefly discussed later.

According to DLVO theory, the primary forces, or potentials, that determine whether two particles in suspension will aggregate when they collide are *van der Waals* attraction potential and *electrostatic* repulsion potential. These potentials operate simultaneously to different degrees depending on the particles and their medium, and both potentials diminish with distance away from the particle surface, each eventually reaching zero potential. The net effect of the two combined forces will often be the critical determinant of whether repulsion or attraction dominates at a given distance between particles and thus whether and how strongly the two colliding particles will aggregate.

A sound understanding of the outcomes of these potentials requires a brief review of their sources.

The van der Waals forces, which are the main attractive forces between particles, arise from transient fluctuations in electron density, which can create instantaneous dipole moments in a molecule. These instantaneous dipole moments can induce polarization of neighboring molecules, resulting in the correlated interaction (attraction) of the dipole and induced dipole. Though these are relatively weak attractive forces between individual molecules, they are fairly strong forces between two particles that possess a very large number of interacting molecules.

The electrostatic repulsive forces between particles depend on the presence and magnitude of surface charge. Particles in an aqueous pharmaceutical dispersion typically develop a surface charge by one of two main mechanisms: (1) ionization of ionizable moieties (e.g., carboxyl or amine) of molecules on the surface of the particle (this degree of ionization and thus the magnitude of surface charge depend on the molecule's pKa and the medium's pH), or (2) adsorption of ions from solution onto an uncharged particle's surface. An example is a surfactant with an ionizable headgroup that resides on the surface of the dispersed particle (the ionization of this type of surfactant is also pH-dependent).

As described in Chapter 9 and illustrated in Figure 8-2, the ionic environment that creates the electrostatic potential around a charged particle can be visualized using the Gouy-Chapman-Stern model of the electrical *diffuse double-layer*, in which a double layer of ions is formed around the particle in order to neutralize its charge. In the first layer surrounding the particle, there is a layer of counterions strongly attached to the oppositely charged particle surface ions. This first layer, called the *Stern layer*, does not have sufficient ions to neutralize the charge of the particle. The second layer—engulfing the Stern layer and the particle—is a *diffuse* layer of more weakly held (*diffuse*) counterions (along with some associated ions that have the same charge as the particle surface). Moving outward from the particle, the concentration of these weakly held counterions in the *diffuse layer* diminishes until it equals their concentration in the bulk of the dispersion's medium. At this point

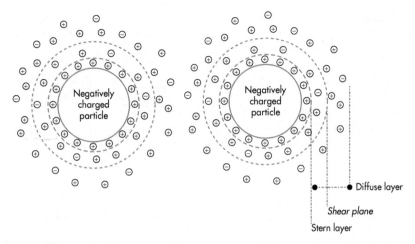

FIGURE 8-2 Two negatively charged particles, each with an electrical double layer (Stern layer and diffuse layer), along with a shear plane that determines the zeta potential.

the diffuse layer ends and the particle's charge is neutralized.

The possible magnitudes of the van der Waals attractive and electrostatic repulsive forces, and their possible net effects, as a function of distance from the particle's surface, are illustrated in Figure 8-3. The repulsive electric potential is represented as a positive value on the curve and is a barrier that must be overcome (e.g., by the kinetic energy of the collision) for the two particles to unite. The attractive van der Waals potential is represented as a negative value, and its depth indicates the strength of the attraction as the particles come closer together. The net interaction potential curve, produced by subtracting the repulsive and attractive potentials at a given distance, shows the positive repulsive energy barrier (where electrical repulsion dominates) that must be overcome, and as the particles get very close, the negative and strong

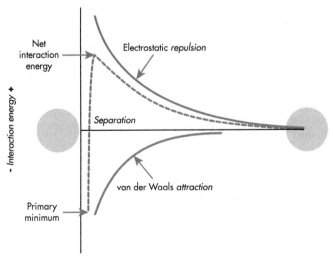

FIGURE 8-3 Possible magnitudes of the van der Waals attractive and electrostatic repulsive forces, and their possible net effects, as a function of distance from a particle's surface.

van der Waals attractive potential—an energy trap known as the *primary minimum* in interaction energy. The curves in Figure 8-3 illustrate one possibility, but for a given dispersion the magnitudes of the attractive and repulsive potentials, and their net effects can differ significantly from this example, and can be modified by the formulator to minimize the chance of reaching the primary minimum. One of the ways to adjust the net forces of interaction is to adjust the particle's *zeta potential*.

The Zeta Potential

An important measure of the charge on a particle's surface is the *zeta potential*. As particles move around in their medium, the bound counterions in the Stern layer stay with the moving particle, as does a certain fraction of the more strongly bound ions in the diffuse layer. Other, more weakly bound ions in the diffuse layer cannot move with the particle and are shorn off during movement. An imaginary *shear plane* (see Fig. 8-2) defines the boundary between the diffuse layer ions that are unaffected by fluid motion and those that are shorn off by fluid motion. The net charge at the shear plane of the diffuse layer is measurable as the *zeta potential*, which can be defined as the electrostatic potential difference between an average point on the shear plane compared to the potential out in the bulk medium. The zeta potential can be seen as an important measure of the electrostatic repulsive forces between particles, because it is the shear planes of the double layers of the colliding particles that meet, overlap, and repel each other because of like charges.* All else being equal, the greater the zeta potential, the greater the repulsion.

Non-DLVO Forces

The DLVO forces of electrostatic repulsion and van der Waals attraction are important forces to be aware of, but they do not completely explain the interaction of all types of dispersed systems. Other forces that may be critical for particle interactions include *hydration forces* and *steric forces*. Along with the DLVO forces, understanding these "non-DLVO forces" helps paint a more complete picture of the possible forces that must be understood for the pharmaceutical scientist and pharmacist to more thoroughly understand dispersion stability.

Hydration forces can be important for hydrophilic colloids, such as proteins, polysaccharides, and liposomes. These are repulsive forces that operate at a lesser distance of particle separation (less than 3 nm) than electrostatic forces and can compete with van der Waals attractive forces to keep hydrophilic particles apart. The origin of hydration repulsion lies in the polarization of water molecules bound to the hydrophilic surfaces. The bound water molecules, being less active than those in the bulk, leads to an accumulation of additional water molecules between the surfaces, thus forcing the surfaces apart.

Steric forces are another type of non-DLVO force than can help keep particles apart. These forces usually result from coating the particles with polymeric macromolecules that are large enough to physically prevent them from getting close enough to experience van der Waals attraction. This is illustrated in Figure 8-4. Therefore, certain macromolecules that have portions that can adhere to the particle and portions that project into the dispersion medium can be used to help stabilize dispersions.

Ostwald Ripening as a Cause of Particle Growth

Another source of particle growth is called *Ostwald ripening*, which results from the preferential creation of large particles at the expense of smaller ones. This process is well illustrated with coarse suspensions,

---

*Another factor that contributes to the repulsion between the two colliding particles is an osmotic effect that derives from the increased ion concentration at the points of diffuse layer overlap. The high concentration of ions osmotically draws in water from the bulk medium, helping to force the two particles apart.

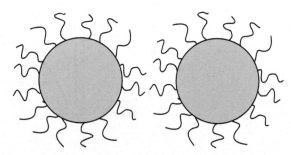

FIGURE 8-4 Illustration of using polymers to promote steric repulsion between particles.

where the dissolution of small particles in the mix occurs more quickly than that of larger particles. As the dissolution occurs, the solutes from the smaller particles diffuse away to the surface of the larger particles, where it deposits, thus enabling the larger particle to grow in size. The smaller particles, now even smaller, continue to dissolve, thereby perpetuating this mechanism of large particle growth, with a consequent increase in average particle size for the dispersion. Ostwald ripening can be accelerated by processes that alter dissolution. For example, storing suspensions at high temperature can increase solubility and dissolution speed and storing at low temperature can have the opposite effects. Thus wide temperature swings during storage can exacerbate Ostwald ripening.

Applying the Basic Knowledge to Product Formulation

With this background knowledge of the forces underlying dispersion uniformity and particle growth, formulators can rationally employ different techniques to minimize particle settling or rising and to keep particles from becoming strongly aggregated.

For example, one of the most common ways to decrease settling is to adjust the viscosity of the dispersion medium. (For a more detailed explanation, see Chapter 10.) Formulators often try to create a *pseudoplastic, thixotropic* system—pseudoplastic so that on the shelf the viscosity can remain high, thereby reducing settling, and off the shelf, when the pharmacist or patient agitates it, flow can increase for ease of pouring. Being thixotropic, the suspension won't revert too quickly back to its more viscous state.

A number of approaches, employing the concepts discussed, can be used by formulators and pharmacists to minimize particle aggregation (or its avidity). For example, the formulation can be adjusted to increase the zeta potential to maximize repulsion.

Another fairly common, seemingly counterintuitive approach to preventing permanent aggregation is to intentionally create weak particle aggregates called *floccules* (from the Latin *flocculus*—to collect together in a loose aggregation like flocks, or tufts, of wool). Particles that are not aggregated are called *deflocculated* particles. The reason that creating flocculated suspensions is reasonable is based on what can often occur in coarse dispersions of deflocculated particles that have sufficient charge to prevent aggregation during collisions. If the particles are large enough to be resistant to Brownian motion and are denser than their medium (which is common with suspensions), they will ultimately succumb to gravitational forces and sediment at the bottom of their container. Over time the sediment becomes progressively more compact as smaller particles in the mixture move to fill voids. In this compact sediment, larger particles can press down upon the smaller ones with enough pressure to overcome the repulsive forces that have kept the particles apart. This leads to strong particle aggregation at the primary minimum, and as the process becomes widespread in the sediment, it leads to the formation of a *cake* that is extremely difficult of not impossible to redisperse.

The purpose of *flocculation* then, is to create a network of weakly held particles that although they sediment quickly because of the size of the aggregate, they do so as a large and loose group of particles that typically resembles a cloud in the liquid. Being weakly held together, the particles are easy to redisperse to uniformity by simple shaking.

One approach to creating floccules is to reduce the electrostatic repulsive barrier to a point where the van der Waals attractive potential is greater than the electrostatic repulsive potential, but at significant distance of particle separation (Fig. 8-5). At a distance in this case, the net interaction potential will be negative (attractive), and strong enough to keep particles loosely held together (i.e., as a flocc), but a weak enough attraction to enable easy redispersion. This net effect is seen as a slight dip at a distance known as the *secondary minimum* in potential interaction energy on the energy of interaction curve.

The most common way to reduce the electrostatic potential, to enable flocculation, is to increase the salt content of the medium, which puts more counterions in the diffuse layer to neutralize the particle's charge, causing charge neutralization at a closer distance to the particle surface (i.e., a thinner double layer). Thus the decay in the electrostatic repulsion occurs faster, allowing the van der Waals

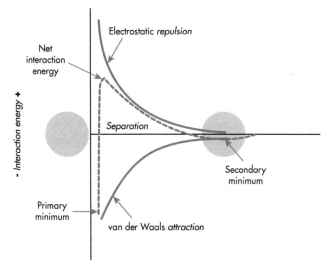

FIGURE 8-5 A potential energy of interaction curve illustrating the role of reducing electrostatic repulsion, to promote the creation of a secondary minimum in interaction energy, which can enable particle flocculation.

forces to become dominant at a distance, thereby creating the weak secondary minimum (Fig. 8-5).

Another approach to creating flocculates is by formulating the dispersion with polymeric macromolecules (discussed previously), such as polysaccharides that can adsorb to a particle's surface to sterically prevent close approach (see Fig. 8-4). Macromolecules can also be used to form weak *bridges* between particles, thereby creating a meshwork of loosely held particles. The bridging approach to flocculation, as illustrated in Figure 8-6, employs long polymers that are used to bridge suspended particles by one of two main mechanisms: (1) single polymers can adhere to more than one particle, or (2) polymers adhered to the surfaces of separate particles can interact to form bridges between the particles.

Coarse emulsions, unlike microemulsions, typically do not form by self-emulsification and they are not thermodynamically stable. To aid the coarse emulsion's formation, surface tension–lowering emulsifiers can be used to reduce the energy needed to create the droplets. To prevent the droplets from uniting, emulsion stabilizers can be used to stabilize the newly formed interface and to reduce droplet aggregation. Some excipients can function to both reduce interfacial tension and stabilize the interface.

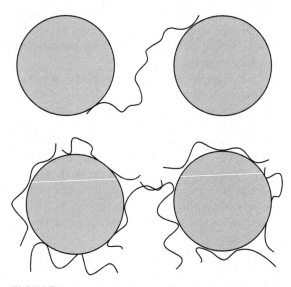

FIGURE 8-6 An illustration of two approaches of using polymers to form bridges to promote particle flocculation.

For example, the natural polysaccharide Acacia gum, has some capacity to reduce interfacial tension thus facilitating emulsification, and to quickly adsorb to the interface during the emulsification process, thereby minimizing intimate contact and subsequent coalescence of newly formed droplets.

SUMMARY

Dispersed systems are commonplace in drug products and preparations. The purpose of this chapter was to introduce the student pharmacist to the different types of dispersed systems and some of the more common physical stability problems that they share. This will serve as foundational knowledge that forms the framework for subsequent learning about the formulation and manufacture of dispersed dosage forms. An awareness of the different systems and their potential problems, coupled with scientific explanations of their causes, will help the pharmacist practitioner to recognize, and prevent or resolve, potential formulation, preparation, or patient use challenges, each of which can affect patient care.

PROBLEMS

1. In which type of dispersion, homogenous or heterogeneous, is the dispersed phase physically distinguishable from the medium in which it is dispersed?
2. List pharmaceutical examples of the following types of dispersions:
   a. solid in liquid
   b. solid in gas
   c. liquid in liquid
   d. gas in liquid
3. Compare the particle size differences of colloidal versus coarse dispersions and briefly state why colloidal dispersions are thought of separately from coarse dispersions.
4. Briefly define and give an example of hydrophilic colloids and association colloids.
5. According to Stoke's equation, predict the effects of an increase in the following properties of a coarse suspension on the sedimentation rate of the suspension:
   a. Crystal particle size.
   b. Crystal density.
   c. Viscosity of the medium.
6. Briefly describe the breaking of a coarse emulsion and the caking of a coarse suspension and comment on the usability of the products when these phenomena occur.
7. Briefly explain how adjusting the salt content of a suspension can be used to create a flocculated suspension of particles. Include terminology found in the potential energy of interaction curve (Fig. 8-5).

ANSWERS

1. Heterogeneous dispersion.
2. a. Solid in liquid: Crystal suspensions (e.g., pediatric suspensions, some antibiotics, and depot injections)
   b. Solid in gas: Aerosolized spray powder, inhalation dry powder.
   c. Liquid in liquid: Emulsions (e.g., mineral oil emulsions, IV lipids, propofol, cream bases).
   d. Gas in liquid: Foams (e.g., contraceptive, corticosteroid).
3. Colloidal dispersions roughly encompass the particle diameter range of 0.1 to 1000 nm (1 µm). Coarse dispersions have larger particle diameters (e.g., 1 to 200 µm). Colloidal dispersions are thought of separately from coarse dispersions because they can exhibit somewhat different behavior (e.g., Brownian motion and light scattering) than particles in coarse dispersions.
4. Hydrophilic colloids are those in which the colloidal particles have a strong affinity with their

aqueous dispersion medium. Examples include dispersions of certain macromolecules, such as proteins (e.g., albumin and gelatin) and polysaccharides (e.g., natural gums and semi-synthetics cellulose derivatives). Association colloids are those formed by the association of dissolved molecules of a substance to form particles of colloidal dimensions. Examples include surfactant micelles, liposomes, and microemulsions.

5. a. Increased crystal particle size increases sedimentation rate.
   b. Increased crystal density increases sedimentation rate.
   c. Increased viscosity of the medium decreases sedimentation rate.
6. For an emulsion, the aggregation of droplets can lead to their coalescence, and as this process continues, the emulsion can ultimately undergo a complete phase separation, known as emulsion breaking, with the oil phase residing on top of the aqueous phase. For a suspension, particles can sediment to the bottom of their container. Over time the sediment becomes progressively more compact and larger particles can press down upon the smaller ones with enough pressure to overcome the repulsive forces that have kept the particles apart. This leads to the formation of a cake that is extremely difficult, if not impossible, to redisperse. With either emulsion breaking or suspension caking, the products should not be used by the patient as they are no longer uniform dispersions.
7. Added salts reduce the electrostatic repulsive barrier by adding counterions to the diffuse layer, thereby causing charge neutralization at a closer distance to the particle surface (i.e., a thinner double-layer). Thus the decay in the electrostatic repulsion occurs faster, allowing the van der Waals forces to become dominant at a distance, thereby creating the weak secondary minimum where particles can weakly attach to each other to create the flocculated suspension.

KEY POINTS

▶ The dispersions discussed in this chapter are those where the dispersed components are physically different from the medium in which they are dispersed. Common examples include suspensions and emulsions.

▶ Dispersions can vary based on the size of dispersed particles or on the nature of the dispersed particles and dispersion medium.

▶ Colloidal dispersions are those that have very small particles (e.g., 1 μm diameter or less). They include hydrocolloids, microemulsions, and surfactant micelles.

▶ Coarse dispersions are those with larger particles and include the common pharmaceutical suspensions and emulsions.

▶ The particle size and uniformity of pharmaceutical dispersions can be critical determinants of drug safety and efficacy.

▶ Most dispersions have a tendency for the particles to aggregate or separate from their medium by settling or rising. This can lead to significant problems in dosage uniformity in the dispersion.

▶ Colloidal dispersions have minimal tendency to rise or settle because of the effects of Brownian motion.

▶ Among the factors that affect aggregation are electrostatic repulsive forces and van der Waals attractive forces.

▶ Among the factors that affect settling or rising in coarse dispersions are dispersed phase particle size and dispersion medium viscosity.

▶ Knowledge of the factors that determine particle behavior in a dispersion can be readily employed to prepare physically stable dispersions.

CLINICAL QUESTIONS

1. A patient returns to your pharmacy with an antibiotic suspension that doesn't appear to be working as it has in the past. Your first thought might be the development of resistance, but when she shows you the bottle, there appears to be sediment in the bottom that is difficult to redisperse when you shake it. Explain how the medication failure may be due to a physicochemical phenomenon in the suspension.

2. A patient brings you a tube of hydrocortisone 1% cream. The expiration date has not been reached yet, but when he squeezes the tube, out comes a mixture of clear oil with some cream. He explains to you that he's kept it in his car all summer just to have it handy. Explain what has occurred to the cream and comment on whether it should still be used.

SUGGESTED READINGS

Birdi KS: *Handbook of Surface and Colloid Chemistry.* Boca Raton, Florida: CRC Press/Taylor & Francis, 2009.

Cosgrove T: *Colloid Science: Theory, Methods and Applications.* Ames, Iowa: Blackwell Pub. Professional, 2005.

Liang Y, Hilal N, Langston P, Starov V: Interaction forces between colloidal particles in liquid: Theory and experiment. *Adv Colloid Interface Sci* 134–135; 151–166, Oct. 31, 2007. Epub Apr. 21, 2007. Review.

Lieberman HA, Rieger MM, Banker GS: *Pharmaceutical Dosage Forms—Disperse Systems.* New York: M. Dekker, 1996.

Nielloud F, Marti-Mestres G: *Pharmaceutical Emulsions and Suspensions.* New York: Marcel Dekker, Inc., 2000.

Sinko PJ, Martin AN: *Martin's Physical Pharmacy Pharmaceutical Sciences: Physical Chemical Principles in the Pharmaceutical Sciences.* Philadelphia: Lippincott Williams & Wilkins, 2006.

Wu L, Zhang J, Watanabe W: Physical and chemical stability of drug nanoparticles. *Adv Drug Deliv Rev* 63(6):456-469, May 30, 2011. doi: 10.1016/j.addr.2011.02.001. Epub Feb. 21, 2011.

9 Interfacial Phenomena

Maria Polikandritou Lambros and Shihong Li Nicolaou

Learning Objectives

After completing this chapter, the reader should be able to:

- Discuss the significance of surface and interfacial phenomena in biological and pharmaceutical systems.
- Measure surface and interfacial tensions of liquids using the appropriate method.
- Estimate the ability of a liquid to spread over a surface.
- Propose ways to enhance the wettability of pharmaceutical powders in liquid media.
- Organize surfactants in different categories.
- Classify surfactants on the basis of their hydrophile-lipophile balance values.
- Evaluate the process of adsorption at the liquid–air interface, with emphasis on the adsorption of surface active agents (surfactants).
- Determine the critical micelle concentration of a surfactant.
- Calculate the surface excess concentrations and molecular areas of adsorbing species by using the Gibbs adsorption equation.
- Discuss the adsorption of gases on solid surfaces and the use of

SURFACE TENSION

In this chapter we discuss the behavior of the molecules at the boundaries of phases. The behavior of molecules at those boundaries is different from their behavior in the bulk of the phases, which has implications for the physiology of the human body as well as for pharmacy. Interfacial phenomena affect drug delivery systems. For example, solubilization and dispersion of drugs, suspension or emulsion stability, and adsorption of drugs on different substrates are all affected by the interfacial properties of drugs and their environment.

The boundaries of solids, liquids, or gases with other solids, liquids, or gases are called *interfaces*. The boundaries of solids or liquids with air are called *surfaces* (Table 9-1).

TABLE 9-1 Surfaces and Interfaces

Surfaces Between	Examples
Liquid and gas	Water surface
Solid and gas	Tablet surface
Liquid and liquid	Oil and water interface
Solid and liquid	The interface between the suspended solid particle and the suspending medium

Definition of Surface Tension

Molecules that reside in the bulk of phases are surrounded by other molecules of the same kind; thus, they experience equal forces from all directions (Fig. 9-1). In contrast, molecules at the interfaces are subjected to differing forces because only some of their neighboring molecules are of the same kind. They experience high attractive forces from molecules of the same kind and lower attractive forces from molecules of the neighboring phase. (If the molecules of the neighboring phase were attracting the molecules of the other phase with stronger attractive forces than their own, the two phases would be miscible and mix together like alcohol and water.)

the Langmuir, Freundlich, and BET adsorption isotherms.

▶ Analyze adsorption isotherm data and calculate the amount of adsorbate required to form a monolayer, and the specific surface area of the adsorbent.

▶ Assess the adsorption at the solid-liquid interface, and discuss the adsorption of proteins onto solid surfaces.

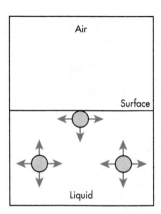

FIGURE 9-1 Molecules at the surfaces experience different forces from the direction of the other phase, whereas bulk molecules experience similar forces from all directions.

Molecules at the surfaces or interfaces are subject to surface tension or interfacial tension, respectively. Surface tension and interfacial tension manifest in many aspects of everyday experience, for example, when a person blows a film of soap to make soap bubbles.

Surface tension plays an important role in the lungs. Lungs have an agent (lung surfactant) that lowers the surface tension of the alveolar membrane; thus, alveoli can inflate easily with each inspiration and do not collapse at the end of each expiration. If there is little or no surfactant in the lungs to assist these processes, the alveoli collapse. This leads to respiratory distress syndrome, which is characterized by atelectasis, or collapse of the air sacks, with a decrease in lung volume and an inability to exchange oxygen and carbon dioxide within the lung.[1] This disease afflicts mainly preterm infants who, because of premature lungs, have a deficiency in the lung surfactant.

Interfacial tension or surface tension causes immiscible phases to resist mixing and shrink their surfaces. Oil and water do not mix because of interfacial tension. To disperse one in the other, it is necessary to introduce another kind of molecule that has an affinity for both oil and water—an amphipathic molecule. This type of molecule is called a surface active agent.[2] These molecules place themselves at the interface of oil and water and reduce the interfacial tension, allowing oil and water to mix. Surface tensions of various liquids and interfacial tensions of various liquids with water are given in Tables 9-2 and 9-3.

It is possible to illustrate surface tension by using a wire frame ABCD (Fig. 9-2). The ABCD part of the frame is rigid,

TABLE 9-2 Surface Tension of Liquids at 20°C (68°F)

Liquid	Surface Tension, dyne/cm
Water	72.8
Benzene	28.9
Oleic acid	32.5
Heptane	19.7
Hexane	18.0
Chloroform	27.1
Isoamyl alcohol	23.7
n-Octanol	26.5

TABLE 9-3 Interfacial Tension Between Various Liquids and Water at 20°C (68°F)[3,4]

Liquid	Interfacial Tension, dyne/cm
Benzene	35.0
Oleic acid	15.6
Hexane	50.8
Chloroform	32.8
Isoamyl alcohol	8.1
n-Octanol	8.5
Hexadecane	52.1

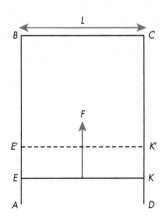

FIGURE 9-2 The frame with the movable bar (EK) shows the pull the soap film exerts on the bar as a result of surface tension.

and only the EK part can slide along the frame sides AB and DC. If a soap solution is placed on the frame, it will create a thin film, and then the film will try to shrink itself, forcing the movable part of the frame (EK) to move closer to the BC side. The new position of the movable part will be E′K′. The force F required to move the EK part is proportional to the surface tension γ times 2 (since the film has two surfaces) times the length L of the EK bar:

$$F = \gamma \times 2 \times L$$

The unit of force is dynes, and the unit of surface tension is dynes/cm in the metric [centimeter = gram = second (CGS)] system or N/M in the Système International (SI) system. Since $1\,J = 1\,N\,m$, surface tension also can be expressed as $J\,m^{-2}$. The surface tension of the water at 20°C (68°F) is 72.8 dynes/cm or 72.8×10^{-3} N/M. At a higher temperature, the surface tension will be lower, since the surface tension of a liquid decreases as temperature rises.

In the previous example, the work (energy) required to move the bar back to the EK location from E′K′ is $w = F \times dx = \gamma \times 2 \times L \times dx = \gamma \times 2 \times L \times EE' = \gamma \times 2 \times$ (area $EKK'E'$), where L is the length of the moving bar and EE' is the distance dx. Thus:

$$w = \gamma \Delta A \text{ and } \gamma = \frac{w}{\Delta A}$$

where ΔA is the area change. Furthermore we can say that the surface tension is the free energy change per unit area change.

EXAMPLE 1 ▶▶▶

The force required to move the movable bar EK to the new position E′K′ is 234 dynes, and the length $L = 3$ cm. The surface tension of the solution γ is

$$\gamma = \frac{F}{2L} = \frac{234\ \text{dynes}}{2 \times 3\ \text{cm}} = 39\ \text{dynes/cm}$$

Spreading Coefficient

Lotions, creams, sunscreens, and many cosmetics have to be spread on the skin to exert their effect. When a liquid spreads over the surface of a substrate, it covers all or a part of the surface. The substrate can be a solid or another liquid that is immiscible with the spreading liquid. In applying lotions, creams, and other materials on the skin, in essence one is spreading them on the sebum, which is a thin liquid film covering the skin. Sebum is composed of triglycerides, free fatty acids, wax esters, squalene, and cholesterol. The main function of sebum is to protect the skin against moisture loss and fungal or bacterial infections.

When a drop of oil is added on the surface of water, three things may happen:

1. The drop may spread as a thin film on the surface of water.
2. It may form a liquid lens if the oil cannot spread on the surface of water (Fig. 9-3a).
3. The drop may spread as a monolayer film with areas that are identified as lenses (Fig. 9-3b).

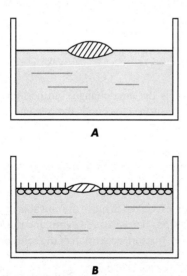

FIGURE 9-3 *A.* Formation of a lens when a drop of oil is added to water. *B.* Monolayer of oil with a lens.

There are several methods of measuring surface tension. These are described in the Appendix of this chapter.

Spreading depends on the surface tension of the liquids involved and on the interfacial tension between them. For a liquid to spread over the surface of a substrate, the spreading coefficient S has to be positive:

$$S = \gamma_s - \gamma_l - \gamma_{sl}$$

where γ_s = surface tension of the substrate, γ_l = surface tension of the spreading liquid, and γ_{sl} = interfacial tension between the liquid and the substrate.

Thus, if oleic acid ($\gamma_l = 32.5$) is placed on top of water ($\gamma_s = 72.8$) and ($\gamma_{sl} = 15.6$), it will spread because the spreading coefficient is positive:

$$S = 72.8 - 32.5 - 15.6 = 24.7$$

Sometimes spreading will occur only initially, and later the spreading liquid will retract and a lens will form. The retraction of the liquid to a lens occurs because the two substances in contact with each other become mutually saturated and their respective surface tensions become different from the initial surface tensions (when the liquids were pure).[1] For example, if benzene is added to water, one will observe an initial rapid spreading of benzene on the surface of the water, but later when benzene and water become mutually saturated with each other, the benzene retracts and forms a lens. The surface tensions of pure water and benzene are 72.8 and 28.9 dyne/cm, respectively. Their interfacial tension is 35 dyne/cm. The initial spreading coefficient of benzene spreading on water $S_{initial}$ indicates spreading because it is positive:

$$S_{initial} = 72.8 - 28.9 - 35 = 8.9$$

The final spreading coefficient S_{final} will be different since the surface tension of water saturated with benzene is 62.2 dyne/cm and the surface tension of benzene saturated with water is 28.8 dyne/cm:

$$S_{final} = 62.2 - 28.8 - 35 = -1.6$$

Since the final spreading coefficient $S_{final} = -1.6$ is a negative number, the benzene will not spread on water.

Initial Spreading Coefficients on Water

Liquid	S	Result
n-Octanol	72.8 − 26.5 − 8.5 = 37.8	Spreading will occur
Hexadecane	72.8 − 30.0 − 52.1 = −9.3	Spreading will not occur

Derivation: Spreading is dependent on the forces of cohesion and adhesion. Cohesive forces correspond to the work required to pull apart a volume of liquid that has a unit cross-sectional area (Fig. 9-4a). In other words, the work of cohesion is the work required to create two new liquid surfaces. Thus the work of cohesion (W_{coh}) is:

$$W_{coh} = \gamma_A + \gamma_A = 2\gamma_A$$

The work of adhesion is the work required to separate two immiscible liquids that form an interface that has a cross-sectional area of one unit (Fig. 9-4b). The work of adhesion follows this equation:

$$W_{adh} = \gamma_A + \gamma_B - \gamma_{AB}$$

where W_{adh} is the work of adhesion, γ_A is the surface tension of liquid A, γ_B is the surface tension of liquid B, and γ_{AB} is the interfacial tension between liquids A and B.

Spreading occurs only when the work of adhesion is greater than the work of cohesion. In mathematical terms:

$$W_{adh} > W_{coh}$$

$$\gamma_A + \gamma_B - \gamma_{AB} > 2\gamma_A$$

Liquid A will spread on liquid B when the spreading coefficient S is positive:

$$S = \gamma_B - \gamma_A - \gamma_{AB} \geq 0$$

A positive S (S > 0) means that spreading will occur spontaneously.

Contact Angle

Many pharmaceutical applications require the dispersion of solids in liquids. The preparation of suspensions, for example, requires that fine solid particles be immersed and subsequently dispersed in a liquid vehicle. The solid particles will be immersed in the liquid if the liquid can replace the air that surrounds the particles. In other words, the solid particles have

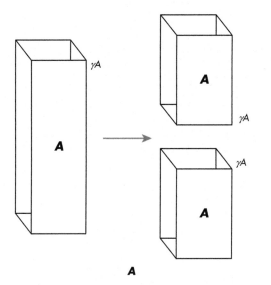

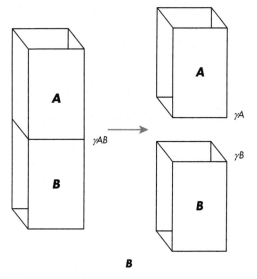

FIGURE 9-4 A. The creation of new surfaces requires work, which is termed the work of cohesion. B. The work required to separate two immiscible liquids is called the work of adhesion.

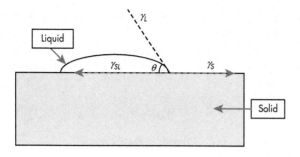

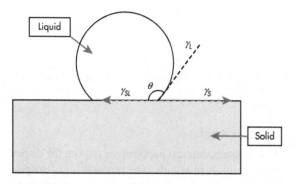

FIGURE 9-5 A solid partially wetted by a liquid.

to be wetted to be dispersed in a liquid. The degree of wetting will determine whether a solid can be dispersed in a liquid. Often, to increase the extent of wetting, surface active agents are used. The extent of wetting is determined by the contact angle, which is the angle between a tangent drawn on the surface of the drop (at the contact point with the solid) and the tangent to the solid surface (Fig. 9-5). Thus, the contact angle measures the tendency of a liquid to wet the solid surface. The smaller the contact angle, the greater the tendency of the liquid to wet the solid. A contact angle of zero degrees signifies complete wetting. A contact angle of 180 degrees signifies no wetting. Contact angles between 0 and 180 degrees signify partial wetting. Solid materials, which form contact angles less than 20 degrees, with liquids, are highly wettable by the liquid. Contact angles more than 90 degrees mean that the solid is not wettable by the liquid. The contact angle between Teflon and water is 112 degrees. Hydrophylic materials have small contact angles with water, for example, the contact angle between lactose and water is 30 degrees, whereas hydrophobic materials have large contact angles, that is, magnesium stearate forms a contact angle of 121 degrees with water. Contact angles depend on the surface tensions of the liquid and the solid and the interfacial tension between them.[1-7]

The Young's equation gives the contact angle θ:

$$\gamma_l \cos \theta = \gamma_s - \gamma_{sl}$$

or:

$$\cos \theta = \frac{\gamma_s - \gamma_{sl}}{\gamma_l}$$

where θ is the angle, γ_s is the surface tension of the solid, γ_{sl} is the interfacial tension between the solid and the liquid, and γ_l is the surface tension of the liquid. For complete wetting, $\theta = 0$ and $\cos \theta = 1$. Surfactants that assist in wetting by lowering the contact angle are called *wetting agents*. They displace the air from the surface of the solid and lower the contact angle between the solid and the liquid by reducing both the liquid surface tension, γ_l, and the interfacial tension, γ_{sl}. The amounts of the surfactant that locate at the solid-liquid interface and the liquid surface are important and affect the wetting efficiency.[8] A drop of water on clean glass will spread, and the contact angle between water and clean glass is zero. A drop of water on dirty glass does not spread. However, the water will spread on the glass if a wetting (surface active agent) is added.

EXAMPLE 2 ▶▶▶

The surface tension of water at 20°C (68°F) is 72.8 dyne/cm. The surface tension of lactose is 71.6 dyne/cm. The interfacial tension between water and lactose is 8.55 dyne/cm. What is the contact angle between lactose and water?

Solution

$$\cos \theta = \frac{\gamma_s - \gamma_{sl}}{\gamma_l}$$

$$= \frac{71.6 - 8.55}{72.8}$$

$$= 0.866$$

$$\theta = 29.99$$

The disadvantage of the Young's equation is that it requires knowledge of the interfacial tension γ_{sl} and the surface tension of the solid γ_s. Neither of these two terms can be measured accurately. As was mentioned, a liquid will wet a solid if spreading of the liquid on the solid occurs. The spreading coefficient measures the tendency of spreading. Thus, substituting the Young equation into the spreading coefficient equation:

$$S = \gamma_l (\cos \theta - 1)$$

This equation is an alternative form of the Young's equation and has the advantage of not requiring γ_{sl} and γ_s. Furthermore, it gives the tendency of spreading wetting. If the contact angle is larger than 0 degrees, the term $\cos \theta - 1$ is negative; thus, the spreading coefficient S will be negative, indicating no wetting.[3,6,7]

The methods of measuring contact angle are described in the Appendix of this chapter.

ELECTRICAL DOUBLE LAYER

At an interface there is an uneven distribution of charges that results in one side of the interface having one type of electrical charge and the other side having the opposite type. However, the net charge of both sides is zero, and overall neutrality is maintained.

Let us take as an example a pharmaceutical solid-liquid interface such as a suspension, in which there is an uneven distribution of electrical charges between the two phases. If in the solid part of the interface there are positive ions, in the liquid part of the interface there will be a dense layer of negative ions (counterions) strongly attached to the solid surface. This layer is called the Stern layer (Fig. 9-6a). Surrounding the Stern layer is a diffuse layer of negative ions that also are trying to approach the solid surface and maintain electroneutrality but are repelled by the Stern layer. This diffuse layer (the Gouy-Chapman layer) can be visualized as an atmosphere of ions surrounding the solid particle. The Stern layer and the Gouy-Chapman layer (diffuse layer) constitute the double layer. The difference between the two layers is that the Stern layer is rigidly attached to the surface of the solid particle, whereas the Gouy-Chapman layer is not. Moreover, the electrical

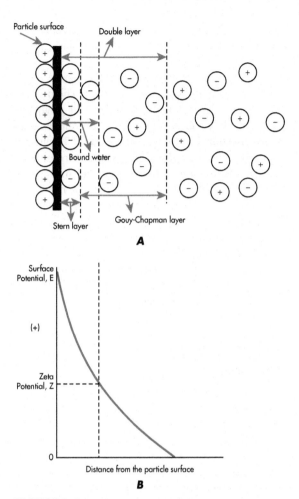

FIGURE 9-6 *A.* The interface of a solid particle inside an aqueous medium and the distribution of charges around it creating different types of layers. *B.* Change in the electrical potential as a function of distance from the solid surface.

potential drops rapidly across the Stern layer and more gradually across the Gouy-Chapman layer.[9] In other words, the charge density is greater near the solid particle and gradually diminishes toward zero at a distance from the particle as the negative and positive ions merge (Fig. 9-6b). When two particles, each having its own double layer, approach each other, electrostatic repulsive forces arise. The zeta potential, which is a measure of magnitude of the electrical charge at the double layer, expresses the magnitude of the repulsive forces between particles. These forces can be affected by the concentration and the valence of the electrolyte in solution. The zeta potential is useful because it can be measured and changed by the addition of electrolytes.

The concept of the zeta potential can be understood by studying particles in solution. The particles of aqueous dispersions that contain polar groups are surrounded by a layer of water like an aqueous shell, which is attached strongly to a particle because of electrostatic forces. This aqueous layer surrounds the surface of the particle like a shell and moves with the solid particle. Even if the particle is moved out of the water, this aqueous layer will move along with the particle. The electric potential at the plane that separates the bound water from the free water (bulk) is called the zeta (electrokinetic) potential (Fig. 9-6b). The zeta potential is related to the mobility of the particle, and its magnitude has implications for the stability of pharmaceutical dispersions.[9,10]

ADSORPTION

The Surfaces

The surface of a solid can hold molecules of other substances that come in contact with it. This also is observed on the surface of a liquid; this phenomenon is called *adsorption* and refers to the process by which a molecule (adsorbate) binds to the surface of the substrate (adsorbent). Adsorption is a surface phenomenon and should be distinguished from *absorption*, in which a molecule enters the bulk of the substrate. For example, water is absorbed by a sponge. Adsorption takes place between a gas and a solid or a liquid and a solid as well as between a liquid and another liquid. Adsorption plays an important role in pharmacy.

Adsorption of Gases or Vapors at Solid Interfaces

Solid materials such as carbon have been used for their gas-adsorbing abilities. Gas masks contain very small particles of carbon (colloidal size) to adsorb poisonous gases. The degree of gas adsorption onto solids depends on the pressure and temperature of the gas. If a gas (adsorbate) adsorbs onto a solid (adsorbent) by utilizing weak forces such as van der Waals forces that keep the adsorbate and adsorbent together, this process is called physical adsorption and is reversible. In contrast, chemisorption requires strong bonding between the adsorbent and the adsorbate, such as covalent or ionic bonding, because there is an exchange of electrons between the adsorbate and the adsorbent. Chemisorption is an irreversible process. However, both physical adsorption and chemisorption (with a few rare exceptions) are exothermic processes and liberate heat. Since adsorption is an exothermic phenomenon and releases heat, according to Le Chatelier's principle, lowering the temperature will increase the amount adsorbed onto the solid.

The extent of the adsorption of a gas onto a solid is described by adsorption isotherms. An *adsorption isotherm* is a graph that shows the amount of adsorbate per gram of adsorbent as a function of the equilibrium pressure or concentration at a certain temperature. In the case of gases adsorbed onto solids, an adsorption isotherm will show the amount of the gas (adsorbate) adsorbed per gram of solid (adsorbent) as a function of pressure at a constant temperature.

Three types of adsorption isotherms can describe the adsorption of gases onto solids: the Langmuir, Freundlich, and BET isotherms. The equations describing each of these isotherms are found in the Appendix of this chapter.

Langmuir Isotherm

A Langmuir isotherm shows that the amount of gas adsorbed onto a solid increases with pressure but reaches a maximum and remains constant thereafter. A Langmuir isotherm indicates the formation of a gas monolayer on the surface of the solid (Fig. 9-7). Langmuir isotherms are often indicative of chemisorption. However, these types of isotherms also have been found for physical adsorption on solids with a very fine pore structure.[5,7,11]

Langmuir isotherms can be used to measure the surface area of solids, since they represent adsorption of a monolayer at the surface of the solid. The point at which the Langmuir isotherm reaches a plateau is used to calculate the amount of the gas adsorbed at the surface. From the amount of gas adsorbed, the number of gas molecules adsorbed is determined, and this is multiplied by

Interfacial Phenomena

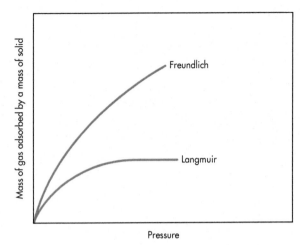

FIGURE 9-8 Langmuir and Freundlich isotherms.

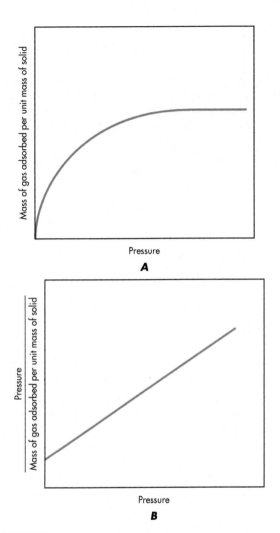

FIGURE 9-7 *A*. The Langmuir isotherm. *B*. The linearized form of the Langmuir isotherm.

the cross-sectional area of each molecule to give the surface area of the solid adsorbent per unit weight. The surface area of the solid per unit weight is called the specific area and plays an important role in the dissolution of drug particles: With increased specific surface area, more molecules or atoms in the crystals will be exposed to the solvent, which will result in a faster dissolution.

Freundlich Isotherm

The Freundlich isotherm shows that the amount of gas adsorbed per mass of the solid increases as the pressure increases.

The Freundlich isotherm indicates multilayer adsorption, in contrast with the Langmuir isotherm, which indicates monolayer adsorption. Furthermore, as Figure 9-8 shows, the Freundlich isotherm is not linear. Most of the time, the Freundlich isotherm represents physical adsorption. Physical adsorption depends on the van der Waals forces of attraction. The Freundlich isotherm, in contrast to the Langmuir isotherm, does not reach a saturation value.

BET Isotherm

BET isotherms, developed by Brunauer, Emmett, and Teller, is a modification of Langmuir's approach. It addresses an inability of the Langmuir isotherm to describe cases where molecules adsorb to other molecules already adsorbed on the surface, thus creating multilayers. BET isotherms have a sigmoidal shape (Fig. 9-9).

Solid-Liquid Interfaces
Adsorption from Solution

Many pharmaceutical systems deal with the adsorption of solutes onto solid surfaces. Drugs or preservatives in solution may adsorb onto containers; thus, the concentration of the drug in the solution is reduced, or with the adsorption of preservative onto the container, the preparation may be left unpreserved and susceptible to microbial attack. Bactericidal compounds such

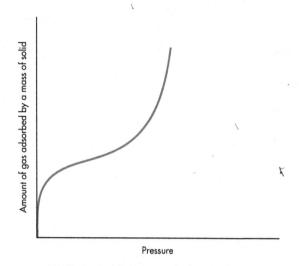

FIGURE 9-9   The BET isotherm.

as cationic surfactants have to be adsorbed on the surface of the bacteria to kill them. Adsorption from solution onto solids also plays an important role in chromatography and water purification.

The amount of material adsorbed onto the solid surface can be estimated by analyzing the liquid and finding the amount of solute adsorbed onto the surface or by removing the solid surface from the liquid and estimating the amount of the material adsorbed onto the surface.

Adsorption from solution is a rather complex phenomenon. It depends on the adsorbate and the nature of the liquid (e.g., the dielectric constant of the solvent and its pH) and the nature of the solid. A polar adsorbent will preferentially adsorb the more polar solute of a nonpolar solution. Likewise, a nonpolar adsorbent will adsorb the nonpolar solute of a polar solution.

Surface active agents may adsorb at the liquid-solid interfaces. Mechanisms responsible for the adsorption of a solubilized surface active agent on solids are as follows:[7]

- *Ion exchange.* This mechanism involves surface active agent ions replacing similarly charged counterions adsorbed onto the substrate.
- *Ion pairing.* This occurs when surfactant ions are adsorbed from the solution onto oppositely charged sites of the solid that are unoccupied by the counterions.
- *Hydrogen bonding.* This occurs when the surfactant adsorbed onto a solid substrate forms hydrogen bonds.
- *Adsorption by dispersion forces.* London forces are responsible for this type of adsorption. Furthermore, this type of adsorption increases, if the molecular weight of the adsorbate increases.
- *Hydrophobic bonding.* This type of mechanism is the result of a very strong attraction between the hydrophobic groups of the surfactant, and their tendency to escape the aqueous environment is so large that they adsorb onto the solid.
- *Adsorption by polarization of π electrons.* This type of adsorption occurs when the adsorbent has positive sites and the adsorbate contains electron-rich aromatic nuclei.

The adsorption from solution onto a solid is commonly described by a Langmuir or Freundlich isotherm.

Case Study: Adsorption to Activated Charcoal

On a hot August day a patient came to the emergency room. He had signs of food poisoning, nausea, vomiting, and diarrhea. The doctors suspected food poisoning because during his lunch 4 hours ago, he consumed shellfish. Shellfish such as mussels, clams, and oysters are filter feeders and accumulate toxins produced by microscopic algae. The toxins are responsible for most shellfish poisoning because they are water-soluble, heat and acid stable, and common cooking methods do not destroy them. The doctors in the emergency room administer to him by mouth activated charcoal at a dose of 1g/kg.

Activated charcoal is a black powder made of wood that has been heated at high temperatures and then reheated again and exposed to an oxidizing gas to create a fine black powder that is highly adsorbant. It is given in the case of drug or toxin poisoning and adsorbs 100 to 1000 mg of drug per gram of charcoal. It does not dissolve in water.

Activated charcoal is administered orally as an antidote to several drugs and poisons. It has an immense specific surface area; depending on the brand, it can have up to 3000 m^2/g. Drugs such as acetaminophen, ibuprofen, phenobarbital, tetracycline, tolbutamide, sedatives, hypnotics, and tricyclic antidepressants adsorb onto the surface of activated charcoal; therefore, their absorption through the gastrointestinal tract is limited. Neutral drugs are adsorbed better to activated charcoal than are their ionic forms because adsorbed neutral molecules do not electrostatically repel each other. However, adsorption of fluoxetine hydrochloride at pH 7.5, where 97.55% of the drug is ionized, is still much larger than expected. This phenomenon is attributed to the ionic strength of the solution: The added salts in the solution help reduce the electrostatic interactions of the adjacently adsorbed ionized fluoxetine molecules.[12]

The functional states of activated carbon are hydroxyls or ethers (C—O), carbonyls (C=O), and carboxylic acids or esters (O—C=O). The primary site involved in the binding of phenobarbital by activated charcoal is the C—O site.[13]

EXAMPLE: ADSORPTION OF PROTEINS TO SURFACES ▶▶▶

Therapeutic proteins are often potent agents with high specificity and activity at low concentrations. Thus, a small loss can have a significant impact on the therapeutic outcome. Protein formulation can be challenging because of physicochemical stability problems.[14] Proteins adsorb to many interfaces, such as air-water interfaces, surfaces of containers, and tubes. Adsorption of a protein to a polymer surface may catalyze protein unfolding and aggregation.[15] Adsorption of proteins onto surfaces is a fast process. Ninety-three percent of human granulocyte colony-stimulating factor is adsorbed onto polyvinyl chloride (PVC) container surfaces within the first 10 min of contact at 22°C (71.6°F).[16]

Concentration, charge, temperature, and hydrophobicity are among the factors that influence the adsorption of proteins onto interfaces and solid surfaces. The administration of low concentrations of interleukin 1β (IL-1β) through a polypropylene syringe pump resulted in losses up to 80% because of adsorption of protein.[17] Even transient association of IL-2 with the surfaces of a clinical infusion pump and its silicon tubing caused irreversible structural changes on IL-2, resulting in a 90% loss of its biological activity.[18] The protein adsorption on surfaces is concentration-dependent and for some proteins may reach saturation.[14,19] Loss of insulin resulting from adsorption to surfaces is a problem with dilute solutions of insulin less than 1 IU/mg.[20] The fraction of insulin adsorbed decreases with increasing insulin concentration and varies with the nature of the surface; however, more insulin is adsorbed onto hydrophobic than onto hydrophilic surfaces.[21] Agitation of insulin in the presence of hydrophobic surfaces (i.e., Teflon) causes insulin to aggregate.[22,23] Insulin aggregation creates serious delivery problems because the biological activity is reduced and routes of drug administration are obstructed. Substances such as albumin and low concentrations of nonionic surface active agents have been added in the formulation to reduce adsorptive losses of insulin and other protein molecules.[22,24,25]

Factors Affecting Adsorption from Solution[2-7]

Nature of Adsorbent and Adsorbate

The natures of the adsorbent and adsorbate are of the most important factors affecting adsorption at the liquid-solid interface. The specific surface area of the adsorbent dictates the extent of adsorption. The effectiveness of the activated charcoal increases as the surface area and the percentage in surface hydroxyl groups increase.[27] Activated charcoal with a surface area of 2773 m²/g can absorb 2.6 times more than another brand with a specific surface area of 1511 m²/g.[28] The smaller the particle size or the more porous the solid, the higher is its capacity of adsorption. Nonpolar compounds adsorb more onto nonpolar adsorbents. Thus, acetic acid is more polar than propionic acid and it will be adsorbed more onto a silica gel that is polar, whereas the propionic acid will be adsorbed more onto charcoal (nonpolar).[29] Electrical charges on both the solid adsorbent and liquid adsorbate play a significant role on the adsorption. Hydrogen bonds and the van der Waals forces that are weak attractive forces are responsible for physical adsorption from solution onto the solids, whereas ion exchange and other strong attractive forces are responsible for chemisorption.

Solubility

The solubility of a solute has an inverse relationship with its adsorption. The lower the solubility of the solute in the solvent, the higher the extent of its adsorption onto the solid adsorbent. For example, the longer the fatty acid chain, the lower its solubility in water. The adsorption of fatty acids from water onto charcoal is higher the longer the fatty acid chain.[4]

pH

The pH of the solution affects the degree of ionization of the drug and its solubility. Unionized drugs in their majority have low solubility, thus the higher the extent of their adsorption compared to their ionized forms.

Temperature

Adsorption is an exothermic process; thus, decreasing temperature will enhance the adsorption process. The effect of temperature can be described by:

$$\log \frac{C_2}{C_1} = \frac{\Delta H^0_{ads}}{2.303R}\left(\frac{T_2 - T_1}{T_2 T_1}\right)$$

where C_1 and C_2 are the amounts adsorbed (per gram of adsorbent) at temperatures T_1 and T_2, respectively, ΔH^0_{ads} is the standard enthalpy change upon adsorption, and R is the gas constant.[2,4–7]

Adsorption at Liquid Interfaces: Surface Active Agents

Adsorption phenomena also occur at the surfaces or interfaces of liquids with other liquids. When dispersed in the bulk of liquids, substances with part of their molecule lipophilic and part hydrophilic move on their own to the surfaces or interfaces of the liquid, where they lower the surface or interfacial tension. These substances are called *surface active agents* or *surfactants*. They also are called *amphiphiles* because of the dual character of their molecule, part of it lipophilic and part of it hydrophilic. Surface active molecules often are represented in a graphic form as a circle with a tail. The circle signifies the hydrophilic, polar part, and the tail signifies the lipophilic, nonpolar part (Fig. 9-10). The surface active agent can be predominantly hydrophilic or lipophilic, depending on the kind and number of lipophilic or hydrophilic groups the molecule carries.

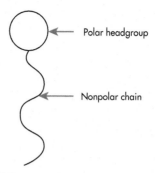

FIGURE 9-10 A surface active molecule.

When surface active molecules relocate at the surfaces after being dispersed in water, the hydrophilic, polar part of the molecule will be in the water, while the lipophilic part will stick out of the surface of the water. In the case of a water-oil interface, the polar part of the surface active molecule will stay in the water side of the interface, while the lipophilic part will locate itself at the oil part of the interface. When the surface active molecules saturate the surface of the water, they form micelles in the bulk of the water. Micelles are aggregates of 50 or more surface active molecules. Micelles are often spherical; however, other shapes may form (Fig. 9-11).

Surface active agents are classified by their hydrophilic groups. Also, they can be classified as anionic, cationic, nonionic, and zwitterionic.[1,7]

Anionic Surface Active Agents

The surface active portion of an anionic surface active agent has a negative charge, for example, sodium lauryl sulfate, $C_{12}H_{25}OSO_3^-Na^+$. Anionic surface active agents are widely used in the pharmaceutical and cosmetic industries; however, they have an unpleasant taste, and some have skin irritation potential. They are not compatible with cationic surface active agents but are compatible with nonionics and zwitterionic surface active agents.

Types of anionic surface active agents include the following:

- **Sodium and potassium salts of straight-chain fatty acids (soaps) ($RCOO^-M^+$ where M^+ is Na^+ or K^+).** The fatty acid chain ranges between 12 and 18. More carbon atoms in the fatty acid chain will render the compound insoluble. Soaps in the presence of divalent or trivalent metallic anions

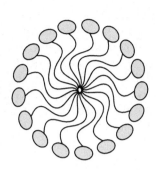
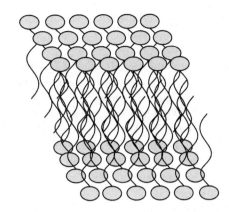

FIGURE 9-11 Different types of micelles.

such as calcium, magnesium, and aluminum form water-insoluble soaps and precipitate. Soaps are unstable below pH 7 and yield water-insoluble free fatty acids.

- **Sulfates ($ROSO_3^- M^+$).** Sodium lauryl sulfate $[CH_3(CH_2)_{10}OSO_3^- Na^+]$ is the most popular representative of this group. It is used in toothpastes, shampoos, and other cosmetic products as well as in fabric detergents.
- **Sulfonates ($RSO_3^- M^+$).** The characteristic of this class is the sulfur atom connected to the carbon atom. Thus, the molecule is less liable to hydrolysis than are sulfates.[10]
- ***N*-Acyl taurines ($RCON(R')CH_2CH_2 SO_3^- M^+$).** These have good skin compatibility, exhibit a good stability over wide ranges of pH, and are compatible with hard water since their Mg and Ca salts are soluble.
- **Monoalkyl phosphates ($ROP(=O)(OH)(O^- M^+)$).** This class has low skin irritation potential and is used in face and body liquid cleansers.
- **Acyl isethionates ($RCOOCH_2CH_2SO_3^- M^+$).** These are used in soaps and shampoos because of their mildness and foaming properties.
- ***N*-Acyl sarcocinates ($RCON(CH_3)CH_2COO^- M^+$).** These produce a rich foam and have excellent skin compatibilities.

Cationic Surface Active Agents

The surface active portion of a cationic surface active agent has a positive charge, for example, dodecyl ammonium chloride, $C_{12}H_{25}NH_3^+Cl^-$. Cationic surface active agents can be used as bactericidal agents. Since they adsorb onto negatively charged surfaces, they also are used as hair conditioners and fabric softeners. Cationic surface active agents, like anionic ones, are electrolytes and they are incompatible with anionic surface active agents. Cationics are compatible with nonionics and zwitterionics. Quatenary ammonium compounds are among the most extensively used cationic surface active agents.

Types of cationic surface active agents include the following:

- **Alkylbenzyldimethyl ammonium salts.** Benzalkonium chloride is a popular cationic surfactant of this class that is used as a germicide.
- **Alkyltrimethyl ammonium salts ($RN^+(CH_3)_3X^-$).** They are used as emulsifiers and in cases where adsorption onto the substrate is desirable. They are also very effective germicides.

Nonionic Surface Active Agents

Unlike anionic and cationic surface active agents, nonionics are not electrolytes, and their surface active portion has no charge. Moreover, they are not affected as much by the presence of salts or changes in pH. Their hydrophylic group may contain hydroxyl groups, polyoxyethylene groups, or saccharides (sugar, sorbitan, etc.).

Types of nonionic surface active agents include the following:

- **Polyoxyethylene alkyl ethers ($RO(CH_2CH_2O)mH$).** These are widely used in the pharmaceutical

and cosmetic industries. The longer the polyoxyethylene chain, the more hydrophilic the molecule and the higher the HLB value.
- **Fatty acid alkanolamides (RCONH(CH$_2$CH$_2$OH)n).** They are used extensively in shampoos as foam stabilizers and viscosity enhancers.
- **Sorbitan fatty acid esters.** They are oil-soluble and form w/o emulsions. They are widely used, also in combination with polyoxyethylene sorbitan fatty acid esters. They also are used in the food industry. SPAN is a commercial name for this type of surface active agents.
- **Polyoxyethylene sorbitan fatty acid esters (TWEEN).** They are hydrophilic and form o/w emulsions. They are used extensively in the pharmaceutical, cosmetic, and food industries.
- **Alkyl polyglucosides.** They are used in dishwashing detergents and shampoos.

Zwitterionic Surface Active Agents

Zwitterionic surface active agents are compatible with all types of surface active agents. Depending on the pH of the medium they are in, they can be anionic, cationic, or zwitterionic. Their main use is as cosurfactants to boost the foaming properties of other surfactants.

At pH 7, N-alkylbetaines (RN$^+$(CH$_3$)$_2$ CH$_2$COO$^-$) lead to minimal skin irritation. Hard water does not affect their foaming properties. The hydrophilic groups of cationic, anionic, amphoteric, and nonionic surface active agents are cations, anions, amphoteric ions, and nonions, respectively.

Hydrophile-Lipophile Balance

Hydrophile-lipophile balance (HLB) is a number that describes the hydrophilic-lipophilic nature of the surface active molecule. It is an arbitrary scale, that made it possible to organize information about surface active agents. Several formulas are used to calculate the HLB value of a surface active agent. Among them, the one that was suggested by Davies and Rideal[29] takes into account the contribution of different chemical groups:

$$HLB = \Sigma \text{ [hydrophilic group numbers]} - \Sigma \text{ [lypophilic group numbers]} + 7$$

Surface active agents with an HLB value less than 9 are lipophilic. The smaller the HLB number, the more the lipophilic the character of the surfactant. A surface active agent with an HLB value less than 9 forms water-oil (w/o) emulsions. An HLB value less than 6 signifies an oil-soluble surface active agent. Surface active agents in the range of 1 to 3 are used as antifoaming agents. Surface active agents with an HLB value above 9 form oil-water (o/w) emulsions. The higher the HLB value of the surface active agent, the more hydrophilic its character.

The HLB value required for emulsification of a particular oil in a water emulsion is determined by trial and error, that is, by mixing different combinations of surface active agents until an optimum (smallest droplets) and a stable emulsion is formed.[1] The combination of the surface active agents has a new HLB value (HLB mixture) equal to the algebraic mean of both HLB values:

$$HLB_{\text{mixture}} = fHLB_1 + (1-f)HLB_2$$

where f is the fraction of surfactant 1 and the fraction of surfactant $-$ is $(1-f)$.

Case Study

Last night Jim and his friends visited their favorite restaurant. Jim ordered the restaurant's signature vegetarian dish of beans and rice. After the dinner, although lactose intolerant, Jim could not resist the banana split ice cream. In a while, he started having cramps, abdominal pain, and stomach distension; furthermore, he did not feel comfortable staying with his friends and left. Returning home he stopped at his local pharmacy and discussed with the pharmacist the discomfort he felt because of the abdominal pain, cramps, and flatulence. The pharmacist gave Jim some tablets and asked him to chew them. The active ingredient in the tablets was simethicone. After he chewed two tablets he felt relief. The bloating and the flatulence were gone.

Question: Describe the action of simethicone in this case.

Resolution: Foods such as beans contain oligosaccharides-stachyose, verbascose, and rafinose. Lactose is an oligosaccharide contained in milk.

More than 50% of the world population is lactose intolerant and cannot digest the lactose contained in milk products. Undigested oligosaccharides are fermented by the intestinal microflora and form foams with high amounts of hydrogen and carbon dioxide gases. Simethicone has foam-inhibiting action and relieves pressure and bloating. Simethicone is not absorbed by the gastrointestinal track. It is a surface active agent with a low HLB value and acts on the foams and bubbles in the intestine produced by the undigested oligosaccharides. Simethicone as a surface active agent reduces the surface tension of the gas bubbles and causes them to collapse and coalesce, releasing the gas and thus reducing pressure and bloating.

Monolayers at the Surfaces

Substances that reduce the surface tension of a liquid when added to it concentrate preferentially onto the surface or interface of the liquid. The number of surface active molecules that reside on the surface can be calculated by using the Gibbs adsorption equation:

$$\Gamma = -\frac{C}{RT}\frac{d\gamma}{dC} = -\frac{1}{RT}\frac{d\gamma}{d \ln C}$$

where Γ is the surface excess or surface concentration in moles per unit area of surface. C is the concentration of the substance, γ is the surface tension, the differential $d\gamma/dC$ shows the change in surface tension with a change in concentration, R is the gas constant, and T is the temperature.

From this equation, when the differential $d\gamma/dC$ is negative, the value of Γ is positive. This indicates surface excess, which means that more molecules of the solute (surface active agent) are on the surface than in the bulk. If the differential $d\gamma/dC$ is positive, the value Γ is negative, indicating surface deficiency of the solute. This form of the Gibbs equation is the most common one and can be applied to a dilute solution of a nonionic surface active agent.

While surface active substances stay preferentially at the surface of the liquid to which they are added and lower its surface tension, surface inactive substances, when added to a liquid, concentrate in the bulk of the liquid and increase its surface tension. For example, the addition of an inorganic electrolyte in water (e.g., NaCl) will cause the surface tension of the water to increase. This indicates a surface inactive substance that prefers to concentrate in the bulk of the water rather than on the water surface.

EXAMPLE 3 ▶▶▶

The concentration of a surfactant in water is 0.01 mole/L, and $d\gamma/dC$ is -5.87 dyne liter mole^{-1} cm^{-1}. What is the surface concentration of the surfactant at 20°C?

Solution

$$\Gamma = -\frac{C}{RT}\frac{d\gamma}{dC} \Rightarrow$$

$$\Gamma = -\frac{0.01 \text{ mole/L}}{8.31 \times 10^7 \text{ erg/deg mole} \times 293°K}$$

$$(-5.87)\frac{\text{dyne liter}}{\text{mole cm}} \Rightarrow$$

$$\Gamma = -\frac{0.01 \text{ mole/L}}{8.31 \times 10^7 \text{ dyne cm/deg mole} \times 293°K}$$

$$(-5.87)\frac{\text{dyne liter}}{\text{mole cm}} \Rightarrow$$

$$\Gamma = 2.41 \times 10^{-12} \text{ mole/cm}^2$$

The surface area per molecule can be found by multiplying the surface concentration by Avogadro's number and dividing 1 by the result:

Area/molecule

$$= \frac{1}{2.41 \times 10^{-12} \text{ mole/cm}^2 \times 6.0221 \times 10^{23} \text{ molecules/mole}}$$

$$= 6.89 \, 10^{-13} \text{ cm}^2/\text{molecule}$$

Figure 9-12 shows that the surface tension decreases with increasing concentrations of the surface active agent; however, after a certain concentration of the surface active agent, the surface tension stops decreasing and reaches a plateau. This concentration is called critical micelle concentration (CMC) because at this concentration the surface is saturated

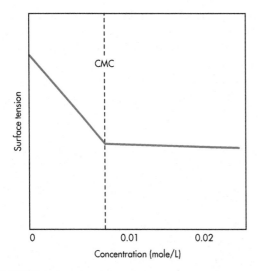

FIGURE 9-12 Changes in surface tension as a function of the concentration of the surface active agent.

with surface active molecules and any increase in their concentration will cause them to form micelles in the bulk to protect their hydrophobic groups from the aqueous environment (Fig. 9-13).

The concentration of the surface active agent also affects other properties of the solution. Figures 9-14 and 9-15 show the effect of the concentration of the surface active agent on other properties, such as interfacial tension, osmotic pressure, detergency, light scattering, and solubility of a drug that is difficult to solubilize. Interfacial tension follows a path parallel to that of surface tension: It decreases with increasing concentration of the surface active agent until the CMC is reached. After that, interfacial tension remains constant. Osmotic pressure increases as the concentration of the surface active agent increases, but at CMC it reaches a plateau. The ability of the solution to remove soil (detergency), the solubility of a drug, and the light-scattering ability of the solution increase sharply when the concentration of the surface active agent increases beyond the CMC concentration.

Micelles are aggregates of surface active agents (see Fig. 9-11). The size of micelles varies, but it is more than 0.1 μm. The number of molecules in micelles is approximately 50 to 100, and micelles are always in equilibrium with monomers of surface active agents in solution. At concentrations of surfactants smaller than CMC, the surfactant monomeric

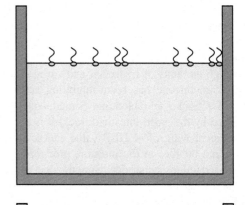

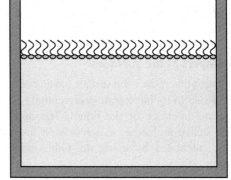

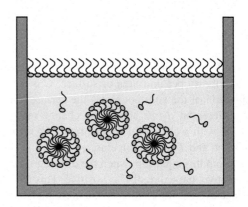

FIGURE 9-13 As more surface active agent is added to water, the surface of the water is covered by molecules of the surface active agent, and then micelles start forming.

numbers are more predominant than the micellar numbers, while the opposite is true at surfactant concentrations above CMC.

In aqueous media, CMC decreases as the number of carbons in the hydrophobic group of the surface active agent increases. Also, in aqueous media the

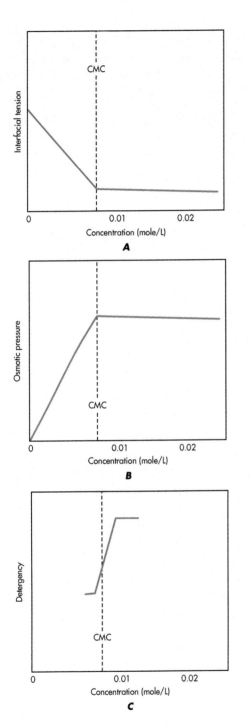

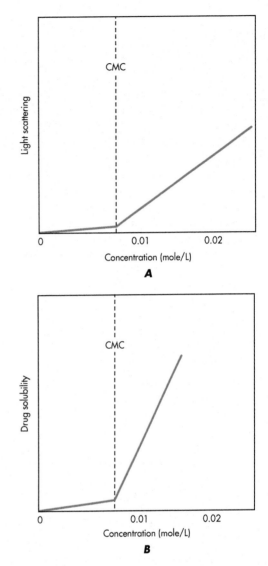

FIGURE 9-15 *A.* Changes in the light scattering of a solution as a function of the concentration of the surface active agent. *B.* Changes in the solubility of a difficult-to-solubilize drug as a function of the concentration of the surface active agent.

FIGURE 9-14 *A.* Changes in the interfacial tension as a function of concentration. *B.* Changes in the osmotic pressure of a solution as a function of the concentration of the surface active agent. *C.* Detergency (cleaning power) of a surface active agent as a function of its concentration.

ionic surfactants have a much higher CMC than do nonionic surfactants containing equivalent hydrophobic groups. Zwitterionic surfactants have a slightly lower CMC than do ionic surfactants with the same number of carbon atoms in the hydrophobic group.[7]

CMC decreases with electrolytes in solution because electrolytes reduce repulsion between micelles. The effect of electrolytes on CMC is more pronouced for anionic and cationic surface active agents than for nonionics.

PROBLEMS

1. A quantity of 10^{-8} M of a surfactant is added to water at 20°C (68°F), and the surface tension of the water-surfactant solution is 2.5 dyne/cm less than that of water. The water surface tension at 20°C (68°F) is 72.8 dyne/cm. What is the surface excess (Γ) in moles of surfactant per unit area of surface?

2. Calculate the work required to increase the surface area of castor oil from 1 cm² to 100 cm² at 20°C (68°F). The surface tension of castor oil at 20°C (68°F) is 39.0 dyne/cm.

3. The surface tension of methanol at 20°C (68°F) is 22.5 dyne/cm. What is the expected height (h) of the capillary rise of methanol inside a capillary tube that has an inside diameter of 0.04 mm? The contact angle is zero, and the density of methanol at 20°C (68°F) is 0.7914 g/cm³.

4. The spreading coefficient of a skin care product is -5 dyne/cm. The contact angle of the product with the skin is 75 degrees. a. Find the surface tension of the product. b. Suggest ways to improve its spreadability.

5. Use the following table to estimate the CMC of an anionic surfactant:

γ dyne/cm	64	56	46.5	42	30	29.2	29
Concentration, $\times 10^{-4}$ M	0.11	0.29	0.99	1.9	9.5	19	28

6. The adsorption of decanol from a solution onto activated carbon particles is as follows:

Equilibrium Concentration, M	Adsorbed, $\times 10^{-4}$ mole/g
0.02	27.0
0.06	53.1
0.08	73.2
0.14	89.3
0.17	94.1
0.21	94.2

Plot the data and show that they fit the Langmuir adsorption isotherm; then evaluate the constants.

ANSWERS

1. $\Gamma = -\dfrac{C}{RT}\dfrac{d\gamma}{dC}$

   $= \dfrac{10^{-8} \text{ M}}{8.3143 \times 10^{7} \text{ erg deg}^{-1} \text{ mole}^{-1} \times 293 \text{ deg}}$

   $\times \dfrac{2.5 \text{ dyne/cm}}{10^{-8} \text{ M}}$

   $\Gamma = 1 \times 10^{-10}$ mole/cm²

   Surface excess, $\Gamma = 1 \times 10^{-10}$ mole/cm²

2. $w = \gamma \Delta A = 39$ dyne/cm $(100 - 1)$cm² = 3861 dynes · cm = 3861 erg

   $w = 3861$ erg

3. $h = \dfrac{2\gamma \cos\theta}{drg}$

   $= \dfrac{2 \times 22.5 \text{ dyne/cm}}{0.7914 \text{ g/cm}^3 \, 0.02 \text{ cm} \, 980 \text{ cm/sec}^2}$

   $= 2.90$ cm

   $= 2.90$ cm

4. $S = \gamma_1 (\cos\theta - 1) \rightarrow \gamma_1 = S/(\cos\theta - 1)$

   $= -5/(\cos 75 - 1) = 64.1$ dyne/cm

   a. $S = 64.1$ dyne/cm

   b. The addition of a surfactant will increase the spreadability of the product.

5. We plot surface tension data versus log concentration

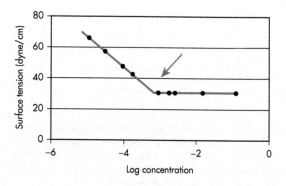

From the figure we see that the line plateaus at log conc = −3.0. Thus CMC = 0.001 M.

6. The Langmuir adsorption isotherm for adsorption from solution is as follows:

$$\frac{x}{M} = \frac{abC}{1+aC}$$

and its linear format is:

$$\frac{C}{x/M} = \frac{1}{ab} + \frac{1}{b}C$$

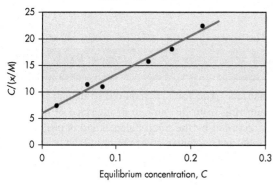

We will use the linear format of the Langmuir equation and we will plot $C/x/M$ versus C:

C	x/M	C/x/M
0.02	2.70E − 03	7.41
0.06	5.31E − 03	11.30
0.08	7.32E − 03	10.93
0.14	8.93E − 03	15.68
0.17	9.41E − 03	18.07
0.21	9.42E − 03	22.29

The slope is $1/b = 74.8$ and thus $b = 0.0133$ moles/g, 0.0133 mole decanol are required to cover 1 gram of charcoal with one monolayer. The intercept is $1/ab = 5.807$ and thus $a = 12.96$.

KEY POINTS

- Interfacial phenomena are those that occur at the boundaries of two or more phases that are different from each other.
- Molecules have lesser attraction to molecules of a different kind at an interface than they do to molecules of the same kind in the bulk medium, away from the interface.
- Because of their greater attraction to molecules of the same kind, the molecules at the interface are drawn inward toward the bulk medium, thereby causing a contraction of the interface that resists external force. This interfacial phenomenon is known as surface (or interfacial) tension.
- Another interfacial phenomenon is exhibited when one substance spread across a substrate, such as a cream over skin or an eyedrop over the eye surface. The greater the interfacial tension between the two surfaces, the less able the substance is to spread across its substrate.
- Spreading is also important for the wetting of a solid, such as water wetting the surface of a solid drug, which is an important first step before the drug can begin to dissolve.
- Another interfacial phenomenon occurs between charged surfaces, where, for example, electrical forces of attraction and repulsion can determine if particles in a suspension are attracted to or are repelled from each other.
- Adsorption to an interface occurs when molecules dispersed in one phase adhere to the surface of another phase. Examples of this phenomenon include the adsorption of drugs to glass, plastic, and activated charcoal.
- Surfactants are amphiphilic chemicals that have properties of both phases, causing them to accumulate at the interface, usually resulting in a lowering of interfacial tension. This allows them, for example, to facilitate spreading, wetting, and emulsion formation. There are a variety of surfactants, usually with a lipophilic portion and a charged or uncharged hydrophilic portion.

CLINICAL QUESTIONS

1. Wetting agents are common excipients in tablets and capsules used to enhance the wetting of drug crystals by gastrointestinal fluids. Explain how they work in terms of their effect on contact angle and explain how this process can potentially improve a drug's bioavailability.
2. Insulin can adsorb to intravenous infusion tubing, so practitioners can choose to flush the tubing with a specific volume of insulin solution or allow the insulin solution to dwell for a specific time period within the tubing, prior to initiating the infusion into the patient. Explain what is accomplished by flushing or dwelling the insulin solution prior to beginning the infusion.

APPENDIX

Measuring Surface Tension

Several methods are used to measure surface tension or interfacial tension. Some of them require more expensive instrumentation than others. Here we will describe a few of the most commonly used methods.[2,5]

Du Nouy Ring Method: This method also is called the detachable ring method and is used to measure both the liquid surface tension and interfacial liquid-liquid tension. It employs a tensiometer that consists of a hanging platinum-iridium ring of defined geometry connected with a microbalance (Fig. 9-16). The instrument also has a mechanism to move the sample liquid in the beaker near the platinum-iridium ring. The surface of the liquid or interface is brought in contact with the platinum-iridium ring. During the measurement, the ring is pulled away from the surface of the liquid or interface. The force F required to detach the platinum ring from the surface or interface of the liquid is recorded by the microbalance and is proportional to the surface tension γ of the liquid:

$$\gamma = kF$$

where k is a proportionality constant that depends on the geometry of the ring.

This method requires a perfectly clean and round ring for accurate estimation of the surface or interfacial tension.

Capillary Rise: If a capillary tube is placed in a liquid that wets the surface of the capillary, the liquid will rise inside the capillary tube and its surface will be concave. For example, water can wet the clean surface of the glass. The height h to which the liquid rises inside the capillary tube depends on the surface tension of the liquid (Fig. 9-17a). The function that connects the capillary rise to the surface tension is:

$$h = \frac{2\gamma \cos\theta}{drg}$$

where γ is the surface tension, θ is the contact angle, g is the acceleration of gravity, r is the internal radius of the capillary tube, and d is the density of the liquid. Because it is difficult to measure the contact angle θ correctly, in practice the capillary rise method is used when the angle is zero or very small, since if $\theta \approx 0$, $\cos\theta = 1$. Zero contact angles for aqueous and other liquid media can be achieved with

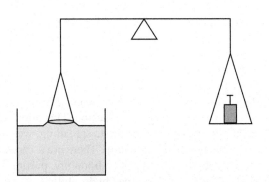

FIGURE 9-16 The Du Nouy ring method of measuring surface tension.

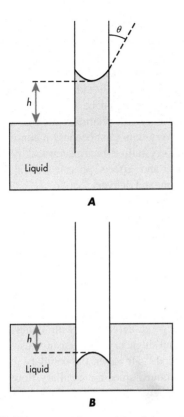

FIGURE 9-17 The capillary rise (A) and depression (B) methods for measuring surface tension.

very well-cleaned glass capillaries. Thus, the previous equation can be simplified:

$$h = \frac{2\gamma}{drg}$$

and the surface tension γ is:

$$\gamma = \frac{1}{2} drhg$$

In some cases (depending on the liquid), if a capillary tube is placed inside a liquid, instead of the surface of the liquid rising inside the capillary, it reaches a level lower than the surface and becomes convex (Fig. 9-17b). This happens when the liquid does not wet the surface of the capillary glass, as is the case with mercury. This treatment also can be used to find the surface tension of the liquid; however, h is now the depth of depression.

The capillary rise (or depression) method cannot be used to measure interfacial tension.

Wilhemy Plate Method: The Wilhemy plate method is a simple and easy way to measure the surface tension of a liquid. It requires hardware similar to that used in the ring method; however, instead of a ring, it uses a thin plate made of mica, glass (such as a microscope cover), or a piece of platinum foil, which is suspended from a balance (Fig. 9-18). The thin plate is placed vertically so that its lower edge nearly touches the surface of the liquid. At this position, a liquid meniscus surrounds the perimeter of the plate. The force (F) required to pull the plate out of the surface depends on the surface tension (γ) of the liquid:[2,29]

$$F = 2(L + T)\gamma \cos \theta$$

where L is the length of the plate. Often the thickness T of the plate is negligible ($T \ll L$). Moreover, because of difficulties in measuring θ, the Wilhemy method is used mostly when $\theta = 0$ so that the $\cos \theta = 1$, and this relationship becomes:

$$F = 2 \times L \times \gamma$$

and surface tension γ is:

$$\gamma = \frac{F}{2 \times L}$$

Measuring the Contact Angle

The contact angle between a liquid and a nonporous solid can be measured using goniometry and tensiometry. In the case of pharmaceutical powders and porous solids the Washburn method is used.

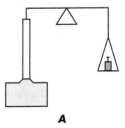

FIGURE 9-18 The Wilhemy plate method for measuring surface tension.

The goniometric method analyzes the sessible (sitting) drop of a liquid on a solid substrate and measures the contact angle formed between the solid and the tangent to the drop. In addition to the goniometer this method makes use of a light source, a sample stage, a lens, and a camera.

Tensiometric methods may employ a Wilhemy plate to measure nonzero contact angles formed by liquids and nonporous solids. The measurement of the contact angle using the Wilhemy plate apparatus is based on the equation:

$$F = \gamma P \cos \theta$$

where F is the wetting force, γ is the surface tension of the liquid, P is the perimeter (length + thickness) of the solid, and θ is the contact angle. This method requires that the surface tension of the liquid, γ, is known. Commercial instruments such as KSV Sigma 70 can be used to measure both the surface tension of the liquid and its contact angle with a solid.

In the processing of suspensions, where a pharmaceutical powder is dispersed in a liquid vehicle, the knowledge of the contact angle of the powder and its liquid vehicle is important. The contact angle of pharmaceutical powders can be determined by the Washburn equation, which relates the rate of penetration of a liquid into a compact powder bed to the contact angle and to the surface tension of the liquid. This method assumes that the pores of the powder bed have uniform diameter. The Washburn equation is as follows:

$$\frac{h^2}{t} = \frac{r\gamma \cos \theta}{2\eta}$$

where η is the viscosity of the liquid, r is the pore radius of the packed bed, h is the height of the liquid penetrating the bed in time, t, and θ is the contact angle. The plot of h^2 versus t is linear. The term, capillary radius in this equation, is difficult to estimate. However this term, r, is eliminated in the following equation, which utilizes the rate of penetration of a liquid (which is perfectly wetting to the solid powder and thus its contact angle with the solid is zero, $\cos \theta = 1$):[30]

$$\cos \theta = \frac{\left(h^2/t\right)_{test} (\gamma)_{test} \eta_{test}}{\left(h^2/t\right)_{wet} (\gamma)_{wet} \eta_{wet}}$$

where $(h^2/t)_{test}$ and $(h^2/t)_{wet}$ are the wetting rates for the test liquid and perfectly wetting liquid, respectively. The term η_{test}, γ_{test}, and η_{wet}, γ_{wet}, is the viscosity and the surface tension of the test liquid and perfectly wetting liquid, respectively.

Contact angle can be used to characterize the surface energy of pharmaceuticals. The surface energy is very important because it controls the interaction of crystalline pharmaceutical powders with excipients, and affects pharmaceutical processing, formulation, and delivery.[31]

A modified sessile drop method was used to measure the contact angle to determine the surface energy of the crystalline L-lysine monohydrochloride dihydrate.[32] The capillary penetration technique was used to determine the wettability and the contact angle of morphine sulfate powders by different liquids.[33] The contact angles of paracetamol particles measured by capillary penetration techniques were found in good agreement with contact angles of paracetamol single crystals measured using the sessile drop method.[34] However, the capillary technique has been questioned since the capillary flow phenomena may differ between large crystals and fine powder.[35]

Calculation of Isotherms
Langmuir Isotherm

The Langmuir equation is based on the following assumptions:

1. All of the sites for adsorption are equivalent.
2. There are no lateral interactions between adsorbate molecules.
3. The solid absorbent is covered by only one layer of the adsorbate.
4. Adsorbed molecules are localized.

The final form of the equation is:

$$\frac{x}{M} = \frac{abP}{1 + aP}$$

where P = pressure, x = mass of gas adsorbed by a mass M of adsorbent, and a and b = constant characteristics of the adsorbent and adsorbate, respectively.

At very low pressures, the Langmuir isotherm equation reduces to:

$$\frac{x}{M} = abP$$

which signifies that the amount of the gas adsorbed x varies linearly with the pressure.

At very high pressures a limiting monolayer is reached:

$$\frac{x}{M} = b$$

The constant b signifies the amount of gas required to achieve complete monolayer coverage of the surface of a unit mass of the adsorbent.

The Langmuir equation also can be written as:

$$\frac{P}{x/M} = \frac{1}{ab} + \frac{1}{b}P$$

The advantage of this form is that when $P/(x/M)$ is plotted versus P, a straight line results. The slope to this line is $1/b$, and the intercept is $1/ab$. Thus, using this form of the Langmuir equation, one can calculate the a and b constants of the adsorbent-adsorbate system (Fig. 9-18).

The Langmuir equation may be based on over-simplified assumptions; however, it can fit many experimental adsorption isotherms reasonably well.

EXAMPLE 4 ▶▶▶

The following experimental data indicate the amount of CO adsorbed on charcoal at different pressures and at 0°C (32°F). Do these data fit a Langmuir isotherm? What is the amount of CO required to cover 1 g of charcoal?

P (mmHg)	x/M (mmole/g)
100	0.15
200	0.27
300	0.38
400	0.46
500	0.55

Solution

The Langmuir isotherm is given by:

$$\frac{x}{M} = \frac{abP}{1+aP}$$

The linearized form of the Langmuir equation is:

$$\frac{P}{x/M} = \frac{1}{ab} + \frac{1}{b}P$$

A plot of $P/(x/M)$ versus P gives a straight line. The slope of the line is $1/b$, and the intercept is $1/ab$.

P (mmHg)	x/M (mmole/g)	P/x/M $\left(\frac{mmHg}{mmole/g}\right)$
100	0.15	666.7
200	0.27	740.7
300	0.38	789.5
400	0.46	869.6
500	0.55	909.1

The data provide a good fit in the Langmuir model ($r = 0.9952$).
The slope of the line ($1/b$) is $(909.1 - 666.7)$ (mmHg/mmole/g)/$(500 - 100) = 0.6$ g/mmole. Since b gives the amount of gas (CO) required to cover by one monolayer the surface of 1 g of adsorbent (charcoal), $1/b = 0.6$ g/mmole and $b = 1.7$ mmole of CO are required to cover 1 g of adsorbent.
The intercept of the line is 601 mmHg/mmole/g.

The Langmuir equation used to describe adsorption from solution onto solids is similar to the Langmuir equation describing the gas adsorption onto a solid discussed previously. The only difference is that the solute concentration, C, is substituted for the vapor pressure term, P, used for solid–gas systems.

$$\frac{x}{M} = \frac{abC}{1+aC}$$

The Langmuir equation for solid–liquid systems is shown above, where x is the amount of solute adsorbed per mass, M, of solid, C is the concentration of solution at equilibrium, b is a constant that describes the amount of solute that has to be adsorbed to cover the solid by one monolayer. The term a is a constant.

Freundlich Isotherm

The Freundlich equation describes the amount of solute adsorbed per gram of adsorbent and is expressed as:

$$\frac{x}{M} = KP^{1/n}$$

where x is the amount absorbed on M grams of adsorbent at STP (standard temperature and pressure), K and n are constants, and P is the equilibrium pressure of the gas.

Another form of this equation is the logarithmic form:

$$\log \frac{x}{M} = \log K + \frac{1}{n} \log P$$

which yields a straight line if log x/M is plotted versus log P. The slope of the line is $1/n$, and the intercept is log K.

The Freundlich equation used to describe adsorption from solution onto solids is as follows:

$$\frac{x}{M} = KC^{1/n}$$

where x is the amount of solute adsorbed per mass, M, of solid. C is the concentration of solute at equilibrium and K and n are constants.

BET Isotherm

The BET equation, considered an extension of the Langmuir equation, addresses cases of multilayer adsorption, which is not addressed with the Langmuir equation. It assumes that the adsorption at one site does not affect adsorption at other sites and that molecules can be adsorbed in a second, third . . . nth layer:

$$\frac{x}{M} = \frac{bcp}{(P_0 - p)\left[1 + (c-1)\frac{P}{P_0}\right]}$$

where x/M is the amount of gas per M mass of solid, b is the amount of gas that has to be adsorbed to cover all of the M mass of the solid by one monolayer, p is the pressure, P_0 is the saturation vapor pressure, and c is a constant.

10 Rheology

Maria Polikandritou Lambros

Learning Objectives

After completing this chapter, the reader should be able to:

- Define the terms Newtonian, plastic, and pseudoplastic flow.
- Discuss examples of Newtonian rheological behavior, with special emphasis on Newtonian pharmaceutical systems.
- Calculate the viscosity of Newtonian solutions by using the appropriate units.
- Identify the units and magnitude of viscosity values of common pharmaceutical liquids.
- Explain the role of temperature in reducing the viscosity of Newtonian fluids.
- Determine the Newtonian viscosity of a pharmaceutical using capillary viscometers and the appropriate experimental methodology and calculations.
- Discuss the non-Newtonian rheology of pharmaceutical systems.
- Explain the significance and calculate the plastic viscosity and yield values of plastic rheograms.
- Interpret pseudoplastic and dilatant rheograms and identify shear-thinning and shear-thickening behaviors.

INTRODUCTION

The term *rheology* is derived from the Greek words *rheo*, "to flow," and *logos*, "science." Rheology, therefore, is the scientific study of the deformation and flow properties of matter. Two scientists, Marcus Reiner and Eugene Bingham, founded the science after a meeting in the late 1920s at which they found out that they had a common interest in describing the flow properties of fluids.[1]

Rheology can be applied to solid, liquid, and gaseous states of matter. From a rheological perspective, solids are classified as being completely resistant to deformation, liquids as being less resistant, and gases as being completely nonresistant. For pharmaceutical applications, however, the discussion of rheology in this chapter focuses primarily on liquids and semisolids (e.g., creams, ointments, and gels). In addition to medical and pharmaceutical applications, rheology is important in many other industries. For example, the rheological property of lubricating fluids such as engine oil is the main determinant of their functional value. The deformation and flow properties of cement, concrete, and paints are important in the construction industry. In the food industry, the rheological properties of ingredients and final products are essential for quality control. Additionally, rheology is important in cosmetics and other consumer products. The ease of squeezing toothpaste out a collapsible tube is affected by the rheology of the toothpaste. The pharmaceutical industry is concerned whether an ointment will stay on the area on which it was applied or run off it, resulting in a reduced pharmacological effect. Moreover, the industries are concerned with the rheological properties of their materials because the rheological properties will affect the processing, the manufacturing, and the characteristics of final product. For example, a powder that does not flow well may not mix homogenously with other powders or produce tablets or capsules with variable weights. Medicine is concerned with the rheology of biological fluids because a change in the rheological properties of blood, saliva, mucus, or synovial fluid may indicate certain diseases.

Rheology itself is used to describe the consistency of different products. The main components of rheology are viscosity and elasticity. *Viscosity* is a measure of resistance to flow or thickness,

- Discuss the importance of thixotropy in pharmaceutical systems.
- Explain the significance of rheology in biological and pharmaceutical systems.
- Identify and discuss the properties of viscosity-inducing agents and their significance in pharmaceutical products.

and *elasticity* refers to stickiness or structure. The higher the magnitude of viscosity, the more resistant the material will be to flow. All liquids are considered viscous; however, the term "viscous" is used commonly to describe materials that are highly resistant to flow. Elastic materials (e.g., rubber bands) can be stretched and will resume their original length when released. Elasticity is primarily a property of solid materials, while viscosity is a property of liquid materials. *Viscoelasticity* is a phenomenon observed in materials that exhibit both elastic behavior and viscous flow. A polymer gel, for instance, will retain its shape and act as a solid (infinite viscosity) at rest. Application of mechanical stress to the gel, however, will result in liquefaction, and the material will begin to flow. Some gels (e.g., gelatin) can undergo a reversible transition from liquid to solid upon a change in an environmental stimulus such as temperature.

NEWTONIAN FLOW

In the field of rheology, flow is typically described by the relationship between viscosity and the applied force: The systems are called Newtonian if viscosity stays constant as the applied force changes, and non-Newtonian if the viscosity changes as some function of applied force.

Newtonian flow is one of the simplest types of flow. Water and low-molecular-weight oils show Newtonian flow. To better understand the concept of the Newtonian flow, let's consider a simple experiment with a deck of cards. If one stacks the cards and blows air horizontally across the top surface of the deck, it is likely that one or two cards at the top will move away. The remaining cards in the deck probably will move a very short distance or not move at all. This creates a gradient in the movement of cards from the top to the bottom of the deck. Air present between the cards creates a *lubricating* layer. That experiment can be repeated by applying a nonsticky substance (e.g., talcum powder) to the surface of the cards and arranging them in a deck. Then, one will notice that the distance moved by blowing the same force of air will be greater or that the cards will move more easily with less resistance. Alternatively, if one puts a sticky substance (e.g., glue) on the cards, the cards will move a much shorter distance or they will not move at all. This simple observation is the basis for Newton's law of rheology.

As is shown in Figure 10-1, the application of horizontal force (F) over a unit area (A) is known as *shear stress*. Newton proposed that the velocity (δv) of the material over the small distance (δx) that it travels is directly proportional to the applied shear stress.[2]

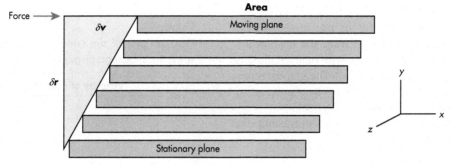

FIGURE 10-1 Diagrammatic representation of the relationship between shear stress and the rate of shear in a Newtonian system.

That is:

$$\frac{F}{A} \alpha \frac{\delta v}{\delta x}$$

The change in velocity over distance is known as the *rate of shear*. To change from proportionality to an equals sign, it is necessary to add a constant. The preceding equation then changes to:

$$\frac{F}{A} = \eta(\delta v/\delta r)$$

where the constant η is known as the *coefficient of viscosity* (or simply viscosity) of the Newtonian fluid. Based on this equation, viscosity is defined as a parameter that measures resistance to flow. Therefore, the higher the magnitude of the viscosity, the more resistant the material will be to deformation and flow. Based on this relationship, when the cards in the deck in the experiment described have no sticky substance on them, they move with little resistance when blown and therefore the system of the cards in this case is analogous to a fluid with lower viscosity. In contrast, the deck of cards with the sticky material on them resembles a fluid with higher viscosity.

EXAMPLE 1 ▶▶▶

Calculate the rate of shear applied by a patient in rubbing a 200-μm-thick film of ointment on the skin surface at a rate of 10 cm/s.

Solution
First convert the units into centimeters:
Since, 1 μm = 10^{-4} cm, 200 μm will be 2 × 10^{-2} cm. The rate of shear ($\delta v/\delta r$) will be

$$\delta v/\delta r = (10)/(2 \times 10^{-2})$$
$$= 500 \text{ s}^{-1}$$

The reciprocal of viscosity is *fluidity* (ϕ), or

$$\phi = 1/\eta$$

Figure 10-2 shows the linear relationship between the rate of shear and shear stress. Note that the graphic representation of the previous equation is opposite to the way the equation is written (i.e., the y variable is rate of shear and the x variable is shear stress). This is the case because in many instruments that measure

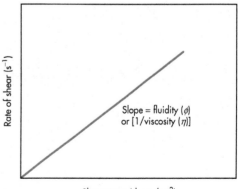

FIGURE 10-2 Linear relationship between the rate of shear (y axis) and shear stress (x axis) in a Newtonian fluid.

viscosity, the shear stress is the independent variable (controlled by the user) and the rate of shear is the dependent variable (obtained from the instrument). The slope of the line therefore will yield the fluidity (or reciprocal of viscosity) value. The graphic representation of the rate of shear versus shear stress at any given temperature is known as a *rheogram*.

The units of shear stress (F/A) in a centipoise (cgs) system are:

$$\text{Units of force/unit of area} = \text{dynes/cm}^2$$

or the same units as pressure.

The units of the rate of shear ($\delta v/\delta r$) in a cgs system are:

$$\text{Units of velocity/units of distance} = \text{(cm/s)/cm} = 1/\text{s, or s}^{-1}$$

Based on dimensional analysis, the units of viscosity in a cgs system are:

$$\text{dynes/cm}^2 = (\text{units of viscosity})(\text{s}^{-1})$$
$$\text{Units of viscosity} = (\text{dynes/cm}^2)\text{s}$$

Since dynes = g(cm/s^2):

Units of viscosity = [1; 2; 3]
or Units of viscosity = g/cm · s

A more convenient way to write the units of viscosity is *poise* in honor of the French physician Poiseuille, who studied the flow properties of blood. Since many liquids, such as water, have relatively low viscosity, cps is also a frequently used unit in pharmaceutical systems:

$$1 \text{ cps} = (1/100) \text{ poise}$$

EXAMPLE 2 ▶▶▶

The following data show the rate of shear as a function of shear stress for a Newtonian fluid at 25°C (77°F). Plot the data and determine the fluidity and viscosity of the fluid.

Rate of shear (s^{-1}): 200 400 600 800 1000

Shear stress
(dynes/cm^2): 50 100 150 200 250

Solution
First, the plot of the rate of shear versus shear stress on a linear graph paper will be as follows:

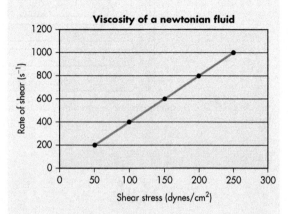

The fluidity and viscosity values are calculated from the slope of the line. Fluidity is the slope, and viscosity is (1/slope). Therefore, fluidity (ϕ) will be equal to slope = $[(y_2 - y_1)/(x_2 - x_1)]$:

$$\phi = [(1000 - 200)/(250 - 50)]$$
$$= 800/200$$
$$= 4.0 \text{ (cm}^2/\text{dynes} \cdot \text{s)}$$

Viscosity (η) is (1/slope):

$$\eta = (1/4)$$
$$= 0.25 \text{ poise}$$

or $\eta = 25$ cps

VISCOSITY OF NEWTONIAN FLUIDS

An important feature of Newtonian rheology is that viscosity is independent of shear stress and remains constant (the slope in Fig. 10-2 is represented by a straight line). This means that regardless of how much force of agitation (or other methods of inducing shear stress) is applied to the liquid of interest, it will have the same viscosity value. This type of viscosity is known as *absolute viscosity*, as opposed to *apparent viscosity*, which is relevant for non-Newtonian fluids. Table 10-1 shows a number of liquids of pharmaceutical interest that exhibit a Newtonian rheological profile. It is important to note that these compounds share the features of low

TABLE 10-1 The Viscosity of Some Newtonian Fluids at 20°C (68.0°F) at 1 atm

Fluid	Viscosity, cps
Acetone	0.34
Water	1.00
Ethanol	1.20
Olive oil	100
Glycerin, 95% (w/w)	545
Castor oil	1000

Source: Adapted from **Martin A, Bustamante P:** *Physical Pharmacy: Physical Chemical Principles in the Pharmaceutical Sciences*, 4th ed. Philadelphia: Lea & Febiger, 1993.

molecular weight relative to polymeric systems and molecular homogeneity.

The United States Pharmacopeia (USP) and other reference texts list kinematic viscosity values as well. *Kinematic viscosity* (κ) is defined as absolute viscosity divided by the density of the substance:

$$\kappa = \eta/\rho$$

where ρ is the density of substance in g/cm³. The units of kinematic viscosity, therefore, will be equal to poise/(g/cm³), conveniently known as a *stoke* (s). As with viscosity, 1/100 of a stoke is a *centistoke* (cs).

The terms *microfluidity* and *microviscosity* are used to describe the fluidity and viscosity, respectively, of microscopic systems, for instance, the microviscosity of cell membrane changes in response to the transport of materials from the extracellular fluid into the cytoplasm. Many anticancer drugs are known to affect the microviscosity of the cell membrane. The physical stability of liposomes (phospholipid vesicles used for drug delivery) also is related to the microviscosity of the bilayer membrane. The addition of cholesterol to the liposome composition is known to increase the microviscosity and thus the physical stability of liposomes. Microviscosity studies are used to determine the hydrophobic core properties of micellar structures formed by the aggregation of surfactants. Microviscosity is determined by using fluorescence or electron paramagnetic resonance (EPR) spectroscopy with probe molecules that selectively localize in the hydrophobic regions of the microstructures. Diphenylhexatriene is a fluorescent probe that is used to determine the microviscosity of phospholipid bilayers. Fluorescence polarization (or anisotropy) studies with diphenylhexatriene are used to measure the *rotational correlation time* (τ) of the probe molecules that are imbibed in the bilayer membrane. When the microviscosity is high, the magnitude of τ will be relatively large as the molecule takes more time to rotate 360 degrees around an axis. Conversely, the τ value will be very small when the microviscosity of the system is low.

EFFECT OF TEMPERATURE ON VISCOSITY

The viscosity of Newtonian fluids *decreases* sharply with increasing temperature. The temperature dependence of viscosity is expressed by the following equation, which is analogous to the Arrhenius equation in Chapter 11:

$$\eta = A \exp^{E_v/RT}$$

The linearized logarithmic form of the equation is:

$$\ln \eta = \ln A + E_v/R(1/T)$$

where A is a constant (has the same units as viscosity), E_v is known as the activation energy (in calories/mole or kilocalories/mole) required to initiate the flow of molecules, and R is the universal gas constant (1.987 cal/mole · K). The units of temperature are in an absolute scale (Kelvin) or 273.15 + °C.

The graph of $\ln \eta$ versus $1/T$ is plotted in Figure 10-3. The slope ($\Delta y/\Delta x$) can be used to determine the magnitude of E_v (slope = E_v/R), and the y-intercept can be used to determine the value of A.

EXAMPLE 3 ▶▶▶

The viscosity of human plasma at 37°C (98.6°F) is 1.2 cps. Assuming that plasma behaves as a Newtonian fluid, determine the viscosity of plasma required for an infusion that is kept at room temperature (25°C, or 77°F). The activation energy of plasma is 4.25×10^3 cal/mole.

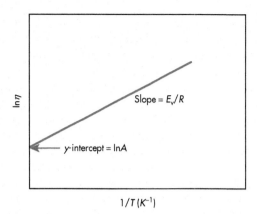

FIGURE 10-3 Linear relationship between ln η and $1/T$.

Solution

First, using the equation $\ln \eta = \ln A + E_v/R(1/T)$, determine the A value by using the parameters at 37°C (98.6°F) (or $T = 310$ K):

$\ln(1.2) = \ln A + [(4.25 \times 10^3/1.987)][(1/310)]$

$0.182 = \ln A + 6.90$

$\ln A = -6.718$

$A = \exp^{(-6.718)}$

$A = 1.21 \times 10^{-3}$ cps

Using the A value, calculate the viscosity at 25°C (or $T = 298$ K):

$\ln \eta_{25} = \ln(1.21 \times 10^{-3}) + [(4.25 \times 10^3/1.987)][(1/298)]$

$= -6.718 + 7.177$

$= 0.460$

$\eta_{25} = \exp^{(0.460)}$

$= 1.58$ cps

NON-NEWTONIAN FLOW

Often pharmaceutical systems such as creams, ointments, or suspensions show non-Newtonian flow. Non-Newtonian flow is observed in complex heterogeneous systems in which the relationship between shear stress and the rate of shear is nonlinear. There are three common types of non-Newtonian rheological profiles of pharmaceutical interest: *plastic flow*, *pseudoplastic flow*, and *dilatant flow*.[3]

Plastic Flow

Plastic flow describes a situation in which no flow occurs in response to shear stress until a certain transition point is reached. This transition point is called the *yield value* (ψ). The yield value is defined as the minimum shear stress required by the system before it deforms and begins to flow. Once the yield value has been reached, the relationship between shear stress and rate of shear becomes linear, as shown in Figure 10-4. The slope of the linear portion of the plastic rheograms is used to calculate the *plastic viscosity* (v) value.

Plastic rheological behavior is observed with many liquid and semisolid dosage forms, including some suspensions, gels, ointments, and creams. It is important to determine the yield value of the final packaged product the patient will use. Since the yield value is a measure of the minimum stress required to induce flow, the magnitude of that stress is critical for the effective use of a pharmaceutical compound. For instance, in the case of a suspension that displays a plastic rheological profile, the yield value is a measure of how vigorously the bottle needs to be shaken before the product flows. For an ointment or a cream, it is a measure of the force required to spread the material on the skin surface. If the product has a very high yield value, it will not rub easily, whereas a product that has a very low yield value may run.

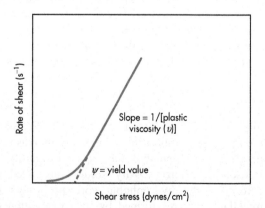

FIGURE 10-4 The relationship between rate of shear and shear stress in a plastic system.

EXAMPLE 4 ▶ ▶ ▶

The following data show the relationship between the rate of shear and shear stress for a topical formulation that exhibits plastic rheology. Plot the data and determine the yield value and plastic viscosity of the formulation.

Rate of shear (s^{-1}): 250   500   750   1000   1500

Shear stress
(dynes/cm^2):   162.5   275   387.5   500   725

Solution
First, plot the rate of shear as a function of shear stress on linear graph paper:

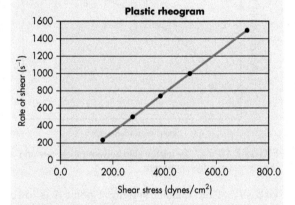

Extrapolating the straight line to the x axis, it is clear that the yield value (ψ) will be approximately 50 dynes/cm^2.

Plastic viscosity (v) = 1/slope or $[(x_2 - x_1)/(y_2 - y_1)]$:
$\qquad v = [(725 - 162.5)/(1500 - 250)]$
$\qquad\quad = 0.45$ poise
or $\qquad v = 45$ cps

Pseudoplastic Flow

Pseudoplastic flow is observed in systems that deform and flow instantaneously with applied stress. The relationship between shear stress and rate of shear, as shown in Figure 10-5, however, is not linear.

In pseudoplastic systems, the slope at point A is lower than that at point B, and since viscosity is inversely proportional to slope, the viscosity at point A is higher than that at point B. Pseudoplastic

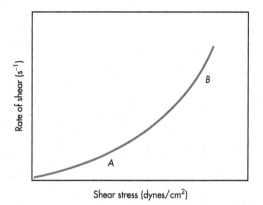

FIGURE 10-5 The relationship between rate of shear and shear stress in a pseudoplastic system.

systems, therefore, also are referred to as *shear-thinning* systems because at increasing shear stress, the viscosity decreases.

Many people are familiar with the frustration of getting ketchup out of a glass bottle. Ketchup (or tomato sauce) is a classic example of a pseudoplastic system that forms a highly viscous material upon rest (no shear stress). To get ketchup out of the bottle, one must shake the bottle several times. Shaking the bottle, or increasing shear stress, causes the three-dimensional structure in ketchup to collapse, and it begins to flow as a result of shear thinning.

Aqueous solutions of certain polymers, such as some of the polysaccharides described later in this chapter, exhibit a pseudoplastic rheological profile. At rest or at low shear stress, the polymers are in a coiled or *globular conformation*, as shown in Figure 10-6, because of intramolecular stabilizing forces.[4] The immobilized water within this globular conformation provides a three-dimensional structure to the system and thus results in higher viscosity. When shear stress is applied (such as agitation), the polymer chain untangles and expands in the direction of flow. The viscosity of the solution decreases sharply as a result of conformation change in the polymer and the release of water, which leads to the breakdown of the structure. This is analogous to the building of house of cards–type structures that collapse upon the application of a small force.

Pharmaceutical products that contain water-soluble polymers as viscosity modifiers, suspending agents, or emulsifiers are examples of products that

FIGURE 10-6 Relaxation of polymer chains with increasing shear stress to show shear-thinning behavior.

often display pseudoplastic rheology. The shear-thinning behavior of a pseudoplastic system can be used to design suspensions and emulsions that have enhanced stability. When used as suspending agents or stabilizers, polymers such as methylcellulose and hydroxypropyl cellulose will decrease the rate of settling and aggregation of particles. When the product is at rest, as on a pharmacy shelf, and there is no shear stress, the viscosity can be very high because of the structure formed by the globular conformation of the polymer. After dispensing the product, the patient shakes the bottle, leading to an increase in shear stress, breakdown of the structure, and flow of the suspension from the bottle. Industrial pharmacists design suspension formulations with polymeric stabilizers that have very high viscosity at rest but flow almost instantaneously after the bottle is shaken a few times. The shear-thinning property of polymeric systems in suspension provides not only stability but also uniformity of drug dose. Pseudoplastic behavior can also be valuable for topical semisolids, because the decrease in viscosity with stress assists both the squeezing of the product out of its container and the rubbing of the product on the skin.

Dilatant Flow

The dilatant rheological profile is exactly opposite to that observed for pseudoplastic systems. In this case also, the deformation and flow occur instantaneously upon the application of stress. As in pseudoplastic systems, the relationship between shear stress and the rate of shear is nonlinear, as shown in Figure 10-7.

However, unlike pseudoplastic systems, the slope at point A is higher than that at point B. As viscosity is inversely proportional to slope, the

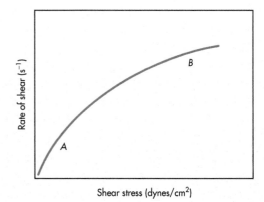

FIGURE 10-7 The relationship between shear rate and shear stress in a dilatant system.

viscosity of the dilatant system at point A is lower than that at point B. Dilatant systems therefore are also known as *shear-thickening* systems.

Whipping cream is an example of a dilatant system. When whipping cream is bought from the supermarket, it comes in the carton as a liquid. Upon the application of shear stress in the form of mixing, the cream begins to thicken (viscosity increases with shear stress) until it solidifies into a foam. A spoonful of solidified whipped cream, when inverted over the bowl, will remain in place.

Dilatant rheology is observed with suspensions that contain a very high (> 40% to 50%—w/w) solid content. In dilatant systems, the solid particles of suspension are loosely aggregated at rest, resulting in low viscosity. Upon the application of stress, the particles begin to break apart and distribute in the medium. The system acquires structure at this point, and the viscosity increases drastically. In some instances the shear-thickening behavior of suspensions can become problematic. For instance, during the industrial preparation of suspensions, the large mixing vessels provide a significant area to the suspension product, which minimizes shear stress (shear stress is *inversely proportional* to area). When the suspension is prepared, it needs to flow through conduits that transport it from the mixing vessels to the packaging area. These conduits have a much smaller area than that of the mixing vessel, and the shear stress increases substantially. Because of high shear stress in these conduits, the viscosity of the suspension increases and the product cannot flow through. Another example where the area of the opening makes a difference in shear stress involves injectable suspensions. The suspension will fill easily into the syringe because the barrel diameter is large; therefore, there is less shear stress. When the suspension is forced through the needle, however, it experiences a much smaller diameter or very high shear stress, and the viscosity increases sharply. A health-care provider injecting this type of suspension will need to use a lot more force, which is bound to inflict some harm on the patient.

THIXOTROPY

In a pseudoplastic system, the shear-thinning behavior at increased stress is observed in curve AB of Figure 10-8. At some point, when the shear rate has reached a maximum and the stress is stopped to allow the material to recover, the downward curve CD does not superimpose over the upward curve AB. The viscosity at point B will be much higher than that at point C even though the shear stress is the same. This property is due to the time required for the recovery of polymer configuration once stress has been applied. It indicates that the structure does not revert back instantaneously when stress is removed. This property of the shear-thinning fluids, called *thixotropy*, is defined as a comparatively slow recovery of the material structure on standing that was lost by shear thinning (analogous to rebuilding the house of cards). The magnitude of the difference between the upward curve and the downward curve is known as the *degree of hysteresis*. Depending on the magnitude of hysteresis, it will take a shorter or a longer time for the pseudoplastic system to acquire structure and become viscous upon standing. Reversal of structure upon standing is dependent on Brownian motion, which is a diffusion-controlled event. Since the diffusion coefficient is directly proportional to molecular weight, high-molecular-weight polymeric materials are slow to recover their structure once shear thinning has occurred.

Shear thinning is a desirable property of pharmaceutical suspensions and emulsions that should have high viscosity upon standing to prevent settling and aggregation but should pour easily from the container once it is shaken. Thixotropy of these shear-thinning systems can be used as a benefit in a well-formulated suspension dosage form. A thixotropic suspension will have high viscosity on standing and is easily pourable after the bottle is shaken, and the structure does not revert quickly so that a dose can be dispensed. The reversal of structure, however, should not be so slow that it does not allow the particles to settle or aggregate. Another desirable feature of thixotropy is in the parenteral administration of shear-thinning suspensions. When the suspension is injected subcutaneously or intramuscularly, it experiences a high shear stress as it passes through the needle (less area), which causes it to liquefy and be easy to inject. At the injection site, the product will form a viscous depot upon rest and release the drug slowly over a long period of time. The simple property of the formulation in this case can be exploited for the development of a sustained-release drug delivery system.

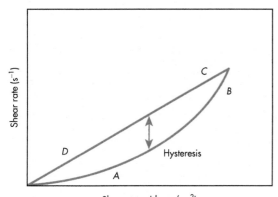

FIGURE 10-8 Upward and downward shear rate versus shear stress curves for a pseudoplastic system exhibiting thixotropic behavior.

VISCOELASTICITY

Some materials exhibit both viscous flow and elastic deformation (such as stretching) when stressed. An aqueous solution of high-molecular-weight poly(ethylene oxide), for instance, will flow slowly from the beaker but also form strings when pulled (like melted mozzarella cheese) that can be cut with scissors. Silly putty is another example of a viscoelastic material in that it deforms when at rest to acquire the shape of the container but will bounce like a rubber ball when stress is applied for a short period. Many types of aqueous gel formulations (e.g., Carpobol) display viscoelastic behavior. When these gels are packaged in collapsible tubes, for instance, they acquire the shape of the tube, but when stress is applied in the form of compression, the material flows from the tube. When a polymeric material remains under shear stress for a long period, the macromolecular chains are uncoiled and disentangled progressively. The chains slip past one another with sufficient ease to cause the system to flow. When the stress is removed, the polymer chains will acquire the structure and show the properties of an elastic material. Permanent disentanglement of the polymer chain results in irreversible deformation of the structure, which does not form a gel upon the removal of stress.

SIGNIFICANCE OF RHEOLOGY

Rheology of Biological Systems

Rheological Properties of Blood

Blood is a complex suspension of cells in plasma. In normal adults, about 40% to 45% of blood composition consists of cells (Table 10-2). The cell density of blood is easily measured as *percent hematocrit* by taking a small volume and centrifuging it to separate the plasma from the cells. The three major types of blood cells are *erythrocytes* (red blood cells), *leukocytes* (white blood cells), and *thrombocytes* (platelets). Erythrocytes make up the majority (~95%) of the cellular component of blood and are responsible for the transport of oxygen and carbon dioxide in the body. Leukocytes are important for defense against infections and generation of antibodies. Platelets, the smallest blood cells, are present in a larger concentration than leukocytes and function in blood coagulation.

TABLE 10-2 Physical Properties of Normal Human Blood

Blood Component	Properties	Values
Whole blood	pH	7.35–7.40
	Viscosity [at 37°C (98.6°F)]	3.0 cps
	Hematocrit: Male	47%
	Female	42%
	Blood volume	78 mL/kg
Plasma or serum	pH	7.30–7.50
	Viscosity [at 37°C (98.6°F)]	1.2 cps
	Osmotic pressure	330 mmH$_2$0
Cells:		
Erythrocytes	pH	7.40
	Count: Male	5.4×10^9 cells/mL
	Female	4.8×10^9 cells/mL
	Diameter	8.4 μm
	Production rate	4.5×10^7 cells/day
Leukocytes	Count	7.4×10^6 cells/mL
	Diameter	7–20 μm
Thrombocytes	Count	2.8×10^8 cells/mL
	Diameter	2–5 μm

Source: Adapted from **Merrill EW:** Rheology of blood. *Phys Rev* 49:863, 1969.

The whole blood viscosity (WBV) depends on the blood cell composition (erythrocytes, leucocytes, and platelets) on plasma proteins (albumin, globulin, and fibrinogen), the shear rate, vessel diameter, cell flexibility, and deformability. The blood is a non-Newtonian fluid. Its viscosity is not constant even when its composition remains unchanged. Blood becomes thinner—less viscous—at higher shear rates in large arteries and veins. Blood becomes thicker—more viscous (WBV increases)—in the small arterioles and capillaries where the shear rate is low. At lower shear rates, the viscosity of blood is high as a result of aggregates of erythrocytes, known as a *rouleaux* formation. Above the shear rate of 100 s^{-1}, the viscosity of blood remains constant (~3.0 cps) and is directly proportional to the hematocrit levels.

The rheological property of blood in normal and diseased states is important for the variety of critical functions that are performed by blood.[5] Blood rheology has been used in diagnosing and treating congenital diseases, infections, rheumatic diseases, malignancy, and inflammatory conditions.

The rheological profile of blood is affected drastically by disease states in which the concentration of cells and proteins changes. In case of polycythemia, where the hematocrit can increase to above 50%, the viscosity of blood increases sharply, resulting in the aggregation of erythrocytes. Above a hematocrit of 65%, the blood rheological profile changes from Newtonian to dilatant flow. The rheological profile is the opposite for patients who have lower erythrocyte counts, such as patients with anemia. In patients with sickle cell anemia, for instance, the viscosity of blood is significantly lower and becomes Newtonian even at lower shear rates.[6]

The viscosity of blood is also dependent on the concentration of plasma proteins such as albumin and fibrinogen. Normal concentrations of albumin and fibrinogen are about 45 to 50 mg/mL and 3 to 5 mg/mL, respectively. In the case of elevated fibrinogen concentration (~6.0 mg/mL), for instance, the transition from non-Newtonian to Newtonian occurs at a much higher shear rate of 300 s^{-1}. Additionally, for patients who suffer from the congenital disease tetralogy of Fallot, there is a concomitant polycythemia and elevated fibrinogen concentration in blood. For these patients, peripheral circulation is completely impaired, and their blood has the consistency of a paste. Other disease states, including diabetes mellitus, have not been shown to affect blood viscosity.[7]

A number of pharmacologic agents can affect the blood rheological profile. Anticoagulants (warfarin, heparin) probably would have the largest effect; however, it is difficult to measure the rheological profile of blood *in vitro* in the absence of anticoagulants. Pentoxifylline, when used to treat peripheral vascular disease, decreases the aggregation of erythrocytes, leading to lower viscosity.

Acetylsalicylic acid (aspirin) and other platelet-acting drugs also have been found to lower blood viscosity. Thrombolytic agents (tissue plasminogen activator, streptokinase, urokinase) and calcium channel blockers (verapamil, diltiazem, nifedipine) also reduce the viscosity of blood. Through stimulation of erythrocyte production, the hormone erythropoetin increases hematocrit and therefore increases the viscosity of blood.

Rheological Properties of Mucus

Mucus is a gel-like material secreted by globular epithelial cells on mucosal surfaces of the respiratory tract, gastrointestinal tract, genitourinary tract, and eyes. Primarily, mucus serves a *protective* function, guarding the underlying mucosal tissue against digestive enzymes and preventing infectious organisms from coming in contact with the host at the first site of exposure. Mucus also has other functions, such as lubrication, and aids in the digestion of food and the migration of spermatozoa.[8]

The mucus gel layer is composed mainly of *mucus glycoprotein* (also known as mucin), which is a high-molecular-weight glycosylated protein. The saccharide content of mucus glycoprotein is very high (> 80%), and this gives the molecule a very high charge density. The glycoprotein is made up of subunits of smaller (molecular weight around 500,000) protein residues that are covalently bonded through disulfide linkages. Based on the number of disulfide bonds, the total molecular weight of mucus glycoprotein can be as high as 1.4×10^7. The mucus gel is a heterogeneous mixture of mucus glycoproteins, phospholipids, and enzymes with imbibed water to form a firm layer on the surface of mucosal tissues. In addition to covalent and ionic forces, several studies have shown that noncovalent forces such as hydrogen bonding, hydrophobic interactions, and physical chain entanglements are necessary to maintain the mucus gel structure *in vivo*. Like other gel-like materials, mucus exhibits viscoelastic rheological properties under normal conditions. The rheological profile of mucus is influenced by the concentration of glycoprotein and other components, the location in the body, and physiologic variables. The normal function of mucus is very much dependent on the rheological properties that change it from a viscoelastic gel-like matrix (infinite viscosity) to a fluid consistency (low viscosity) upon an increase in the shear rate.

Bronchial mucus protects the lining of the respiratory tract by filtering inhaled dust particles and microorganisms. The viscosity of bronchial mucus at the epithelial surface is relatively lower because of the rapid movement of the cilia, which produces a shear rate of about 100 s^{-1}. The mucus layer farther away from the epithelial surface is more viscous and better able to trap particles and microorganisms. The cough reflex allows people to expel excessive mucus from the respiratory tract. Cystic fibrosis, a congenital disease caused by a lack of the cystic fibrosis transmembrane conductance regulator (CFTR) gene, affects about 1 in 2000 live births in the United States. One of the complications of cystic fibrosis is abnormality in the secretion and consistency of bronchial mucus, leading to childhood disorders of the respiratory tract. The decreased ionic content and water lead to mucus that is extremely viscous. Obstruction of airflow into the lungs as a result of mucus plugging will lead to upper respiratory infections, emphysema, and eventually lung collapse. The inhalation form of *N*-acetylcysteine, a mucolytic agent that breaks disulfide bonds in mucus, has been found to be effective in reducing the viscosity of mucus and in allowing these patients to expel it. Other medications, including pulmozyme, a proteolytic enzyme that degrades mucus glycoprotein, and rhDNAse, a DNA-degrading enzyme, also work by decreasing mucus viscosity to enable the secretions to be cleared from the respiratory tract.

The mucus layer in the gastrointestinal tract covers all of the area from the mouth to the rectum. The mucus layer of the stomach protects the tissue from pepsin and hydrochloric acid. It has been suggested that *Helicobacter pylori*–secreted phospholipase A$_2$ may be involved in decreasing the viscosity of stomach mucus and therefore allowing the enzyme and acid to diffuse into the tissue, ultimately resulting in ulcer formation. Mucosal protective agents such as misoprostol (prostaglandin E$_2$) and sucralfate increase the production and viscosity of mucus. Acetylsalicylic acid and other nonsteroidal anti-inflammatory compounds, in contrast, decrease the mucus viscosity to break down the mucosal barrier (Table 10-3).[9]

TABLE 10-3 Effect of Various Agents on Stomach Mucus Viscosity

Agents	Viscosity, %[a] Increase	Viscosity, %[a] Decrease
Prostaglandin E$_2$ analog	78	
Sucralfate	107	
Acetylsalicylic acid		75
Lysolecithin		74

[a]Percent increase or decrease in viscosity relative to untreated control.
Data from **Slomiany BL, Murty VLN, Piotrowski J, Slomiany A:** Effect of antiulcer agents on the physicochemical properties of gastric mucus, in *Mucus and Related Topics*, E Chantler, NA Ratcliffe (eds). Cambridge, UK: The Company of Biologist, 1989.

The mucus layer in the female urinogenital tract is formed as a result of secretion from the uterine wall. The viscosity of cervical mucus is affected markedly by the menstrual cycle. During the period of ovulation, the viscosity of mucus is low to allow the penetration and migration of spermatozoa. At other times, the mucus is highly viscous and impenetrable to spermatozoa. Progesterone and similar contraceptive medications increase the viscosity of cervical mucus. A study by Chantler and associates[10] showed that cationic compounds such as chlorohexidine and other biguanides also increase the viscosity of cervical mucus by ionic interaction with the negatively charged mucus glycoprotein. Increasing the viscosity of mucus in the urinogenital tract with drugs may prove to be an effective strategy for contraception and the prevention of sexually transmitted diseases.

Rheological Properties of Synovial Fluid

Synovial fluid is a viscous (egg-white consistency) liquid present in all the skeletal joints of the body. It is composed mainly of hyaluronic acid, proteins such as albumin, and water. The volume of synovial fluid is about 0.2 to 0.5 mL even in large joints such as the knee. Hyaluronic acid is a mucopolyssacharide that forms highly viscous aqueous solutions.[11] The function of synovial fluid is to protect the skeletal joints through lubrication and provide a medium that acts as a shock absorber. The rheological property of synovial fluid is dependent on the structure of hyaluronic acid. The molecular weight of hyaluronic acid in normal synovial fluid is around 1×10^7. In inflammatory conditions such as rheumatoid arthritis, the molecular weight and concentration of hyaluronic acid decrease sharply, and the viscous consistency of the fluid is lost. To restore the joint function in arthritis, viscosupplementation therapy with intraarticular injections of hyaluronic acid has been found to be effective.[12]

Rheology of Pharmaceutical Systems

Rheology of Mixing

For both Newtonian and non-Newtonian fluids, the shear forces generated during the mixing process by the interactions between the fluid and the surfaces of vessels can have profound implications for the quality of the products formed. For Newtonian systems, the viscosity of the product does not change with shear stress. However, for non-Newtonian systems, the apparent viscosity is related to the shear stress and can either decrease (shear thinning) or increase (shear thickening) with mixing. Many solution dosage forms are examples of Newtonian fluids in which the mixing of solute and solvent occurs at the molecular level to form a homogeneous system. Since the shear forces of mixing do not affect the viscosity of the product, solutions tend to be uniformly mixed and provide dose uniformity. Suspensions and emulsions, however, tend to exhibit a non-Newtonian rheological profile, and the mixing process can affect the viscosity of a formulation. In shear-thinning systems, for instance, the apparent viscosity will be low during mixing, but the product may thicken upon standing. Similarly, for shear-thickening systems, the apparent viscosity will be high during mixing, but the product will become less viscous upon standing. For shear-thickening systems, high viscosity can have a significant effect on the content uniformity of drug in the product. Drug incorporation into ointments and other semisolid dosage forms also is affected by the shear stresses induced during mixing. Depending on the method used for drug incorporation (e.g., roller mills versus a hand mixer), the content uniformity of the drug may be affected.[13]

Rheology of Suspensions and Emulsions

Rheological studies are very important in the manufacturing of suspension and emulsion dosage forms. From the pharmaceutical manufacturing perspective, settling and aggregation of solid particles in suspensions are viscosity-dependent (based on Stoke's law). In addition, the final suspension should be pourable or injectable, based on the ultimate site of administration. Emulsion stability is also dependent on the viscosity of the formulation. The rate of creaming and coalescence of oil droplets in oil-in-water emulsions, for instance, is inversely proportional to viscosity (based on Stoke's law). Industrial scientists use apparent viscosity values of suspensions and emulsions to design systems that can provide optimum stability and dose uniformity upon administration. An advantage of using polymeric emulsifiers such as poly(acrylic acid) copolymer resins is that they provide both emulsification and viscosity enhancement to the final product.[14]

Removal and Spreading of Ointments

Incorporation of drugs and other additives in semi-solid dosage forms affects the apparent viscosity of the base. A compounding pharmacist should be especially concerned with the viscosity issues related to the ointments and other semisolid dosage forms that are prepared routinely in community and institutional pharmacies. The increase in apparent viscosity upon the addition of drugs and excipients in the dosage form can affect the removal of the product from the container and spreading on the skin surface. The problem of the ointment removal container can be quite serious with elderly patients who may have difficulty using enough force to extrude the product, especially from a collapsible tube. If the product cannot be spread well on the skin surface, it will be less efficacious or may fail to perform as intended.

Bioavailability of Ophthalmic Drops

Ophthalmic dosage forms such as eye-drops are applied topically to the eye to treat surface and intraocular conditions such as infections, inflammation, and intraocular pressure. The volume of the tear fluid inside the cul-de-sac of the eye is about 7 to 8 µL. An eye that is not blinking can hold up to 30 µL. A blinking eye can only hold up to 10 µL. When a drop, usually 50 µL, is applied to the eye, most of it is lost mainly due to solution drainage and tear turnover and only 10 µL are retained. From this small volume that remains after blinking, only a very small percentage is absorbed. The bioavailability of the drug in the liquid can be increased by decreasing the drainage of the liquid and increasing its contact time with the eye. Increasing the viscosity of the ophthalmic liquid can retard drainage and can increase the contact time, and therefore, can increase bioavailability. The viscosity of the ophthalmic solution can be increased by including hydrophilic polymeric vehicles such as polyvinyl alcohol (PVA) and hydroxypropylmethyl cellulose (HPMC) in the formulation.

Polymer Gels for Drug Delivery

Water-soluble polymers are used in almost all aspects of drug administration in the body. For instance, polymers are used in solid dosage forms as binders and disintegrants, in ointments and suppositories as a water-soluble base, and in suspensions and emulsions as suspending agents and viscosity modifiers. Polymeric gels, as defined by the USP, are semisolid systems made up of large organic molecules (polymers) interpenetrated by a solvent (water). Gels can be formed by physical cross-linking of the polymers through chain entanglement or noncovalent forces such as hydrogen bonding and hydrophobic interactions. Chemically cross-linked gels are formed by the addition of a cross-linking agent, which could result in either ionic or covalent bond formation. For consumer appeal, gels should have clarity and sparkle. Gels exhibit a viscoelastic property in that they are rigid materials at rest but will flow when shear stress is induced. Gel formulations are prepared for oral cavity, nasal, rectal, and topical administration of drugs.

There is significant interest in developing drug delivery systems that gel *in situ* (i.e., after administration to the patient). For injectable systems, the gelling formulation allows the product to be administered as a liquid (consistency of dose, ease of injection), followed by the formation of a gel depot in the body (*in situ*) to provide sustained delivery of the drug, preferably at the disease site (e.g., intratumoral injection). With ophthalmic administration, the liquid drug formulation can be administered as a drop in the cul-de-sac, where it will congeal to form a gel that increases the residence time and provides prolonged release. Based on these and other applications, several polymeric systems have been investigated as potential vehicles for *in situ* gelling.

Temperature-sensitive polymeric formulations change from a liquid to a gel consistency when the critical transition temperature has been reached. Aqueous gelatin (Jello) and agar solutions are examples of natural polymers that change from a liquid to a gel consistency when the solution is cooled. Other polymeric systems work in the opposite direction, forming gels when the temperature is *increased* beyond a certain point. An example is a 20% to 25% (w/w) aqueous solution of Pluronic F-127 (poloxamer 407), which will change from a liquid to a gel consistency when the temperature is increased above 21°C (69.8°F). Based on this thermo-gelling property, Pluronic F-127 solutions have been studied extensively as an *in situ* gel-forming drug delivery system, because they can be liquids during storage and administration and form gels at body temperature following administration. Another example of this phenomenon is found with some synthetic water-soluble copolymers based on poly(D,L-lactic-co-glycolic acid) and poly(ethylene glycol), which show thermo-gelling properties and biodegradation in a physiologic environment.

Polymeric systems that change from a liquid to a gel consistency as a result of ionic interactions have also been investigated and used for drug delivery. Gellan gum, for instance, changes from a liquid to a gel consistency in the presence of monovalent cations (e.g., Na^+, K^+). Timoptic XE (Merck & Company), an ophthalmic formulation of timolol maleate for the treatment of glaucoma, contains gellan gum to provide longer residence time of the formulation in the eye by increasing viscosity (after administration) upon contact with cations in tear fluid. Sodium alginate is another polymeric material that can form a cross-linked gel structure in the presence of divalent cations (e.g., Ca^{2+}). Chitosan, a linear polymer of D-glucosamine, adheres to biological tissues and forms gels in the presence of divalent and polyvalent anions (e.g., SO_4^{2-}, heparin).

VISCOSITY MODIFIERS

Cellulose Derivatives

Cellulose is the most abundant natural polymer and is obtained mainly from cotton and wood pulp fibers. It is made up of repeating units of D-glucose that are bonded by a β,1-4 glycosidic bond. Cellulose is

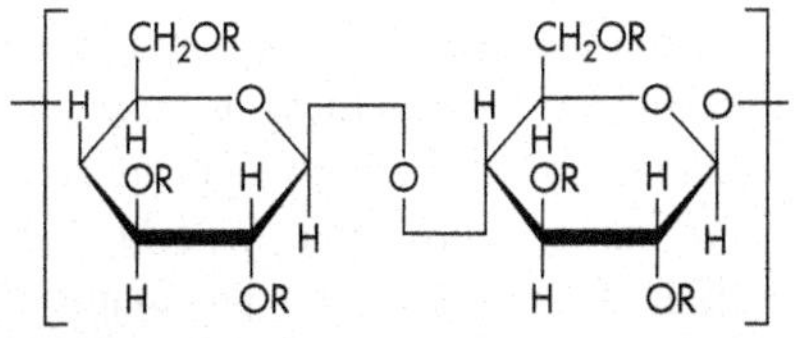

FIGURE 10-9 The chemical structure of cellulose and cellulose derivatives.

Cellulose	R = H
Methylcellulose	R = H, CH$_3$
Sodium carboxymethylcellulose	R = H, CH$_3$C = O$^-$ Na$^+$
Hydroxyethyl cellulose	R = H, CH$_2$CH$_2$OH
Hydroxypropyl cellulose	R = H, CH$_2$CHOHCH$_3$
Hydroxypropyl methylcellulose	R = H, CH$_3$, CH$_2$CHOHCH$_3$

insoluble in water and common organic solvents such as ethanol. Several water-soluble derivatives of cellulose, however, have been synthesized, based on substitution of the hydroxyl groups in glucose residue (Fig. 10-9). These water-soluble derivatives are used frequently as viscosity modifiers.[15]

Methylcellulose

Methoxy-substituted cellulose is the simplest water-soluble cellulose ether derivative. It is formed by reacting cellulose in sodium hydroxide with either methyl chloride or methyl sulfate to yield a product with about 30% methoxy substitution. Dow Chemical Company (Midland, MI) is one of the major suppliers of Methocel brand methylcellulose for pharmaceutical applications in the United States. Methylcellulose is used for oral and topical formulations as a viscosity modifier for both aqueous and hydroalcoholic mixtures. It is a nonionic polymer that dissolves by hydrogen bonding interaction. For dissolution, the methylcellulose powder is dispersed in hot water and agitated to prevent aggregate formation. Cold water (or ice) is then added, and the material dissolves relatively quickly. Ethyl alcohol or propylene glycol may be used to prewet the powder before dissolution. Methylcellulose is available in several different viscosity grades: 15, 400, 1500, and 4000 cps. The viscosity grade refers to the viscosity of 2% (w/v) aqueous solution at 20°C (68°F).

Sodium Carboxymethylcellulose

The sodium salt of carboxymethylcellulose is available in low-, medium-, and high-viscosity grades from several commercial manufacturers. A 1.0% (w/v) aqueous solution of the low-viscosity grade (LF) has a viscosity of 25 to 50 cps, the medium-viscosity grade (MF) has a viscosity of 400 to 800 cps, and the high-viscosity grade (HF) has a viscosity of 1500 to 3000 cps. The polymer powder dissolves in water by hydrogen bonding and ion-dipole interactions. As a result of the presence of an ionizable functional group, the viscosity of the solution is dependent on pH. The viscosity decreases sharply at a pH below 5 or above 10. Being an anionic polymer, sodium carboxymethylcellulose does interact with a number of drugs and other additives (e.g., preservatives) by hydrogen bonding and through ionic forces. It therefore can decrease the efficacy of other compounds in the formulation.

Hydroxyethyl Cellulose

Hydroxyethyl cellulose is a water-soluble cellulose derivative formed by reacting with ethylene oxide in the presence of sodium hydroxide. Hydroxyethyl cellulose is a nonionic polymer and therefore dissolves in water by hydrogen bonding. Various viscosity grades of hydroxyethyl cellulose, based on molecular weights, are commercially available (e.g., Natrosol, Hercules). Unlike sodium carboxymethylcellulose, the viscosity of hydroxyethyl cellulose is not affected by the presence of cations in solution.

Hydroxypropyl Cellulose

Hydroxypropyl cellulose is synthesized by reacting cellulose with propylene oxide in the presence of an alkaline agent such as sodium hydroxide. Hydroxypropyl cellulose is also a nonionic polymer that is soluble in water and polar organic solvents. The solubility in water is temperature-dependent in that hydroxypropyl cellulose is not soluble above 40°C. The Klucel brand of hydroxypropyl cellulose is available from Hercules for pharmaceutical requirements that meet USP guidelines.

Hydroxypropyl Methylcellulose

Hydroxypropyl methylcellulose is synthesized by reacting methylcellulose with propylene oxide in the presence of sodium hydroxide. The typical substitution is 15% to 30% methoxy groups and 4% to 30% hydroxypropyl groups. Like methylcellulose and other cellulose derivatives, hydroxypropyl methylcellulose is nonionic and therefore does not have incompatibility issues when used with cationic compounds. It is an additive in many contact lens solutions and artificial tear fluids.

Natural Gums

Acacia

Also known as gum arabic, acacia is obtained from the gummy extract of the stem of *Acacia senegal* and other African species of *Acacia*. Acacia consists of potassium, calcium, and magnesium salts of polysaccharide arabic acid. An aqueous solution of acacia therefore tends to be acidic. Being an anionic polymer like Na-CMC, acacia also interacts with a number of cationic compounds, including preservatives. The cations (e.g., calcium) in acacia also can interact with anionic compounds. In addition, acacia contains a peroxidase that can oxidize a number of drugs, including morphine and scopolamine. Acacia mucilage, which is prepared by dissolving acacia in water, requires the addition of a preservative (e.g., 0.1% benzoic acid) to prevent microbial growth.

Tragacanth

Tragacanth, which is obtained from the plant *Astragalus*, contains a complex mixture of methoxylated acids (bassorin) that form a gel upon hydration. Tragacanth usually is used as an auxiliary viscosity modifier with another agent. Like acacia, tragacanth mucilage requires a preservative (e.g., 0.1% benzoic acid) to prevent spoilage.

Xanthan

Xanthan, a high-molecular-weight polysaccharide prepared by fermentation of *Xanthomonas campestris*, is a polymer of glucose, mannose, sodium, and potassium or a calcium salt of glucuronic acid. It is largely a neutral polymer with a partial negative charge resulting from glucuronate that dissolves in water and ethanol. It is compatible with almost all univalent and divalent cations. There are possible incompatibilities with high-molecular-weight cations such as benzalkonium chloride.

Carrageenan

Carrageenan is a hydrocolloid extracted from red seaweeds of the class Rhodophyceae. It consists mainly of potassium, sodium, magnesium, and calcium sulfate esters of galactose or 3,6-anhydrogalactose copolymers. The three main types of copolymers are *kappa* carrageenan, *iota* carrageenan, and *lambda* carrageenan, which differ in the composition and manner of linkages of monomeric units and the degree of sulfation. Being a strong anionic polymer, carrageenan interacts with cationic compounds. Those compounds hydrate rapidly in water, but only lambda carrageenan dissolves completely. The other gelling fractions (kappa and iota) require heating to about 80°C (176.0°F) for dissolution.

Case Study

George was very happy to start his third rotation as a pharmacy student at the only compounding pharmacy in his town. A couple of weeks later the pharmacist asked him to prepare soft lozenges containing acacia and phenol 1.4% for sore throat relief. George used vanillin as a flavor. When the patient came to pick up the lozenges, George noticed that their color had changed.

Question: What are the possible reasons for the discoloration of the lozenges?

Resolution: Acacia contains peroxidase, which acts as an oxidizing agent and produces color derivatives of phenol, vanillin, and others such as thymol, tannin, antipyrine, physostigmine, and scopolamine. The lozenges became discolored because of the peroxidase in acacia discolored by phenol and vanillin. To avoid discoloration, George must heat the solution of acacia before mixing with phenol or vanillin because the heat will destroy the peroxidase and thus the discoloration will be avoided.

Sodium Alginate

Alginate, which is extracted from brown seaweed, is a random copolymer of D-glucuronic and D-mannuronic acids (Fig. 10-10). It is produced all over the world, and there are major distributors in the United States (e.g., Pronova Biomedical, Raymond, WA). Sodium alginate is a yellowish white, tasteless, practically odorless fibrous powder. Based on the average molecular weight of the material and the purity, several different grades of alginate are commercially available. It is used at a concentration between 1% (w/v) and 5% (w/v) as a viscosity-inducing agent. Cross-linking of

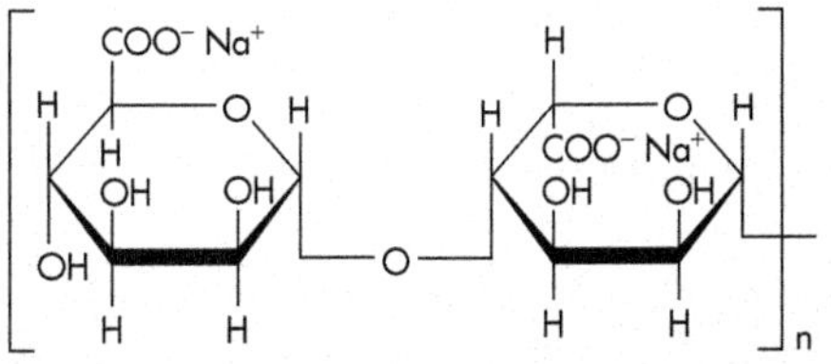

FIGURE 10-10 The chemical structure of sodium alginate.

alginate with calcium or other divalent cations results in gel formation. The gel that is formed is unstable in the presence of phosphate ions, which can react with calcium.

Poly(Acrylic Acid) Resins

B.F. Goodrich's Specialty Polymers and Chemicals Division (Cleveland, OH) manufactures cross-linked poly(acrylic acid) (Carbopol or carbomer) resins for use as viscosity-enhancing and gel-forming agents. Carbopol resins were first described in the scientific literature in 1955. Today, they are used as inactive ingredients in a variety of pharmaceutical products. Since it is a synthetic polymer, several different forms of poly(acrylic acid) resins are available with variation in molecular weight, degree of cross-linking, homo- or copolymer form, and availability as a sodium salt form of acrylate (Fig. 10-11). These are fluffy white dry powders with large bulk densities (208 kg/m^3). The pK_a of the carboxylic acid group of poly(acrylic acid) is between 6.0 and 6.5. The pH of a 1.0% (w/v) aqueous dispersion is between 2.5 and 3.0. Carbopol resins increase the viscosity of the medium when they are neutralized by the addition of an organic (e.g., triethanolamine) or inorganic (e.g., NaOH) base. The reason for the high viscosity at increasing pH is that the COOH group of acrylic acid residue ionizes into the COO— form. Charge repulsion and the presence of Na$^+$ or other counterions lead to extensive swelling of the resin particle and the uptake of water to form a gel. Carbopol resins do not support microbial growth; thus, a gel formulation will not require the addition of preservatives.

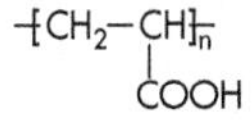

FIGURE 10-11 The chemical structure of poly(acrylic acid).

Homopolymer Resins

These are high-molecular-weight poly(acrylic acid) resins cross-linked with either allyl sucrose or allyl-pentaerythritol. Many homopolymer resins are available in National Formulary (NF) grade for pharmaceutical applications (Table 10-4). Carbopol 910 is very effective at low concentrations when low viscosity, as in suspension formulations, is desired. It is the least ion-sensitive Carbopol resin. Carbopol 934 is used when highly viscous gels, suspensions, and emulsions are desired. Carbopol 934P is a highly purified form of 934 for pharmaceutical applications. Carbopol 940 forms crystal-clear gels with water or hydroalcoholic solvents.

Copolymer Resins

These are polymeric resins of acrylic acid and long-chain (C_{10} to C_{30}) alkyl acrylates that are cross-linked with allylpentaerythritol. In addition to their function as rheology modifiers, these resins serve as oil-in-water (o/w) emulsifiers. The copolymer resins stabilize emulsions by an electrosteric mechanism that involves adsorption onto an oil droplet and prevention of aggregation by the swollen resin particle.

Sodium Salt Resins

These are Carbopol resins in which the acrylate groups have been neutralized with sodium hydroxide. They swell rapidly in water and can be used as viscosity modifiers without the need to add a base to the final composition.

Poly(N-vinylpyrrolidone)

Poly(N-vinylpyrrolidone) (PVP) is a synthetic polymer that is obtained by free radical polymerization of N-vinylpyrrolidone (Fig. 10-12). Its water solubility is based on hydrogen bonding interactions between nitrogen and oxygen atoms in the polymer chains with the hydrogen atoms of water. BASF Corporation (Parsippany, NJ) supplies a variety of pharmaceutical-grade linear PVPs under the trade name Kollidon. The polymers are supplied with different K values, which correspond to the average molecular weight. For instance, Kollidon 12PF is a polymer with a viscosity average molecular weight of 12,000 and is supplied as a pyrogen-free (PF)

TABLE 10-4 Properties of Selected Carbopol Resins for Pharmaceutical Applications

Resin Type	Designation	Viscosity,* cps	Suggested Uses
Homopolymer resins	910,NF	3,000–7,000	Topical formulations, solutions, and suspensions
	934P,NF	30,500–39,400	Topical formulations, stable at high viscosity; used for gels, emulsions, and suspensions
	940,NF	40,000–60,000	Topical formulations, clear gels with water or hydroalcoholic solvents
	941,NF	4,000–11,000	Topical formulations, emulsion stabilization, low-viscosity gels
	974P,NF	29,400–39,400	Oral, topical, and controlled-release formulations; novel drug delivery systems
	980,NF	40,000–80,000	Topical formulations, clear gels with water or hydroalcoholic solvents
Copolymer resins	1342,NF	9,500–26,500	Topical formulations; excellent viscosity modifier in the presence of ions; polymeric w/o emulsifier
	1382	25,000–45,000	Topical formulations; excellent viscosity modifier in the presence of ions; polymeric w/o emulsifier
Sodium salt resins	EX-161 (salt of 934P)	8,000–25,000	Oral, topical, and novel delivery systems; rapid dispersion viscosity modifier
	EX-214 (salt of 974P)	8,000–25,000	Oral, topical, and novel delivery systems; rapid dispersion viscosity modifier

*Viscosity of 1.0% (w/v) aqueous gel of Carbopol resins neutralized to pH 7.0.
Data from **the Lubrizol Corporation / Technical Data Sheets**.

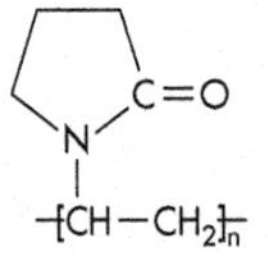

FIGURE 10-12 The chemical structure of poly-(N-vinylpyrrolidone).

liquid. Polymers with a molecular weight greater than 25,000 are solids. USP and other international pharmaceutical standards also list PVPs as povidones, polyvidones, and poly(1-vinyl-2-pyrrolidone). Since PVP is a neutral polymer, it does not interact with cationic or anionic compounds. The viscosity of PVP solutions in water increases with its molecular weight (or K value). PVP does form organic complexes with a number of compounds, including acetaminophen, aspirin, benzocaine, benzoic acid, iodine (Povidone), sulfathiazole, and trimethoprim. These complexes can serve as slow-releasing formulations for antibacterial or topical use. However, complexation also can decrease the bioavailability of these drugs from a suspension or emulsion formulation administered orally.

Pluronic Copolymers

Pluronics or poloxamers are ABA-type triblock copolymers of poly(ethylene oxide) and poly(propylene oxide) (Fig. 10-13). BASF Corporation's Performance Chemicals Division (Parsipanny, NJ) supplies about 30 different types of Pluronic copolymers. They vary in the chain length of ethylene oxide and propylene oxide residues. Commonly used Pluronics include F-68, F-108, and F-127, which are freely soluble in water. The "F" designation refers to the flake form of the product. A popular viscosity modifier is Pluronic F-127 (molecular weight 11,500), which has 96 residues of ethylene oxide and 69 residues of propylene oxide. Pluronic F-127 forms a clear gel at concentrations of 20 to 25% (w/w) in water. The polymer powder is added to water, and the mixture is placed in the refrigerator for about 48 h for the powder to hydrate

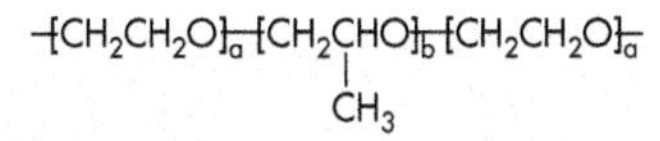

$$\text{-[CH}_2\text{CH}_2\text{O]}_a\text{-[CH}_2\text{CHO]}_b\text{-[CH}_2\text{CH}_2\text{O]}_a\text{-}$$
$$\qquad\qquad\qquad |$$
$$\qquad\qquad\qquad \text{CH}_3$$

FIGURE 10-13 The chemical structure of Pluronic (poloxamer) triblock copolymers.

and dissolve. The cold polymer solution is liquid and changes to a gel consistency when warmed to room temperature. A Pluronic F-127/lecithin combination is marketed as Organogel for extemporaneous compounding of topical gel formulations. As a result of the temperature-induced gelling behavior, drugs can be added and dispersed uniformly or dissolved in liquid Pluronic F-127 at lower temperature and subsequently can form a gel when warmed up to room temperature.

Clays

Bentonite

Bentonite is a naturally occurring clay named after the location where it can be found near Fort Benton, Wyoming. Natural clays are mainly aluminium phyllosilicates, along with varying amounts of other minerals. Bentonite can be used as a viscosity inducer and to prepare gels by sprinkling the solid powder on the surface of hot water and allowing the mixture to stand at room temperature or in a refrigerator for 24 h, to allow for hydration of the clay. Once the bentonite particles are wetted thoroughly, the viscosity of the mixture increases. The final concentration can be adjusted for particular end uses but is usually between 2% (w/v) and 10% (w/v). Glycerin and other nonaqueous liquids can be used to prewet the bentonite particles before they are mixed with water. Aqueous bentonite suspensions retain their viscosity above pH 6.0.

Veegum® (magnesium aluminum silicate) is a brand of a purified form of bentonite that is available in different grades to serve different purposes for the formulator. For example, the standard grade of Veegum can be used at concentrations of about 10% (w/v) to form firm thixotropic gels. Being composed of silicates, Veegum (and bentonite) are anionic, and thus are typically incompatible with cationic compounds, including cationic drugs.

PROBLEMS

1. Ketchup and soda are both available in plastic bottles with narrow openings. Inverting these bottles causes the fluids to flow, but at significantly different rates. Why? What type of flow pattern is exhibited by each fluid?
2. Whipping cream is available as a liquid that converts to a gel/foam structure when it is agitated at a high shear rate. Draw the type of rheological profile displayed by whipping cream and indicate the point where viscosity will be maximal.
3. The following table shows the change in the viscosity of a Newtonian fluid as a function of temperature.

Temperature, °C	Viscosity, cps
25	1.228
30	1.130
35	1.043
40	0.965

   a. Plot the viscosity as a function of temperature by using the linearized form of the equation:

   $$\eta = A \exp^{E_v/RT}$$

   b. From the data, calculate the activation energy (E_v) and the constant A.

4. Poly(ethylene glycol) with a molecular weight of 400 g/mole (i.e., PEG-400) and a density of 1.127 g/cm³ is a Newtonian liquid at 25°C with a viscosity of 7.35 cps. Calculate the time (in minutes) required for PEG-400 to flow through a glass capillary viscometer if the time required for water is 25.4 s (ρ and η of water are 0.997 g/cm³ and 1.002 cps, respectively, at 25°C, or 77°F).

ANSWERS

1. Although ketchup and soda are both sold in plastic bottles with narrow openings, ketchup is a *pseudoplastic fluid* and does not flow right away when the bottle is inverted. At rest, pseudoplastic fluid will form a three-dimensional structure, which is highly viscous. When the product is agitated (e.g., ketchup bottle is shaken several times), the viscosity decreases with increasing shear stress and the material is able to flow out of the bottle. Soda, on the other hand, is a Newtonian fluid and the viscosity remains constant regardless of shear stress.

2. Whipping cream is a dilatant fluid whose viscosity increases with shear stress. As shown Figure 10-7, the viscosity of whipping cream will be highest at higher shear stresses (point B).

3. a. The linearized form of the viscosity-temperature relationship is:

   $$\ln \eta = \ln A + (E_v/R)(1/T)$$

   Therefore, for the plot of viscosity as a function of temperature, the values will be:

Temp. (°C)	Temp (K)	1/T (1/K)	Viscosity (cps)	ln(Viscosity)
25	298	3.36 × 10⁻³	1.228	0.2053
30	303	3.30 × 10⁻³	1.130	0.1222
35	308	3.25 × 10⁻³	1.043	0.0421
40	313	3.20 × 10⁻³	0.965	−0.0356

   The plot of ln viscosity as a function of $1/T$:

   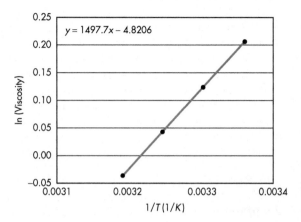

   b. Using the equation of the straight line ($y = 1497.7x - 4.8206$), the slope = E_v/R and the y-intercept is $\ln A$.

   $E_v/R = 1497.7$

   $E_v = (1497.7)(1.987)$

   $= 2975.93$ cal/mole (or ~3.0 kcal/mole)

   $\ln A = -4.8206$

   $A = 8.06 \times 10^{-3}$ cps

4. Using the equation, $\eta_2/\eta_1 = t_2\rho_2/t_1\rho_1$

   $(7.35/1.002) = [(t_2)(1.127)/(25.4)(0.997)]$

   $t_2 = 164.8$ s (~2.75 min)

KEY POINTS

▶ Rheology is the study of flow, which has widespread importance in pharmacy ranging from the flow of powders into capsule shells, to the rubbing of an ointment on the skin, to the drainage of an eyedrop, to the pushing of a fluid through a syringe, to the settling of crystals in a suspension.

▶ Viscosity is a measure of resistance to flow and can be defined as the ratio of shear stress to shear rate. Shear rate is proportional to flow rate. Shear stress is proportional to the applied force, which includes the forces of shaking, rubbing, and pouring.

- In rheology, fluids are characterized as being Newtonian or non-Newtonian. For Newtonian fluids, the flow rate changes in direct proportion to the force applied, that is, the viscosity remains constant. Examples of these include water, glycerin, and olive oil. For non-Newtonian fluids, the flow rate does not change in direct proportion to the applied force, that is, the viscosity can change with different applied forces.
- Among the important types of non-Newtonian systems are plastic systems and pseudoplastic systems. For plastic systems, flow does not begin until enough force is applied to cause flow. Pseudoplastic systems are similar, except that flow does occur as soon as force is applied, but at a relatively low rate. As greater force is applied, viscosity decreases and flow rate increases. In other words, viscosity decreases with increasing shear stress, a process known as shear thinning.
- Thixotropic systems are a type of shear-thinning system that takes a finite time to revert to their more viscous state. This type of system can be valuable for suspensions, as the fluid can be viscous when on the shelf and thereby resist particle-settling. After shaking that product becomes less viscous and easier to pour, long enough to use before reverting to its more viscous state.
- There are a variety of polymers, including natural and synthetic polysaccharides, that can be useful for adjusting the viscosity of drug products as needed.

CLINICAL QUESTIONS

1. Ophthalmic ointments are typically prepared using the rather viscous base of white petrolatum (e.g., Vaseline). Mineral oil is usually included in the formulation to facilitate the application of the ointment and the spreading of the ointment over the eye surface after its application. Conjecture how the mineral oil functions to improve these two processes.
2. An elastomeric infusion pump is an infusion device in which a drug solution is filled into an elastic reservoir, such as an elastic bladder. The infusion flow occurs by the recoil of the bladder, which forces the fluid through a restrictor that controls the flow rate. A drug in a solution of 0.9% sodium chloride was infused into a patient and the nurse noticed that the pump was emptied faster than when the drug was delivered in a fluid of 5% dextrose. Given that 5% dextrose is more viscous than 0.9% sodium chloride, explain the likely reason for the faster emptying.

APPENDIX

Determination of Viscosity

Newtonian Viscosity

Capillary Viscometers: Ostwald, Cannon (www.cannon-ins.com), and Ubbelohde manufacture glass capillary viscometers in which the viscosity of the liquid is dependent on the rate of flow through the capillary, as shown in Figure 10-14. The viscosity is measured at one shear stress only. The capillary method, therefore, is applied to the measurement of the viscosities of Newtonian fluids. A known volume of liquid is placed in the tube and drawn into the upper reservoir bulb (above point 1 in Fig. 10-14) by suction. Then the suction is removed and the liquid is left to flow out of the upper reservoir to the lower reservoir, and the time for complete transfer from the upper to lower reservoirs is measured. A glass capillary viscometer can be placed in a temperature-controlled water bath to minimize temperature-induced variation in viscosity.

Based on Poiseuille's law, the volume of liquid (V) flowing per unit time (t) in a capillary is expressed as

$$V/t = (\pi R^4 \Delta P)/8L\eta$$

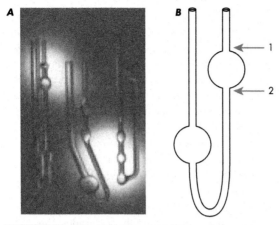

FIGURE 10-14 Some examples of glass capillary viscometers (*A*) and a schematic diagram showing measurement of time of flow between two points (1 and 2) in a viscometer (*B*).

where R is the radius of the capillary in centimeters, ΔP is the pressure difference in dynes/cm^2, L is the length of the capillary in centimeters, and η is the viscosity of the liquid in poise. Rearranging the equation by replacing ΔP with $h\rho g$ where h is the height of the liquid in the capillary in centimeters, ρ is the density in g/cm^3, and g is the acceleration caused by gravity (i.e., 980 cm/s^2), the viscosity of the liquid is determined as:

$$\eta = K\rho t$$

where K is considered an instrumental constant ($K = \pi R^4 hG/8LV$). The viscosity of the liquid of interest is dependent on the time of flow through two points of the capillary viscometer and the density. Calibration of the instrument consists of measuring the instrumental constant (K) with a liquid of known viscosity.

Alternatively, the viscosity of the liquid of interest (η_2) can be calculated from the following equation by comparing the flow time with a reference standard:

$$\eta_2/\eta_1 = t_2\rho_2/t_1\rho_1$$

or

$$\eta_2 = (t_2\rho_2/t_1\rho_1)\eta_1$$

where η_1 is the viscosity, t_1 is the time of flow, ρ_1 is the density of the reference standard, and t_2 and ρ_2 are the time of flow and the density of the liquid of interest, respectively. The η_2/η_1 ratio is known as the relative viscosity of the liquid of interest.

EXAMPLE 5 ▶▶▶

The time taken for a 10% (w/v) aqueous glycerin solution ($\rho = 1.20$ g/cm^3) to pass through a glass capillary viscometer at 25°C (77°F) is 1.24 min. If it takes pure water ($\rho = 1.00$ g/cm^3) 20 s to pass through the same distance and the viscosity of water is 1.00 cps, calculate the viscosity of a 10% (w/v) glycerin solution at 25°C (77°F).

Solution
Using the equation $\eta_2 = (t_2\rho_2/t_1\rho_1)\eta_1$:

$$\eta_2 = \{[(74.4)(1.2)]/[(20)(1.00)]\}(1.00)\}$$
$$= 4.46 \text{ cps}$$

Falling Sphere Viscometers: The viscosity of a liquid is measured from the terminal velocity of a falling or sliding sphere. Many variations of this setup that have a falling needle or a rising air bubble are also available. The falling sphere method is designed for measuring the absolute viscosity of Newtonian fluids, since the value can be obtained only at one shear stress. For the best measurements, the magnitude of the time of fall over the distance of the viscometer should not be less than 30 s.

Based on Stoke's law, the rate of sedimentation of a sphere (v) is inversely proportional to the viscosity of the liquid:

$$v = \{[2R^2(d_2 - d_1)g]/9\eta\}$$

where R is the radius of the sphere, $(d_2 - d_1)$ is the difference in the densities of the sphere and the liquid, g is the acceleration caused by gravity (i.e., 980 cm/s^2), and η is the viscosity of the liquid.

Rearranging the preceding equation, the viscosity of the liquid will be determined as:

$$\eta = \{[2R^2(d_2 - d_1)g]/9v\}$$

EXAMPLE 6 ▶▶▶

Using a falling sphere viscometer, the terminal velocity of a sphere with a radius of 0.05 cm in a Newtonian liquid at 25°C (77°F) was determined to be 15.4 cm/s. Determine the viscosity of the liquid (in centipoise) if the density difference $(d_2 - d_1)$ was 1.65 g/cm³ and the acceleration caused by gravity was 980 cm/s².

Solution
Using the equation $\eta = \{[2R^2(d_2 - d_1)g]/9v\}$, viscosity will be determined as

$$\eta = \{[2(0.05)^2(1.65)980]/[(9)(15.4)]\}$$
$$= 0.058 \text{ poise}$$
or $\eta = 5.80$ cps

Non-Newtonian Viscosity

Dynamic (or apparent) viscosity takes into account the effect of shear rate and time and is therefore the only type of viscosity relevant for non-Newtonian systems. Dynamic viscosity is measured with dynamic instruments, either rotating (shearing) or oscillating. An instrument capable only of measuring shearing viscosities is called a *viscometer*, and the oscillating type is called a *rheometer*. Rotational viscometers are very popular for measurements of dynamic viscosity because they can measure at different shear stresses to account for shear-thinning or shear-thickening behaviors. The shear stress exerted by the torque of a solid rotating body creates viscous drag, which is proportional to the viscosity of the material. Modern viscometers are interfaced with computers to automate the analysis and obtain the apparent viscosities at various shear stresses, the viscosity at zero shear stress, and the yield value directly from the instrument. Temperature control of viscometers is also an important variable, and many newer instruments have built-in temperature settings. The majority of rotational viscometers fall within two main categories: those where the two concentric cylinders rotate relative to one another around a common axis and those consisting of a cone and a plate.

Coaxial-Cylinder Viscometers: Laboratory models of coaxial-cylinder viscometers use the sample in the beaker as an outer cylinder and place another rotating inner cylinder centrally within it. The shear stress is applied by the torque of the rotating cylinder (also called a *spindle*), as shown in Figure 10-15, which can be varied by the speed. The instrument then provides a dial reading of the apparent viscosity value at that shear stress. Many companies in the United States and abroad manufacture laboratory- and industrial-grade coaxial-cylinder viscometers. Cannon Instruments (www.cannon-ins.com), Brookfield Engineering (www.brookfieldengineering.com), and Thermo-Haake (www.haakeusa.com/index.htm), are a few examples.

Cone-and-Plate Rotational Viscometer: These instruments, as shown in Figure 10-16, consist of a rotating cone with a very obtuse angle and a stationary flat plate. The plate is raised until it just touches the apex of the cone. The liquid or semisolid material fills the triangular space between the cone and the plate. Surface tension prevents the fluid from spreading on the plate. The viscous drag created by the rotating cone on a stationary plate is proportional to the apparent viscosity of the material. The speed of the cone can be changed to achieve variation in the shear stress values, and the magnitude of apparent viscosity at different shear stresses can be obtained.

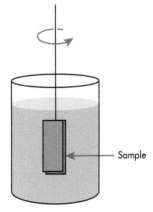

FIGURE 10-15 Schematic diagram of a coaxial-cylinder viscometer.

Ferranti Electric, Inc. (Ferranti-Shirley viscometers), Brookfield Engineering (www.brookfieldengineering.com), and Thermo-Haake (www.haakeusa.com/index.htm) are major manufacturers of cone-and-plate viscometers.

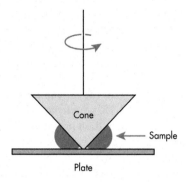

FIGURE 10-16 Schematic diagram of a cone-and-plate viscometer.

11

Chemical Kinetics of Pharmaceuticals

Thomas J. Cook

Learning Objectives

After completing this chapter, the reader should be able to:

- Describe the impact of reaction kinetics in the context of drug formulations.
- Describe the impact of reaction kinetics in the context of pharmacotherapy.
- Indicate the units of the reaction rate constants for zero-order and first-order reactions.
- For zero-order and first-order reactions, define the rate equations, the integrated rate equations, and the mathematical expressions for half-life ($t_{1/2}$) and shelf life (t_{90}).
- Describe the concept of pseudo reaction order.
- From tabulated reaction kinetic data, create a graph of drug concentration versus time and use the graph to determine the rate constants, half-life and shelf-life values, and apparent reaction order.
- Calculate the shelf life (t_{90}) of drugs from mathematical equations and from graphical representations of reaction kinetic data.
- Describe the influence of temperature on reaction rates.

Pharmacists encounter the impact of the chemical degradation of pharmaceuticals in the course of their everyday activities. Proper storage of drug products, providing beyond use dates for prescriptions, and the preparation and storage of sterile products are some examples of common scenarios that are dependent on the knowledge of the chemical kinetics of pharmaceuticals. Therefore, for both safety and economic reasons, it is advantageous for pharmacists to be knowledgeable about the topic of chemical kinetics. The basis for understanding chemical kinetics is the knowledge of the processes involved in chemical degradation and the time dependence on chemical degradation. Understanding the mathematical principles describing the degradation processes is necessary to make sound professional decisions. Fortunately, the primary mathematics are common to other pharmacy topics, particularly pharmacokinetics and biopharmaceutics. In this chapter, we investigate the chemical kinetics of pharmaceuticals focusing on first-order and zero-order reactions, which are the most commonly encountered categories of pharmaceutical degradation reactions.

To ensure that the patient receives the correct dose of a drug, whatever form it is carried in (liquid medicines, tablets, gel caps, etc.), the rate of degradation of a drug must be known. The length of time a drug remains stable is determined by inherent and external factors. Inherently, a drug may be changed or decomposed at a molecular level by the changing of energy and the breaking of molecular bonds. This can result from the temperature at which a drug is stored or by other chemicals with which an active drug is compounded during its formulation as a dosage form (i.e., excipients). Pharmaceutical manufacturers determine the optimum conditions for the long-term stability of a drug product during the drug development process. External factors include such physical properties as pH, pK_a, solubility and dissolution, excipient interactions, and temperature.

The most common degradation reactions that occur with small molecule drugs are solvolysis, oxidation, photolysis, dehydration, and racemization. Of these, solvolysis (the splitting of a drug molecule via a solvent molecule) is the most common

- Estimate the shelf life (t_{90}) and expiration time of drugs at storage temperature by analyzing accelerated stability-testing data.
- Describe the impact of pH and solvent on the catalysis of reactions.
- Evaluate the log k versus pH profile of drugs and identify the pH at which a drug is most stable.
- Apply first-order reaction rate concepts to complex reactions such as reversible, parallel, and consecutive (series) reactions.
- Relate reaction kinetic principles to pharmacokinetics.

reaction type with hydrolysis (where water is the solvent molecule) being the predominant solvolysis reaction.

Chemical kinetics, then, deals with the stability of drugs and the mode of action of their degradation through the examination of rates of reaction and reaction mechanisms. How these rates change and why they change, how the change is measured, and external factors that cause wanted or unwanted effects are discussed in this chapter. In all the discussions, it must be realized that the issue involves the changes of the active drug and how to maintain the medicinal viability of the drug under reasonable conditions.

COMMON DRUG DEGRADATION REACTIONS

Of the common degradation reactions affecting pharmaceuticals, the three most common mechanisms (for small molecular pharmaceuticals) are solvolysis, oxidation, and photolysis; we will focus on solvolysis and oxidation. Since water is the most common solvent encountered by pharmaceuticals (whether in the body or in a formulation), hydrolysis is the primary solvolysis reaction.

Hydrolysis

An hydrolytic reaction occurs when a water molecule, with the hydroxyl ion (OH-) acting as a nucleophile, interacts with a functional group to degrade the drug. A major target of hydrolytic reactions is carbonyl functional groups, which contain C=O in the chemical structure. Examples of carbonyl groups are carboxylic acids, esters, and amides. One well-known example of this type of reaction is the hydrolysis of aspirin (acetylsalicylic acid):

Acetylsalicylic acid + H_2O → Salicylic acid + Acetic acid

In a simplification of this reaction, water reacts with the ester group of aspirin resulting in the production of salicylic acid and acetic acid. The distinctive, vinegary odor of acetic acid may be found in bottles of aspirin that are stored for long periods of time under nonideal conditions (e.g., high temperature and humidity).

Hydrolysis reactions may be accelerated (i.e., catalyzed) by either acids or bases, thus making the pH of the aqueous environment a critical factor in stability. Acid and base catalyses are discussed later in this chapter. The analysis and interpretation of pH rate profiles will graphically illustrate the impact of acid-base catalysis on degradation processes.

Oxidation

Oxidation reactions occur simultaneously with reduction reactions and are generally referred to as redox reactions. These reactions occur together because when one molecule loses electrons (oxidation) another molecule must gain electrons (reduction). More precisely, redox reactions refer to processes where the oxidation state increases (oxidation) or decreases (reduction). While oxygen is not required for an oxidation reaction, oxygen is the primary oxidizing agent of concern for most pharmaceutical degradation reactions. An example of this type of reaction includes the oxidation of morphine, which results in the production of a dimer known as pseudomorphine.

Several approaches can be used to inhibit oxidation reactions in pharmaceutical formulations. Limiting oxygen may be achieved by packing products in an inert atmosphere (e.g., flushing the headspace of liquid parenterals with an inert gas) or, in the case of solid oral dosage forms, sealing the bottle with foil to exclude atmospheric exposure prior to dispensing. Since oxidation reactions of pharmaceuticals may be dependent on pH, the control of pH via buffering may aid in the inhibition of such degradation reactions. In other instances, the use of chelating agents and antioxidants may be warranted.

ORDER OF REACTION

Understanding the rate of reaction, reaction order, and how rate constants are determined will assist the pharmacist in determining beyond use dates, maintaining proper storage conditions, and developing stable products. An effective approach to achieving this understanding is via the mathematical analysis of the degradation process.

Using conventional terminology and symbols, mass action for a degradation reaction can be illustrated as follows:

$$D + W \rightarrow Degradation\ Products \qquad (1)$$

where D represents the drug molecule undergoing degradation and W represents a reacting molecule of water (a common scenario).

Considering the reaction (1) as written, the rate at which the chemical reaction proceeds is proportional to the drug concentration and the concentration of water:

$$\frac{-d[D]}{dt} \propto [D][W] \qquad (2)$$

where $d[D]$ is a very small change in the drug concentration, $[D]$, over a very small change in time, dt. The negative sign indicates that the drug concentration is decreasing with respect to time.

A proportionality is helpful to understand how the degradation process proceeds in a qualitative way. To quantitatively express the degradation process and to permit comparisons between different drugs, the proportionality relationship expressed in (2) must be converted to an equation. To do so, we invoke the rate constant, k, a concrete parameter that represents the speed of the degradation process.

$$\frac{-d[D]}{dt} = k_2[D][W] \qquad (3)$$

Reactions such as those described are often classified by the number of reacting species. In this generic example, the reaction order is determined to be second order. This is also determined by summing up the exponents of the concentration terms (i.e., [D] and [W]) on the right side of equation (3). The values of the exponents of [D] and [W] are each "1" and by convention are not explicitly written. The sum of the exponents is, of course, equal to 2, thus, this reaction would be classified as a second-order reaction. Rate constants, k, from different drug degradation reactions of the same reaction order can be compared to assess the relative stability of drug.

Zero-Order Reactions

A zero-order reaction is one in which the rate is independent of the concentrations of the reactants. The rate of this type of reaction can be determined by a decrease in the drug concentration, $[D]$. The velocity of the decrease in concentration is constant, and the rate equation can be derived for the initial assumption of a decrease of concentration of reactants with time (equation (3) again):

$$-\frac{d[D]}{dt} = k_0 \quad (4)$$

where $[D]$ is the drug concentration and k_0 is the rate constant for a zero-order reaction.

We can integrate equation (4) to generate a linear equation:

$$\int_{[D]_0}^{[D]_t} d[D] = -k_0 \int_0^t dt \quad (5)$$

that upon integration becomes:

$$[D]_t - [D]_0 = -k_0 t \quad (6)$$

And upon rearrangement:

$$[D]_t = [D]_0 - k_0 t \quad (7)$$

where $[D]_0$ is the initial drug concentration and $[D]_t$ is the drug concentration after a specified time, t.

When this linear equation (7) is plotted as drug concentration, $[D]$, against time, t, the slope is equal to $-k_0$ (Fig. 11-1).

Units for a zero-order reaction are obtained by rearranging the rate equations to express the rate constant in terms of the variables. When rearranging equation (4), one can identify the proper units for k_0:

$$k_0 = -\frac{d[D]}{dt} \quad (8)$$

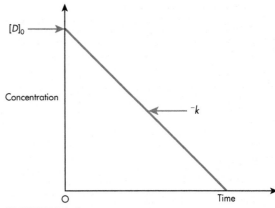

FIGURE 11-1 Zero-order plot of concentration (absorbance) against time, giving a negative gradient equal to k.

From equation (8), one can see that the units for k_0 are ($\frac{concentration}{time}$), where the concentration units in k_0 are the same units for $[D]$.

In addition to the rate constant, the half-life ($t_{1/2}$) is another important parameter for describing drug stability. As the name indicates, the half-life ($t_{1/2}$) is the time required for one-half of the drug to decompose. The equation for $t_{1/2}$ can be derived from equation (8) by setting $[D]_t = \frac{1}{2}[D]_0$ and $t = t_{1/2}$:

$$\frac{1}{2}[D]_0 = [D]_0 - k_0 t_{1/2} \quad (9)$$

Rearrange equation (9):

$$t_{1/2} = \frac{1}{2}\frac{[D]_0}{k_0} \quad (10)$$

The half-life is commonly used to compare the relative stability of different compounds or of the same compound in different formulations.

EXAMPLE 1 ▶▶▶

In a zero-order reaction defined by:

$$-\frac{d[D]}{dt} = k_0$$

at ambient temperature, k_0 was determined to be 0.380 $\left(\frac{mg}{mL} \cdot days^{-1}\right)$. If the initial drug concentration, $[D]_0$, was 200 mg/mL, how long will it take for the initial concentration to decrease by 50%?

Solution

Using equation (10),

$$t_{1/2} = \frac{1}{2}\frac{[D]_0}{k_0}$$

$$= \frac{200\frac{mg}{mL}}{2 \times 0.380\left(\frac{mg}{mL} \cdot days^{-1}\right)}$$

$$= 263 \text{ days}$$

First-Order Reactions

A first-order reaction is one where the rate of reaction is directly proportional to the concentration of one of the reactants. A first-order reaction may be depicted as:

$$D \rightarrow Products$$

And the rate equation that describes this reaction would be:

$$-\frac{d[D]}{dt} = k[D] \quad (11)$$

where $[D]$ is the concentration of drug at time t, $d[D]/dt$ is the rate at which the drug concentration decreases, and k is the first-order degradation constant.

As done with the zero-order process, we integrate equation (11) to derive equations for analyzing kinetic data.

To derive the equation for a first-order reaction, first rearrange equation (11):

$$-\frac{d[D]}{[D]} = k\,dt$$

Integrate the equation:

$$\int_{[D]_0}^{[D]_t}\frac{d[D]}{[D]} = -k_0\int_0^t dt \quad (12)$$

Upon integration, it becomes:

$$\ln[D]_t - \ln[D]_0 = -kt \quad (13)$$

And upon rearrangement we get:

$$\ln[D]_t = \ln[D]_0 - kt \quad (14)$$

This equation can be expressed in its exponential form:

$$[D]_t = [D]_0 e^{-kt} \quad (15)$$

Equation (14) can be used as is with the natural logarithm, but conversion to log to the base 10 is common:

$$\log[D]_t = \log[D]_0 - \frac{kt}{2.303} \quad (16)$$

$$[D]_t = [D]_0 \, 10^{-\frac{kt}{2.303}} \quad (17)$$

Equation (16) is written in the form of a straight line. Since the equation involves the logarithm of the dependent variable, [D], there are two ways to plot first-order degradation data to generate a straight line. When first-order drug degradation data is plotted on semi-logarithmic graph paper (concentration vs. time) a straight line is obtained (Fig. 11-2).

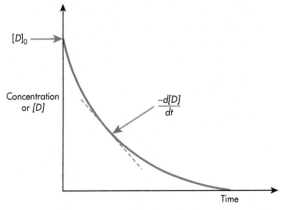

FIGURE 11-2 A first-order plot of concentration against time, showing natural decrease or decomposition of drug.

TABLE 11-1 Equations for Rate Constants, Units of Rate Constants, and Equations for Half-Life ($t_{1/2}$) and Shelf Life (t_{90}) for Zero-Order and First-Order Reactions

	Zero Order	First Order
Equation for k	$k_0 = -\dfrac{[D]_0 - [D]}{t}$	$k_1 = -\dfrac{\ln[D]_0 - \ln[D]}{t}$
Units for k	$k_0 = \dfrac{concentration}{time}$	$k_1 = \dfrac{1}{time}$
Equation for $t_{1/2}$	$t_{1/2} = \dfrac{0.5[D]_0}{k_0}$	$t_{1/2} = \dfrac{0.693}{k_1}$
Equation for t_{90}	$t_{90} = \dfrac{0.1[D]_0}{k_0}$	$t_{90} = \dfrac{0.105}{k_1}$

Also from equation (16), one can derive an expression for the half-life of a reaction ($t_{1/2}$) by setting $[D]_t = \frac{1}{2}[D]_0$ and $t = t_{1/2}$:

$$\log \frac{[D]}{2} = \log[D]_0 - \frac{kt_{1/2}}{2.303} \qquad (18)$$

Rearrange equation (18) and use the laws of logarithm.

$$t_{1/2} = \frac{2.303}{k}\log 2 = \frac{0.693}{k} \qquad (19)$$

From equation (19), one can see that the units for the first-order rate constant are $time^{-1}$. Table 11-1 lists the important equations for zero-order and first-order reactions.

One way to obtain a value for the first-order rate constant k is to plot the log of concentration against time (Fig. 11-3). Notice that plotting log concentration produces a straight line that gives a negative slope (gradient) of $-k/2.303$ from which k can be obtained.

Pseudo Reaction Order

As stated, many drug degradation reactions occur with water as reactant. In such reactions, the "concentration" of water can be considered constant since it is present in vast excess compared to the drug concentration. (Note: Students can work this out on their own to convince themselves of the validity of the assumption.) We can refer back to the equation for a second-order reaction, equation (3).

$$\frac{-d[D]}{dt} = k_2[D][W]$$

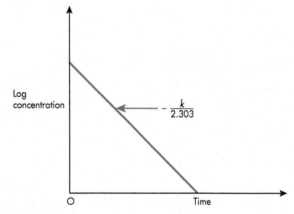

FIGURE 11-3 First-order plot of log concentration against time.

By considering [W] as a constant, the value of the concentration of water can be combined with the other constant on the right-hand side of the equation, k_2. When we combine these constants, we create a new constant, k_1, which replaces "$k_2[W]$" in equation (3) to form:

$$\frac{-d[D]}{dt} = k_1[D] \quad \text{where } k_1 = k_2[W] \qquad (20)$$

Equation (20) now only has one concentration of reactants, [D], on the right side of the equation and the exponent of that concentration is 1, which is indicative of a first-order reaction. This reaction, however, is now considered to be pseudo first order, because while the actual reaction has two reactants, one is in vast excess and the observed reaction appears to be first order both mathematically and graphically.

We can take this pseudo reaction order exercise a step further. If we can then fix the concentration of drug by preparing a suspension of the drug (described in more detail in the pseudo-zero-order reactions-suspensions section of this chapter), the concentration of drug in solution, $[D]$, can now be considered to be constant and combined with k_1 to form a new constant, k_0. Substituting k_0 for $k_1[D]$ in equation (20) gives us:

$$\frac{-d[D]}{dt} = k_0 \quad (21)$$

In equation (21), there are no concentration expressions on the right side of the equation and, thus, the reaction is now considered to be pseudo zero order.

Pseudo reaction order, therefore, refers to the concept whereby a higher-order reaction appears to be a lower-order reaction (mathematically and graphically) because one or more reactant concentrations are constant or in great excess compared to another reactant. This concept is important since many degradation reactions are often at least second order but since one of the reactants is typically in excess, the observed reaction order is pseudo first order or even pseudo zero order.

EXAMPLE 2 ▶▶▶

The stability of stampidine was studied at pH 6 in citrate buffer (D'Cruz et al., Biology of Reproduction, **69**, 1843–1851 2003). The starting concentration of stampidine is 10 mg/mL. After 50 days, the concentration of stampidine in the solution is 6.5 mg/mL. (Note: These are estimates based on the figures in the citation.)

a. Calculate the pseudo-first-order rate constant, k, by using equation (16).
b. Calculate the half-life, $t_{1/2}$, of stampidine at pH 6.

Solution

a. $$\log[D]_t = \log[D]_0 - \frac{kt}{2.303}$$

$$\log\left(6.5\frac{mg}{mL}\right) = \log\left(10\frac{mg}{mL}\right) - \frac{k \times 50 \text{ days}}{2.303}$$

$$k = \frac{2.303}{50 \text{ days}} \log\frac{10}{6.5} = 0.00862 \text{ days}^{-1}$$

b. $$t_{1/2} = \frac{0.693}{k} = \frac{0.693}{0.00862 \text{ days}^{-1}} = 80.4 \text{ days}$$

Pseudo-Zero-Order Reactions—Suspensions

Liquid suspensions fall into this special zero-order situation when the particles of drug in suspension are dissolving continuously to keep the concentration in equilibrium with the degraded drug that is going into solution. Thus, the concentration of drug remains constant for a period of time. The expression k_{app} signifies the apparent constant for the reaction:

Solid drug → drug → saturated solution [A]
$\downarrow k$
degraded drug [B]

$$\frac{-d[B]}{dt} = k[A] \quad (22)$$

A solid drug acts as a drug reservoir, and $[A]^n$ remains constant:

$$-\frac{d[B]}{dt} = k_{app} \quad (23)$$

where $k_{app} = k[A]$.

DETERMINATION OF THE ORDER OF A REACTION

Graphical Method

One can collect data by studying certain reactions and plotting the data on graph paper. The rate of a particular reaction can be found by observing the concentration of one of the reactants or products at given time intervals during the reaction. Various parameters can be measured, such as colorimetric changes, the refractive index, optical rotation, pressure, volume changes, and volumetric analysis. More commonly for pharmaceutical degradation processes, the loss of reactants and/or appearance of products is measured via an HPLC or LC/MS/MS analytical technique. These techniques provide a

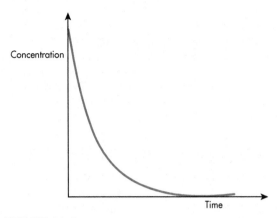

FIGURE 11-4 First-order plot of concentration against time, showing degradation of drug.

TABLE 11-2 Decomposition of Tertiary Butyl Alcohol with Sulfuric Acid at 200°C (392°F)

Concentration (a − x), %	Time, min
100	0
80	10
65	15
50	20
40	25
35	30
30	35
28	40
26	45
25	50

quantitative measure of the concentration of analytes in a sample. These changes over time can be tabulated and graphed, providing information such as rate constants and half-lives.

First-Order Reactions

When plotting the change in concentration of reactants against time, a decreasing curve indicates that the reactants are being consumed in their transition to products (Fig. 11-4).

Consider the reaction where tertiary butyl alcohol undergoes decomposition to form cyclobutane and water:

$$C_4H_9OH = C_4H_8 + H_2O$$

If a is the initial concentration of reactant and $(a − x)$ is the concentration after time t, in a first-order reaction, the rate dx/dt is proportional to the concentration of reactant at time t when the concentration is $(a − x)$.

If the data from Table 11-2 are used to construct a graph of percentage concentration remaining after time, it will be seen to resemble Fig. 11-4. If the log percentage concentration remaining after time is plotted, a negative straight-line relationship will be observed similar to that in Fig. 11-6, where the rate constant k can be obtained from the slope:

$$\text{Rate of reaction} = \frac{dx}{dt} = k(a-x) \text{ (general equation)}$$

and:

$$\frac{dx}{(a-x)} = k\, dt$$

$$\int_0^x \frac{dx}{(a-x)} = k \int_0^t dt \tag{24}$$

$$-\ln(a-x) = kt + c \tag{25}$$

When $x = 0$ and $t = 0$ and there is a substitution indicating c:

Integration constant $= -\ln a \therefore \ln a - \ln(a-x) = k_t$

$$\tag{26}$$

and

$$k_t = \ln\frac{a}{(a-x)} = 2.303 \log\frac{a}{(a-x)} \tag{27}$$

This follows a straight-line equation of the form $y = mx + c$. If $\ln a/(a − x)$ is plotted against t, a straight line is obtained through the origin from which k, the rate constant, can be obtained from the gradient (Fig. 11-5).

That is one way to plot the data for a first-order reaction. Another method is to plot log concentration on the y axis against time to achieve a graph, as in Figure 11-6. In this case k will be a negative number because the gradient is negative. However, for practical purposes, the negative sign may be ignored because it cancels with the negative gradient.

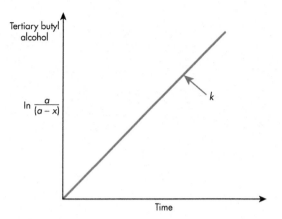

FIGURE 11-5 Plot of tertiary butyl alcohol concentration against time, giving a positive gradient equal to the rate constant k.

To obtain an equation for the half-life ($t_{1/2}$), one can substitute $x = a/2$ in equation (27). The half-life is the time taken for 50% of the reactant to be used:

$$k_{t_{1/2}} = \ln\frac{a}{a/2} = \ln 2$$

$$\therefore t_{1/2} = \frac{\ln 2}{k} = \frac{0.693}{k} \qquad (28)$$

STABILITY AND SHELF LIFE OF DRUGS

Obviously, it is necessary for pharmacists to know how long a drug product remains viable. The manufacturer helps by placing an expiration date on each

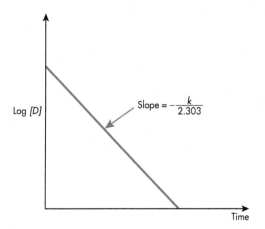

FIGURE 11-6 A plot of log concentration against time, giving a negative gradient from which k may be calculated.

batch of product. This date is obtained by keeping drugs under extreme conditions to accelerate their rate of decomposition.

First, let us consider how to arrive at the shelf life for a drug. Unlike the half-life, where 50% of the drug is decomposed, shelf life is indicated by t_{90}, where greater than 90% of drug remains viable.

Suppose that for a first-order reaction, one had to start with 1.2 mg/mL of drug. To obtain 90% of 1.2, multiply by 0.9:

$$1.2 \text{ mg/mL} \times 0.9 = 1.08 \text{ mg/mL}$$

$[A_t]$ = concentration after time t
(i.e., when 90% is left)

$[A_0]$ = initial concentration

$$[A_t] = [A_0] - k_{t_{90}} \qquad (29)$$

and $\qquad k = 0.03$ mg/mL/h

$$1.08 \text{ mg/mL} = 1.2 \text{ mg/mL} - 0.03 \text{ mg/mL/h}$$

$$1.2 - 1.08 = 0.12$$

$$t_{90} = \frac{0.12}{0.03} = 4 \text{ h}$$

Because the rate constant k is part of the equation, the reaction rate and order must be known to use this approach. t_{90} also can be found graphically by using the Arrhenius equation.

The Effects of Temperature

This topic is vast and can be covered only briefly here. To explain the use of an Arrhenius plot in the previous section, one has to begin with the molecular theory of collision.

Collision Theory of Reaction Rates

The mechanism of a reaction depends on two factors:

1. A frequency factor
2. Activation energy

The larger the frequency factor, the faster a reaction will proceed. However, not all molecules collide with sufficient energy to contribute to a reaction. Only those molecules possessing energy above a

TABLE 11-3 Summary of Order of Reaction

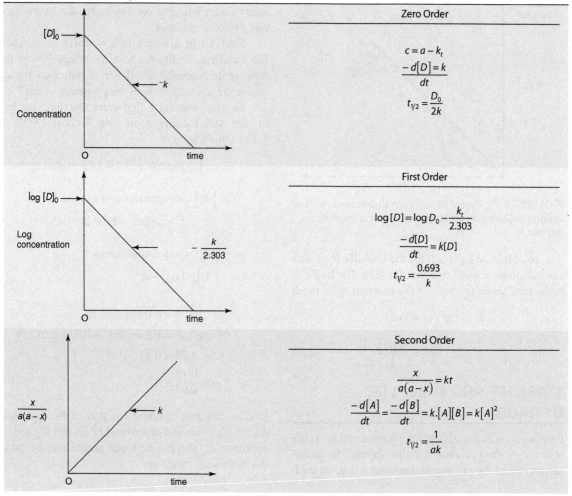

certain value are capable of reacting. Arrhenius stated that "molecules must have a certain energy to react."

The energy of activation (Ea) is the energy possessed by a limited number of reactant molecules in excess of the average molecular energy. The number of activated molecules increases as the temperature is raised:

$$k = Ae^{-Ea/Rt} \quad (A = \text{frequency factor}) \quad (30)$$

The Activated Complex

Consider the reaction:

$$A + B \rightleftarrows (A - B)* \rightarrow P$$

reactants → activated complex (or transition state) → products

The difference in potential energy between the reactants and the products is the heat of reaction.

Figure 11-7 is an energy diagram that shows a transition state where the activated complex (molecular energy) must reach a certain value to proceed to products under constant pressure. Energy is plotted against reaction coordinates in the figure:

$$Ea^1 = \text{energy lost}$$

$$Ea = \text{activation energy}$$

$$\Delta H = \text{heat of reaction}$$

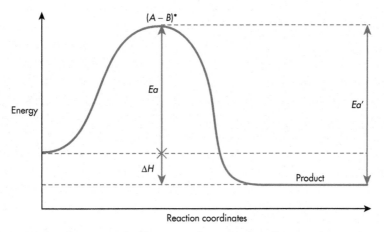

FIGURE 11-7 Representation of energy required for two reaction molecules to form a product.

This is an exothermic reaction; that is, the products possess less energy than the reactants by an amount ΔH.

Temperature and Reaction Rate

The Arrhenius equation can be used to indicate the effect of temperature on the specific rate constant for a reaction. Thus:

$$k = Ae^{-Ea/RT}$$

$$\log k = \log A - \frac{Ea}{2.303\, RT} \quad (31)$$

where k = specific reaction rate constant, A = Arrhenius factor or frequency factor, Ea = energy of activation, R = gas constant (1.987 cal/deg mole), and T = absolute temperature.

Figure 11-8 shows an Arrhenius plot used to determine the activation energy Ea from the negative gradient of the resulting data when $\log k$ is plotted against the reciprocal of absolute temperature. As can be seen from the figure, by applying logarithms to equation (30), one obtains equation (31) in the form of $y = mx + c$, a straight-line equation.

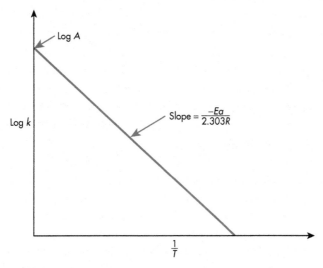

FIGURE 11-8 Arrhenius plot used to determine activation energy.

If log k is plotted against the reciprocal of absolute temperature, an Arrhenius plot can be drawn. The slope of such a plot is equal to:

$$\frac{-Ea}{2.303\,R}$$

from which Ea, the activation energy, can be found readily:

$$Ea = \frac{y_2 - y_1}{x_2 - x_1} \cdot 2.303\,R$$

As k increases, the reaction rate increases. As the activation energy Ea increases, k decreases. As the temperature increases, k increases. A can be found by taking the antilog of the intercept on the y axis.

Care should be taken to realize that when $1/T$ is increasing, the temperature is decreasing, meaning that temperature increase goes from right to left on the x axis.

One also can obtain Ea by considering equation (31) for a temperature T_2:

$$\log k_2 = \log A - \frac{Ea}{2.303\,RT_2}$$

and then considering the same equation for a temperature T_1:

$$\log k_1 = \log A \frac{-Ea}{2.303\,RT_1}$$

After subtracting these equations:

$$\log \frac{k_2}{k_1} = \frac{Ea}{2.303\,R}\left(\frac{(T_2 - T_1)}{T_2 T_1}\right)$$

from which Ea can be found.

EXAMPLE 3 ▶▶▶

A formulation for an analgesic is found to degrade at 110°C (230°F; 383°K) with a rate constant of $k_1 = 2.0\,h^{-1}$ and k_2 at 150°C (302°F; 423°K) of 3.8 h^{-1}.

Calculate the activation energy and the frequency factor A ($R = 1.987$ caldeg^{-1} mole^{-1}).

Solution

$$\log\frac{3.8}{2.0} = \frac{Ea}{2.303}\left(\frac{423-383}{423\times 383}\right)\cdot\frac{1}{1.987}$$

$$\log 1.9 = \frac{Ea}{4.576}\times 2.47\times 10^{-4}$$

$$0.278\times 4.576 = 2.47\times 10^{-4}\,Ea$$

$$1.275 = 2.47\times 10^{-4}\,Ea$$

$$Ea = \frac{1.275}{2.47\times 10^{-4}}$$

$$Ea = 5.164\times 10^{3}\,\text{cal mole}^{-1}$$

Using equation (31) at 110°C (230°F) and converting k_1 from hours to seconds:

$$\frac{1}{2.0\,h} = 0.5\,h \times 3600$$

$$= 1800\,s$$

$$= 5.5\times 10^{-4}\,s^{-1}$$

$$\log(5.5\times 10^{-4}\,s^{-1}) = \log A\frac{-5.164\times 10^{3}}{2.303\times 1.987\times 383}$$

$$-3.26 = \log A - 2.946$$

$$\log A = 2.946 - 3.26 = 0.3135$$

$$A = 4.8\times 10^{-1}\,s^{-1}$$

Accelerated Stability Testing

It is not practical to wait for years to observe how long it takes for a drug to decompose. Therefore, various types of stress are applied to drug compounds to speed the process. These methods include testing by an increase in temperature, humidity, and light.

Temperature increase can accelerate the rate of reaction; therefore, drugs are stored at a variety of higher than ambient temperatures. High humidity increases the decomposition process by hydrolysis. This occurs most readily if drugs are not kept in suitable containers and are left open to the atmosphere. Artificial light is used to increase the effect of daylight or sunlight on a drug product.

At various time periods, samples are taken and analyzed for the viability of the active

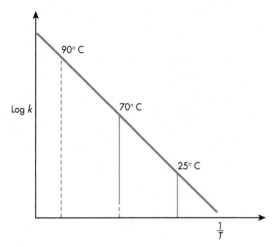

FIGURE 11-9 Arrhenius plot to determine the shelf life of a drug.

contents. Practically, by using elevated temperatures, the k values for a drug can be found under decomposition in solution by plotting a function of the concentration against time. The logarithms of the different k values are then plotted against the reciprocals of the absolute temperature in an Arrhenius plot (discussed in the next section) (Fig. 11-9).

If the plot is extrapolated to ambient (room) temperature, $k25°C$ ($k77°F$) can be used to determine the shelf life of the drug. To determine the expiration date of a drug, the rate equation (assuming a first-order reaction) and equation (16) can be rearranged to give:

$$t = \frac{2.303}{k} \cdot \log \frac{C_0}{C} \quad (32)$$

EXAMPLE 4 ▶▶▶

For a first-order reaction, the initial concentration of a drug is 200 mg/mL. The specific rate constant k is obtained from an Arrhenius plot and is equal to 3×10^{-5} h^{-1} at 25°C (77°F). The limit of viability of the drug is 150 mg/mL, at which stage its shelf life has expired.

Solution
The expiration date is therefore:

$$t = \frac{2.303}{3 \times 10^{-5}\,h^{-1}} \cdot \log \frac{200\,mg/mL}{150\,mg/mL}$$

$$= 95,000\,h \approx 1.1\,year$$

EXAMPLE 5 ▶▶▶

During accelerated stability testing, a medicine was analyzed to have 3.2 mg/mL of active drug, and its stability constant k was found to be 0.05 mg/mL/h. How long will it take before the drug decomposes by 10%?

Solution
Using equation (29), where $[A_0]$ = initial concentration and $[A_T]$ = concentration after time t, 90% of 3.2 mg/mL = 2.88 mg/mL:

$$2.88 = 3.2 - 0.05\,t_{90}$$

$$t_{90} = \frac{3.2 - 2.88}{0.05} = 6.4\,h$$

Effects of pH on Reaction Rates

The magnitude of the rate of a reactions catalyzed by H$^+$ and OH$^-$ ions can vary considerably with pH. For the determination of pH on degradation kinetics, decomposition is measured at several H$^+$ concentrations. The pH of optimum stability can be determined by plotting log k against pH (Fig. 11-10).

In such a plot, when log k increases with a decreasing pH, this is indicative of acid catalysis. When log k increases with increasing pH, it is indicative of base catalysis. The point of the inflection of such a plot represents the pH of optimum stability. These plots, also know as pH-rate profiles, are very useful in guiding the formulation of stable dosage forms. However, the student pharmacist should be aware that different drugs can have different shapes

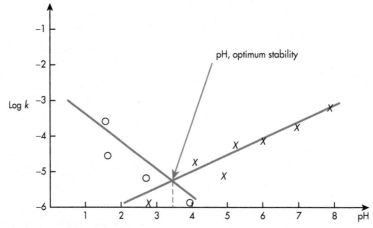

FIGURE 11-10 Determination of the optimal pH for a drug.

of pH-rate profiles than that depicted in Fig. 11-10, and that formulators must also consider the effects of pH on solubility and patient comfort when choosing the appropriate pH to formulate a product.

ENZYME CATALYSIS REACTIONS

Generally speaking, the principles of chemical reaction kinetics can be stated to be true for enzyme-catalyzed reactions. However, enzyme-catalyzed reactions show the characteristic of saturation of substrate, which is not seen in nonenzymatic reactions. Figure 11-11 shows the effect of the substrate concentration [S] on the rate of reaction, with reactants going to products. V_0 is the initial reaction velocity, which at low substrate

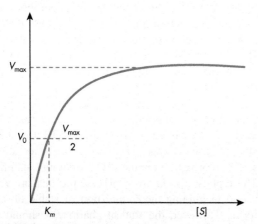

FIGURE 11-11 Effect of substrate concentration on the rate of an enzyme-catalyzed reaction.

concentrations shows a quasi-proportionality between V_0 and [S], and the reaction approximates the first-order reaction with respect to the substrate.

As the substrate concentration increases, the initial velocity V_0 decreases and mixed-order reactions occur. As [S] increases further, the rate (or velocity V_0) becomes independent of [S] and approaches a constant asymptotically. This is where the reaction is of zero order with respect to the substrate, and the enzyme is saturated with its substrate.

The saturation effect indicates that the enzyme and the substrate react reversibly to form a complex. This is an important step in a catalyzed reaction.

In 1913, Michaelis and Menten postulated a general theory of enzyme action and kinetics that basically extends to all aspects of enzyme kinetics and inhibition. If one considers the simple case of a reaction where there is only one substrate and the enzyme E combines with the substrate S to form an enzyme-substrate complex ES, the complex then breaks down to give the product P and the free enzyme E:

$$E + S \underset{k_2}{\overset{k_1}{\rightleftharpoons}} ES \quad (33)$$

$$ES \underset{k_4}{\overset{k_3}{\rightleftharpoons}} E + P \quad (34)$$

where k_1 and k_3 are the rate constants for the forward reaction and k_2 and k_4 are the rate constants for the reverse reaction. To find a general expression for V_0, the initial velocity, assume that enzyme-catalyzed reactions occur in two stages, as shown in equations (33) and (34).

The initial velocity is equal to the rate of breakdown of the enzyme-substrate complex, according to equation (34). The first-order rate equation may be written as:

$$V_0 = k_3[ES] \tag{35}$$

Neither k_3 nor $[ES]$ can be found directly.

To assist in this problem, the second-order rate equation can be utilized. The second-order equation is:

$$\frac{d[ES]}{dt} = k_1([E_T] - [ES])[S] \tag{36}$$

where E_T is the total enzyme concentration and k_1 is the second-order rate constant. Further, the equation for the breakdown of ES by the sum of two reactions is for the forward reaction yielding the product and for the reverse reaction yielding $E + S$:

$$\frac{-d[ES]}{dt} = k_2[ES] + k_3[ES] \tag{37}$$

In the steady-state condition, when the rate of formation is equal to the rate of breakdown:

$$k_1([E_T] - [ES])[S] = k_2[ES] + k_3[ES] \tag{38}$$

Rearranging equation (38):

$$\frac{[S]([E_T] - [ES])}{[ES]} = \frac{k_2 + k_3}{k_1} = K_m \tag{39}$$

where K_m replaces $k_2 + k_3/k_1$ and is known as the Michaelis-Menten constant. From equation (38), solving for $[ES]$ gives:

$$[ES] = \frac{[E_T][S]}{K_m + [S]} \tag{40}$$

and from equation (35), $V_0 = k_3[ES]$, one can substitute for the term $[ES]$. Thus:

$$V_0 = k_3 \frac{[E_T][S]}{K_m + [S]} \tag{41}$$

When the enzyme is saturated with substrate, the maximum initial velocity is reached, V_{max}, which is given by:

$$V_{max} = k_3[E_T] \tag{42}$$

Substituting for $k_3[E_T]$ from equation (49):

$$V_0 = \frac{V_{max} \cdot [S]}{K_m + [S]} \tag{43}$$

which is the Michaelis-Menten equation, the rate equation for a one-substrate enzyme-catalyzed reaction.

The Michaelis-Menten relationship can be transformed algebraically into other useful forms. If the reciprocal of equation (43) is taken:

$$\frac{1}{V_0} = \frac{K_m + [S]}{V_{max} \cdot [S]} \tag{44}$$

Rearranging:

$$\frac{1}{V_0} = \frac{K_m}{V_{max}[S]} + \frac{[S]}{V_{max}[S]} \tag{45}$$

which reduces to:

$$\frac{1}{V_0} = \frac{K_m}{V_{max}} \cdot \frac{1}{[S]} + \frac{1}{V_{max}} \tag{46}$$

This is the Lineweaver-Burk equation. When $\frac{1}{V_0}$ is plotted against $\frac{1}{[S]}$, a straight line is obtained that has a slope of K_m/V_{max} and an intercept of $\frac{1}{V_{max}}$ on the y axis and an intercept of $-\frac{1}{K_m}$ on the x axis (Fig. 11-12). The Michaelis-Menten constant K_m

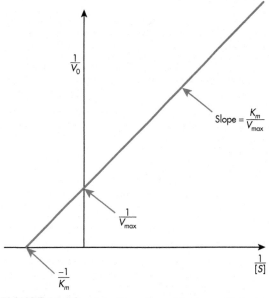

FIGURE 11-12 A double reciprocal Lineweaver-Burk plot.

can be calculated from the gradient of the resulting positive graph. One can obtain V_{max} from the y axis and K_m/V_{max} from the gradient. Thus, K_m can be calculated.

PHARMACOKINETICS: AN EXTENSION OF REACTION KINETICS

It is not the object of this chapter to discuss pharmacokinetics in a comprehensive manner. However, the chapter is aimed at students who will be studying pharmacokinetics at a future date during a pharmacy degree course. Therefore, to try to connect the subjects of pharmacokinetics and reaction kinetics will be profitable to those students. For ease of understanding and to avoid confusing the reader, these connections, with respect to pharmacokinetics, are presented in this section.

Pharmacokinetics is concerned with what happens to drugs in the body and how they are metabolized and excreted. It also involves the extended study of drug absorption and drug distribution. Mathematical models are used to predict the response that a drug will elicit and to analyze the data. Usually the rate of reaction is followed and determined by sampling and measuring the amount of drug degraded from the plasma and urine at certain time periods.

Zero-order absorption and first-order absorption will be discussed with examples. Second-order reactions in regard to both absorption and elimination are extremely difficult to describe and may occur under circumstances such as when a drug and its metabolite are both viable at the same rate. They of course may be absorbed at different rates and show complications of estimation as a result of competition.

Zero-Order Absorption

In zero-order absorption, the rate is independent of concentration. The gastrointestinal tract (GIT) absorption of nutrients and drugs is complex, and molecules go through a variety of processes to be released into the blood. Because numerous variables accompany drug input, including intravenous (IV) infusion, bolus injections, and transdermal patches, which mimic IV infusion, and because the mathematical models proposed to explain these processes are adequate but simplified, a paradox exists in the classification. Some drugs show a pure zero-order absorption rate, such as hydroflumethazide HCl and hydrochloromethazide HCl. Most input processes are classified as first-order reactions and as apparent zero-order reactions. Elimination is wholly a first-order process, although there is a constant rate of diffusion.

The amount of drug in the body at any one time and its rate of change are dependent on both absorption and elimination. Of course, other factors are involved, such as the half-life of drug, bioavailability, and lipid complexation. Let us assume, notwithstanding the extent of the variables and the inadequacy of modeling to provide an accurate account of minute-by-minute drug metabolism, that $-dB/dt$ is the rate of change of a drug in the body. The rate of drug disposition in the body is therefore:

$$\frac{dB}{dt} = \frac{dGI}{dt} - \frac{dE}{dt} \quad (47)$$

where dGI/dt is equal to the amount of drug absorbed from the GIT and dE/dt is the drug eliminated for an oral input.

During absorption, more drug is taken up per unit time than is eliminated, as shown in Figure 11-13 and given by:

$$\frac{dGI}{dt} \geq \frac{dE}{dt} \quad (48)$$

Figure 11-13 also indicates that at the maximum plasma concentration of drug, the rate of absorption equals the rate of elimination:

$$\frac{dGI}{dt} = \frac{dE}{dt} \quad (49)$$

When the peak plasma concentration is passed, drug elimination exceeds drug absorption:

$$\frac{dE}{dt} \geq \frac{dGI}{dt} \quad (50)$$

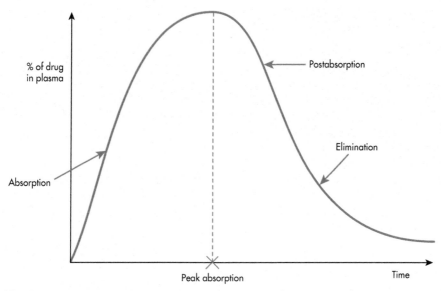

FIGURE 11-13 Percentage of plasma drug against time, giving a profile of drug absorption, peak concentration, and elimination.

When the rate of drug absorption is zero:

$$\frac{dGI}{dt} = 0 \quad (51)$$

The elimination, then, is the only dynamic under consideration and is a first-order process that is described as:

$$\frac{dB}{dt} = -kB \quad (52)$$

where k is a first-order elimination constant.

An apparent zero-order reaction is explained in pharmacokinetics textbooks as a one-compartment model. (It is suggested that the student refer to a pharmacokinetic textbook to obtain the full details.)

Ingested drug from the GIT is absorbed at a constant rate k_0. Eliminated drug follows a first-order process, having a rate constant k. The rate of elimination is therefore equal to $E \cdot k$, whereas the rate of absorption is given by k_0.

The net change per unit time is:

$$\frac{dE}{dt} = k_0 - E \cdot k \quad (53)$$

First-Order Absorption

First-order absorption applies to the oral administration of solutions and instantaneous drug delivery, such as regular tablets, capsules (after exposure to active drug in the GIT), and intramuscular (IM) injection in aqueous solutions.

Action at the absorbtive site is a first-order process in the case of an oral delivery from the GIT, and the rate of absorption can be defined as:

$$\frac{dGI}{dt} = -k_a GI_s F \quad (54)$$

where k_a is the first-order rate absorption constant, GI_s is the amount of drug in solution in the GIT, and F is the fraction absorbed.

Without going into the mathematics, it may be shown that the amount of drug in the body and its rate of change are the rate of absorption minus the rate of elimination:

$$\frac{dB}{dt} = k_a \cdot d\, GI \cdot F - kdE \quad (55)$$

where F is the fraction of drug absorbed in the systemic system.

EXAMPLE 6 ▶▶▶

To titrate a lower dose of digoxin to a patient showing signs of tachycardia and to avoid digoxin toxicity, it is deemed necessary to determine the first-order rate constant k_a and the half-life $t_{1/2}$ of the drug in plasma. By a close examination and analyses of samples of stomach contents, plasma, and urine taken at regular intervals, it is found that the rate of drug absorbed from the GIT (dGI/dt) is 0.0133 mg h^{-1}. The amount of drug in solution in the GIT (GI$_s$) at a given time in the 24-h regimen (drug is administered orally at 0.25 mg once a day) is 0.10 mg after 2 h, and the fraction absorbed (F) is 0.15 mg.

Solution

Plasma analysis, assuming a one-compartment model, is shown to follow a first-order rate process. Using equation (63) to find k_a:

$$k_a = \frac{-0.0133 \, \text{mg h}^{-1}}{0.015 \, \text{mg}}$$

$$= 0.88 \, \text{h}^{-1}$$

as dGI/dt is always decreasing after the peak plasma level has been obtained:

$$t_{1/2} = \frac{0.693}{0.88 \, \text{h}^{-1}} = 0.787 \, \text{h}$$

Radioactivity

The Decay Law

The rate of decay of a radioactive isotope is proportional to the number of undecomposed atoms (N):

$$\text{Rate} = \lambda N \tag{55}$$

where λ = disintegration constant and is a first-order process with a half-life of

$$t_{1/2} = 0.693/\lambda \tag{56}$$

Radioactive decay follows a first-order reaction rate and is illustrated by the following example.

EXAMPLE 7 ▶▶▶

After 32 h, a radioactive isotope decays to a point where 15.7 millicuries of radioactivity remains from its starting activity of 200 millicuries. Find (a) the disintegration constant λ and (b) the time after which only 50% remains.

Solution

Using an equation for radioactive decay:

$$\ln \frac{N}{N_0} = -\lambda t$$

where N = number of remaining atoms after time t, N_0 = number of original atoms, and λ = disintegration/decay constant.

a. $\ln \dfrac{15.7}{200} = -\lambda \cdot 32$

$$\lambda = \frac{2.5446}{32} = 0.0795 \, \text{h}^{-1}$$

b. Using equation (19):

$$t_{1/2} = \frac{0.693}{0.0795} = 8.72 \, \text{h}$$

PROBLEMS

1. First-order half-life is equal to:
   a. $1/k$
   b. k
   c. $0.693/k$
   d. $2k + 1$
   e. None of the above

2. Radioactive decay follows a:
   a. Mixed-order rate
   b. Fractional-order rate
   c. Zero-order rate
   d. First-order rate
   e. Second-order rate

3. A pseudo-first-order reaction is one in which:
   (1) The concentration of only one reactant [A] or [B] is considered important.
   (2) [B] is so large that it can be ignored and only [A] is considered important.
   (3) The speed of reaction depends on the concentrations of A and B.
   a. All the above statements are true.
   b. Only 2 and 3 are true.
   c. Only 2 is true.
   d. Only 3 is true.
   e. All the above statements are false.
4. For a zero-order reaction having $A_0 = 0.580$ absorbance units and $k = 7.6 \times 10^{-4}$, the half-life is:
   a. 0.38×10^2 h
   b. 0.38×10^2 mole/h
   c. 3.8×10^2 mole/h
   d. 3.8×10^{-2} h
   e. 3.8×10^2 h
5. Figure 11-3 shows which of the following?
   a. A linear plot of log c versus time for a first-order reaction
   b. A linear plot of log t versus concentration for a first-order reaction
   c. A graph that follows a second-order decrease in initial concentration given by:
   $$\text{Rate} = \frac{-d[C]^2}{dt} = k[C]^2$$
   d. $\frac{x}{a(a-x)}$ plotted against t
6. An example of a pseudo-first-order reaction is one in which:
   a. The progress of decomposition of a chemical substance is shown by an increasing exponential curve.
   b. $\frac{-d[A]}{dt} = \frac{-d[B]}{dt}$
   c. Either [A] or [B] is in excess, and its decomposition can be ignored with respect to the other.
   d. One in which all the reactants are used in order to form a product
7. The decay law may be expressed as $\ln \frac{N}{N_0} = -\lambda t$. λ is which of the following?
   a. The number of original atoms
   b. The number of atoms at time t
   c. Equal to a constant 0.693
   d. The disintegration constant
8. If λ is found to be 0.0527 min^{-1} from question 7, what is the half-life of the material?
   a. 13.1 h
   b. 13.1 min^{-1}
   c. 26.2 min
   d. 13.1 min
9. The following table shows the decomposition of n moles of X at 25°C (77°F) in an aqueous solution:

t, min	a − x	k, min^{-1}
0	57.90	
5	50.40	0.0278
10	43.90	0.0277
25	20.10	0.0275
45	16.70	0.0276
65	9.60	0.0276
∞	0	

Plotting the data from the table would lead to the observation of which order of reaction?
   a. Zero order
   b. First order
   c. Pseudo first order
   d. Second order
10. In a zero-order reaction, if $A_0 = 0.47$ mole/L and $k = 8.2 \times 10^{-4}$ L/mole/h, what is the half-life of the drug?
    a. 12.08 days
    b. 24.16 days
    c. Just under 3 weeks
    d. Cannot be calculated without the initial time in which the drug began to decrease
11. The energy of activation is the energy possessed by which of the following?
    a. A limited number of reactant molecules in excess of the average molecular energy
    b. An excess of reactant molecules
    c. Product molecules on completion of a reaction
    d. The heat of the reaction
12. What is the predicted shelf-life of a drug that undergoes first-order degradation and has a half-life of 7 days?
13. A liquid medicine having an initial drug concentration of 1000 mg/mL undergoes zero-order kinetics with a rate constant of 100 mg/mL per day.

a. What is the $T_{1/2}$ for this product?
b. How long will it take before the product is completely degraded?

14. From the information in Table 11-1, construct graphs of concentration against time and log concentration against time.
    a. What order of reaction is this?
    b. What is the rate constant for degradation of tertiary butyl alcohol in sulfuric acid?

15. A formulation of use in a nebulizer begins its existence at 100 mg/mL of active substance. Ten months later its concentration has fallen to 75 mg/mL. Calculate the drug's half-life ($T_{1/2}$). What amount is left after 18 months if the order of reactions is (a) zero order and (b) first order?

16. The following table gives the data for a chemical substance that undergoes decomposition at 25°C (77°F) in an aqueous solution:

Concentration, mg/mL	Time, min	Absolute Temperature, °K	k, min^{-1}
540	0	450	0.0356
526	5	400	0.0354
505	10	376	0.0353
500	15	368	0.0353
470	20	361	0.0352
427	25	356	0.0351
401	30	352	0.0349
360	35	348	0.0348
344	45	340	0.0347
298	55	334	0.0346
290	65	330	0.0345
280	75	326	0.0344

a. Plot a graph of log concentration against time.
b. State what order of reaction this represents. If, on integrating the rate equation for this reaction, one obtains the expression:

$$\log C = \log C_0 - kt/2.303$$

c. Find the rate constant k from the gradient of the graph.
d. Find the value of C_0 where the y axis is cut.
   This particular reaction is temperature-dependent, and it is found necessary during preformulation of a tablet containing the active chemical substance to obtain the activation energy Ea. Recalling that the Arrhenius equation for such a reaction is:

$$K = Ae\frac{-Ea}{RT}$$

e. Plot a graph of log (k + 2) against the reciprocal of absolute temperature (T) from the table.
f. Determine the energy of activation Ea. (Take $R = +\cdot 9872$ caldeg^{-1} mole^{-1})

17. The active moiety in an antihistamine formulation has a degradation rate constant of $k_1 = 4.4 \times 10^{-10}$ s^{-1} at 23°C (73.4°F). The amount of energy required to activate the reaction is $Ea = 20.5k$ cal/mole.
    a. What order of reaction does this represent?
    b. Calculate the rate constant (k_2).
    c. Calculate the value of A at 55°C (131°F).

18. An amphetamine was observed to decompose from its original concentration of 0.06 M according to the time given in the following table. The elevated temperature was 120°C (248°F), and the medium was an acidic aqueous solution.
    a. What order of reaction does this reaction follow?
    b. What is its half-life?

Time, h	Amphetamine remaining, mole/L × 102
0.5	5.80
1	5.60
2	5.30
3	5.00
4	4.72
5	4.61
7	4.30
9	4.22
11	4.10
13	4.02

19. The first-order rate constant for the degradation of an antibiotic is 3×10^{-6} s^{-1}, and the solubility is 0.8 g/100 mL. A formulation of 2.0 g/100 mL is required.
    Calculate (a) the zero-order rate constant (k_0) and (b) the shelf life (T_{90}).

ANSWERS

1. c
2. d
3. b
4. e
5. a
6. c
7. d
8. b
9. a
10. a
11. a
12. $T_{90} = 25.45$ h
13. a. 5 days
    b. 10 days
14. a. First order

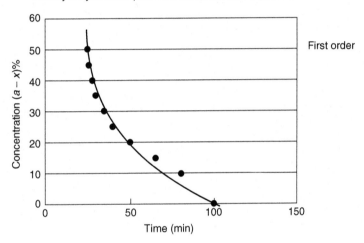

b. $k = 92.12$ h^{-1}

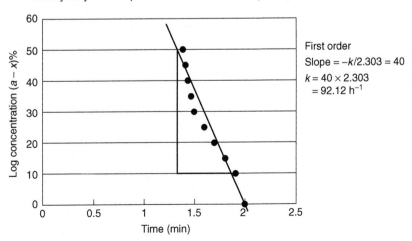

15. a. $T_{1/2}$ = 20 months; 55 mg/mL
b. $T_{1/2}$ = 24.06 months; 59.55 mg/mL
16. a.

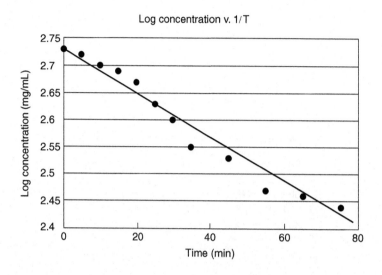

b. First order
c. $k = 1.036 \times 10^{-2}$ min^{-1}
d. C_0 = 549.5 mg/mL
e.

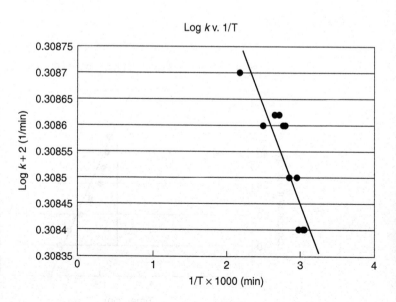

f. $Ea = 1.987k$ cal/mole^{-1}
17. a. First order
   b. $k_2 = 1.3 \times 10^{-8}$ s^{-1}
   c. $A = 6.1 \times 10^5$ s^{-1}
18. a. A possible first-order reaction where $k = 0.043$ L^{-1}
   b. $T_{1/2} = 16.1$ h

19. a. $k_0 = 2.4 \times 10^{-6}$ gdL^{-1} s^{-1}
   b. $T_{90} = 23.14$ h

CLINICAL QUESTIONS

1. Beta lactam antibiotics are highly susceptible to hydrolytic degradation. Explain why most beta-lactam products intended to be administered as liquids are only available as dry powder mixtures, as freeze-dried powders, or if in the liquid form, as frozen premixes.

2. When possible, drug products formulated as solutions are formulated at the pH of maximum stability, but often, compromises must be made because of the effects of pH on other parameters. Name two of those parameters.

SUGGESTED READINGS

Carstensen JT, Rhodes CT: *Drug Stability: Principles and Practices.* New York: Marcel Dekker, 2000.
Rowland M, Tozer TN: *Clinical Pharmacokinetics Concepts and Applications* 3rd ed. Philadelphia: Lea & Febiger, 1995

Yoshioka S, Valentino JS: *Stability of Drugs and Dosage Forms.* New York: Kluwer Academic, 2002.

Appendix: Basic Mathematical Concepts

This appendix covers briefly the basic mathematical procedures that are important for carrying out applications in physical pharmacy. Refer to calculus and mathematical textbooks for a more in-depth discussion.

EQUATIONS

In any equation in physical pharmacy, two main components must be considered first: the dependent variable and the independent variable(s). Consider for instance the following equation for the calculation of the buffer capacity (see Chapter 4):

$$\beta = 2.3\ C\ \{(K_a\ [H_3O^+])/(K_a + [H_3O^+])^2\}$$

To identify the dependent variable, first look at what you are solving for. In the equation, (β) is the independent variable. However, β depends on C (the total molar concentration of the buffer), K_a the acid dissociation constant, and $[H_3O^+]$ the hydronium ion concentration in the buffer solution. Therefore, three *independent* variables can influence the value of β. For a given buffer system, the value of K_a is constant, and thus β is dependent only on C and $[H_3O^+]$ in this case. A change in the value of C and/or that of $[H_3O^+]$ results in a change in the value of β.

In any equation, two sides are clearly identified, usually separated by an equals sign. Whenever a number is added or subtracted from either side of the equation, the other side should be modified by the same value. For example, consider the following equation:

$$Y = 50\ X + 5$$

Adding +10 to the left side of the equation requires adding +10 to the right side, so that the equation remains balanced. The same rule applies to dividing or multiplying by a number. Both sides of the equation must be divided or multiplied by the same number. However, remember that division by zero is not allowed.

LOGARITHMIC FUNCTION

The logarithmic function is a useful arithmetic transformation in physical pharmacy. Often pharmacists deal with variables that have small numeric values. For example, the acid dissociation constant for acetic acid is 1.74×10^{-5}. Converting this value to a logarithmic number results in -4.76; multiplying this number by -1 yields 4.76, which is the pK_a value (defined as the negative logarithm of K_a) for acetic acid. The same principle applies to the concentration of hydronium ion in water ($[H_3O^+]$). Normally, the molar concentration of $[H_3O^+]$ is a small number. Taking its log value and multiplying it by -1 results in the pH of the solution. Two types of logarithmic expressions exist: the logarithm of base 10 (log) and the natural logarithm (ln). The relationship between these two forms is:

$$\ln A = 2.303 \log A$$

The following are some basic logarithmic rules:

1. $\log A + \log B = \log (A \times B)$
2. $\log A - \log B = \log (A/B)$
3. $-\log A = \log (1/A)$
4. $\log (A + B) \neq \log A + \log B$
5. $\log (A - B) \neq \log A - \log B$
6. If $A = B$, $\log (A/B) = 0$
7. $Y = e^K \rightarrow \ln Y = K$ or $\log Y = K/2.303$
8. $A = \log (1/B) \rightarrow B = 10^{-A}$

Also, remember that the exponential equation shown in rule 7 becomes a linear expression when it is converted to its logarithmic form.

245

EXPONENTIAL FUNCTION

The exponential function is encountered frequently in physical pharmacy expressions. It is often cumbersome to use it in calculation, and thus it commonly is converted to a linear function by using logarithmic rules.

DIFFERENTIATION AND INTEGRATION FUNCTIONS

Differential equations are encountered frequently in physical pharmacy applications. Some of the integration rules important for physical pharmacy applications are as follows (q and p are functions in variable x, and C and k are constants):

1. $\int q^n \, dq = q^{(n+1)}/(n+1) + C$ with $n \neq -1$
2. $\int dq = q + C$
3. $\int d \ln q = \ln q + C$
4. $\int e^q \, dq = e^q + C$
5. $\int kq \, dx = k \int q \, dx$
6. $\int (q + p) \, dx = \int q \, dx + \int p \, dx$
7. $\int dq/q = \int d \ln q$
8. $\int \ln q \, dq = q(\ln q - 1) + C$

These rules are *indefinite* integrals because no limit was imposed on the integration and a constant C was used. When specific limits are defined for the integration, the integral is known as a *definite* integral. When definite integrals are used, the constant C can be defined from the limits specified.

Consider the following equation describing the rate of degradation of a drug product following zero-order kinetics:

$$\frac{dA}{dt} = -K_0$$

where dA/dt is the amount of drug A disappearing per unit time t, and K_0 is the zero-order degradation rate constant (units = mass/unit time or concentration/unit time). The negative sign indicates that the amount of the drug is *decreasing* over time. The symbol d signifies that for any infinitesimal time t, an infinitesimal amount of drug A is disappearing. Also, it shows that the process is a *continuous* one.

Integration of this differential equation yields the following:

$$\int dA/dt = -\int K_0$$
$$\int dA = -\int K_0 \, dt$$
$$A - A_0 = -K_0 (t - t_0)$$

When $t_0 = 0$:

$$A = A_0 - K_0 t$$

A more complex example of differentiation and integration is that of a second-order reaction of the type:

$$dA/dt = dB/dt = -K_2 [A][B]$$

where A and B are the concentrations of the two components A and B, and K_2 is the second-order degradation rate constant. Assume that the initial concentrations are A_0 and B_0 for substances A and B, respectively, with $A_0 \neq B_0$. We also assume that z is the concentration of each substance reacting in time t.

The integration proceeds in the following fashion:

$$dz/dt = K_2 [A][B]$$
$$dz/[A][B] = K_2 \, dt$$

Applying a partial fractions method (for more complete information on this method, review a basic calculus textbook) on the left side of the equation:

$$1/[A][B] = X/[A] + Y/[B]$$

where X and Y are constants. Multiplying both sides of the equation by $[A][B]$:

$$1 = X[B] + Y[A]$$

Setting: $X = 1/\{[B_0] - [A_0]\}$ and $Y = 1/\{[A_0] - [B_0]\}$,

$$1/[A][B] = (1/\{[B_0] - [A_0]\})/[A] + (1/\{[A_0] - [B_0]\})/[B]$$

Integrating:

$$\int dz/[A][B] = 1/([B_0] - [A_0]) \int dz/[A] + 1/([A_0] - [B_0]) \int dz/[B]$$

$$\int dz/[A][B] = -1/([B_0] - [A_0]) \ln [A] - 1/([A_0] - [B_0]) \ln [B] + C$$

where C is an integration constant.

$$\int dz/[A][B] = 1/([A_0] - [B_0]) \ln [A] - 1/([A_0] - [B_0]) \ln [B] + C$$

$$K_2 t = 1/([A_0] - [B_0]) \ln [A]/[B] + C$$

At $t = 0$:

$$0 = 1/([A_0] - [B_0]) \ln [A_0]/[B_0] + C$$

$$C = -1/([A_0] - [B_0]) \ln [A_0]/[B_0]$$

$$C = 1/([A_0] - [B_0]) \ln [B_0]/[A_0]$$

$$K_2 t = 1/([A_0] - [B_0]) \ln [A]/[B] + 1/([A_0] - [B_0]) \ln [B_0]/[A_0]$$

$$K_2 t = 1/([A_0] - [B_0]) \{\ln [A]/[B] + \ln [B_0]/[A_0]\}$$

$$K_2 t = 1/([A_0] - [B_0]) \ln \{[B_0][A]/[B][A_0]\}$$

$$\ln \{[B_0][A]/[B][A_0]\} = K_2 \{[A_0] - [B_0]\} t$$

or

$$\log \{[B_0][A]/[B][A_0]\} = [K_2 \{[A_0] - [B_0]\}/2.303] t$$

If one plots $\{[B_0][A]/[B][A_0]\}$ versus t on a semi-logarithmic graph paper, the slope of the straight line equals $K_2 \{[A_0] - [B_0]\}/2.303$. Since A_0 and B_0 are known initial concentrations of the two reacting compounds, the second-order degradation rate constant K_2 can be estimated readily from the slope.

BIBLIOGRAPHY

Anton H: *Calculus with Analytical Geometry.* New York: Wiley, 1980.

Keedy ML, Bittinger ML: *Fundamental College Algebra.* Reading, MA: Addison-Wesley, 1977.

McAloon K, Tromba A: *Calculus of One Variable,* vol. 1BC. New York: Harcourt Brace Jovanovich, 1972.

MATHEMATICAL TABLES

Mathematical Constants

$\pi = 3.1416$
$e = 2.72$
1 atm = 760 mmHg = 1.01×10^6 dynes cm^{-2}
Avogadro number = 6.03×10^{23} mole^{-1}
1 calorie = 4.184 joules
$R = 1.987$ cal °K^{-1} mole^{-1}

Metric Conversion Table

1 inch = 2.54 centimeters
1 mile = 1.6 kilometers
1 pound = 453.6 grams
1 liquid ounce (U.S.) = 29.57 milliliters
1 liquid quart = 0.946 liter
1 angstrom = 1×10^{-8} cm

Metric System

Metric Linear Measure

1000 nanometers (nm)	= 1 micrometer (μm)
1000 millimicrons (mμ)	= 1 micron (μ)
1000 micrometers	= 1 millimeter (mm)
10 millimeters	= 1 centimeter (cm)
10 centimeters	= 1 decemeter (dm)
10 decimeters	= 1 meter (M)
1000 meters	= 1 kilometer (km)

Metric Weight

1000 picograms (pg)	= 1 nanogram (ng)
1000 nanograms (ng)	= 1 microgram (mg, meg)
1000 micrograms (pg, meg)	= 1 milligram (mg)
10 milligrams	= 1 centigram (cg)
10 centigrams	= 1 decigram (dg)
10 decigrams	= 1 gram (g)
1000 grams	= 1 kilogram (kg)

Metric Liquid Measure

1000 picoliters (pl)	= 1 nanoliter (nl)
1000 nanoliters (nl)	= 1 microliter (ml)
1000 microliters (μl)	= 1 milliliter (ml)
1000 milliliters	= 1 liter (l)
10 liters (l)	= 1 dekaliter (dl)
100 liters (l)	= 1 hectoliter (hl)
1000 liters (f)	= 1 kiloliter (kl)

English Systems

English Linear Measure

12 inches (in)	= 1 foot (ft)
3 feet	= 1 yard (yd)
5½ yards	= 1 rod (rd)
320 rods	= 1 mile (mi)
3 miles	= 1 league
4 inches	= 1 hand

Avoirdupois Weight

437.5 grains (gr)	= 1 ounce (oz)
16 ounces	= 1 pound (lb)
7000 grains	= 1 pound

Apothecary Measure (U.S. Fluid System)

60 minims (mm)	= 1 fluid dram (fz)
8 fluid drams	= 1 fluid ounce (fz)
16 fluid ounces	= 1 pint (0.)
8 pints	= 1 gallon (Cong)
128 fluid ounces	= 1 gallon
31 gallons	= 1 barrel

Temperature

$°C = 5/9 (F - 32)$

$°F = 9/5 C + 32$

$(5F - 9C = 160)$

degrees Kelvin $(°K) = °C + 273$

Apothecary Weight

20 grains (gr)	= 1 scruple (z)
3 scruples	= 1 dram (z)
8 drams	= 1 ounce (z)
12 ounces	= 1 pound (lb)
480 grains	= 1 ounce (z)
5760 grains	= 1 pound (lb)

Imperial Measure (British)

160 fluid ounces	= 1 gallon

Endnotes

REFERENCES

Chapter 1

1. **Overington JP, Al-Lazikani B, Hopkins AL:** How many drug targets are there? *Nat Rev Drug Discov* 5(12):993–6, 2006.
2. **von Zastrow M, von Zastrow M:** Chapter 2. Drug receptors & pharmacodynamics, in BG Katzung, SB Masters, AJ Trevor (eds). *Basic & Clinical Pharmacology*, 12th ed. New York: McGraw-Hill, 2012. www.accesspharmacy.com.lp.hscl.ufl.edu/content.aspx?aID=55820126. Accessed August 1, 2013.
3. **Charmot, D:** Non-systemic drugs: A critical review. *Curr Pharm Des* 18(10):1434–45, 2012.
4. **Shargel L, Wu-Pong S, Yu AB:** Chapter 14. Biopharmaceutic considerations in drug product design and *in vitro* drug product performance, in L Shargel, S Wu-Pong, AB Yu (eds). *Applied Biopharmaceutics & Pharmacokinetics*, 6th ed. New York: McGraw-Hill, 2012. www.accesspharmacy.com.lp.hscl.ufl.edu/content.aspx?aID=56604846. Accessed August 1, 2013.
5. **Giacomini KM, Sugiyama Y:** Chapter 5. Membrane transporters and drug response, in LL Brunton, BA Chabner, BC Knollmann (eds). *Goodman & Gilman's The Pharmacological Basis of Therapeutics*, 12th ed. New York: McGraw-Hill, 2011. www.accesspharmacy.com.lp.hscl.ufl.edu/content.aspx?aID=16658945. Accessed August 1, 2013.
6. **Buxton IL, Benet LZ:** Chapter 2. Pharmacokinetics: The dynamics of drug absorption, distribution, metabolism, and elimination, in LL Brunton, BA Chabner, BC Knollmann (eds). *Goodman & Gilman's The Pharmacological Basis of Therapeutics*, 12th ed. New York: McGraw-Hill, 2011. www.accesspharmacy.com.lp.hscl.ufl.edu/content.aspx?aID=16658120. Accessed August 1, 2013.
7. **Zamek-Gliszczynski MJ, Hoffmaster KA, Nezasa K, Tallman MN, Brouwer KL:** Integration of hepatic drug transporters and phase II metabolizing enzymes: Mechanisms of hepatic excretion of sulfate, glucuronide, and glutathione metabolites. *Eur J Pharm Sci* 27(5):447–86, 2006. Epub February 10, 2006.
8. **Shargel L, Wu-Pong S, Yu AB:** Chapter 10. Physiologic drug distribution and protein binding, in L Shargel, S Wu-Pong, AB Yu (eds). *Applied Biopharmaceutics & Pharmacokinetics*, 6th ed. New York: McGraw-Hill, 2012. www.accesspharmacy.com.lp.hscl.ufl.edu/content.aspx?aID=56603200. Accessed August 1, 2013.
9. **Luc EC De Baerdemaeker, LEC, Mortier, EP, Struys MMRF:** Pharmacokinetics in obese patients. *Contin Educ Anaesth Crit Care Pain* 4(5):152–5, 2004.
10. **Prouillac C, Lecoeur S:** The role of the placenta in fetal exposure to xenobiotics: Importance of membrane transporters and human models for transfer studies. *Drug Metab Dispos* 38(10):1623–35, 2010.
11. **Correia MA:** Chapter 4. Drug biotransformation, in BG Katzung, SB Masters, AJ Trevor (eds). *Basic & Clinical Pharmacology*, 12th ed. New York: McGraw-Hill, 2012. www.accesspharmacy.com.lp.hscl.ufl.edu/content.aspx?aID=55820515. Accessed August 1, 2013.
12. **Gonzalez FJ, Coughtrie M, Tukey RH:** Chapter 6. Drug metabolism, in LL Brunton, BA Chabner, BC Knollmann (eds). *Goodman & Gilman's The Pharmacological Basis of Therapeutics*, 12th ed. New York: McGraw-Hill, 2011. www.accesspharmacy.com.lp.hscl.ufl.edu/content.aspx?aID=16659426. Accessed August 1, 2013.
13. **Shargel L, Wu-Pong S, Yu AB:** Chapter 11. Drug elimination and hepatic clearance, in L Shargel, S Wu-Pong, AB Yu (eds). *Applied Biopharmaceutics & Pharmacokinetics*, 6th ed. New York: McGraw-Hill, 2012. www.accesspharmacy.com.lp.hscl.ufl.edu/content.aspx?aID=56603636. Accessed August 1, 2013.
14. **Shargel L, Wu-Pong S, Yu AB:** Chapter 6. Drug elimination and clearance, in L Shargel, S Wu-Pong, AB Yu (eds). *Applied Biopharmaceutics & Pharmacokinetics*, 6th ed. New York: McGraw-Hill, 2012. www.accesspharmacy.com.lp.hscl.ufl.edu/content.aspx?aID=56602254. Accessed August 1, 2013.

Chapter 2

1. **Lennard-Jones JE:** On the determination of molecular fields. *Proc R Soc Lond A* 106 (738):463–77, 1924.
2. **Wada A:** The alpha-helix as an electric macro-dipole. *Adv Biophys* 1–63, 1976.
3. **Alberts B, Johnson A, Lewis J, Raff M, Roberts K, Walter P:** Chapter 3: Proteins. *Molecular Biology of the Cell*, 4th ed. New York, NY: Garland Science, 2002.
4. **Watkins PB:** Noninvasive tests of CYP3A enzymes. *Pharmacogenetics* 4:171–84, 1994.

5. **Bauer J, Spanton S, Henry R, Quick J, Dziki W, Porter W, Morris, J:** Ritonavir: An extraordinary example of conformational polymorphism. *Pharm Res* 18:859–66, 2001.
6. **Stahly GP:** Diversity in single- and multiple-component crystals. The search for and prevalence of polymorphs and cocrystals. *Crystal Growth & Design* 7:1007–26, 2007.
7. **Remenar JF, Morissette SL, Peterson ML, Moulton B, MacPhee JM, Guzman HR, Almarsson O:** Crystal engineering of novel cocrystals of a triaxole drug with 1,4-dicarboxylic acids. *J Am Chem Soc* 125:8456–57, 2003.
8. Guidance for Industry. Q1A (R2) Stability Testing of New Drug Substances and Products. Food and Drug Administration and International Convention on Harmonization of Technical Requirements for Registration of Pharmaceuticals for Human Use (ICH). Nov 2003.
9. **Bak A, Gore A, Yanez E, Stanton M, Tufekcic S, Syed R, Akrame A, Rose M, Surapaneni S, Bostick T, King A, Neervannan S, Ostovic D, Koparkar A:** The co-crystal approach to improve the exposure of a water-insoluble compound: AMG 517 sorbic acid co-crystal characterization and pharmacokinetics. *J Pharm Sci* 97:3942–56, 2008.
10. **Rohrs BR, Thamann TJ, Gao P, Stelzer DJ, Bergren MS, Chao RS:** Tablet dissolution affected by a moisture mediated solid-state interaction between drug and disintegrant. *Pharm Res* 16(12):1850–56, 1999.

Chapter 3

1. **Flynn GL:** Isotonicity—colligative properties and dosage form behavior. *J Parent Drug Assoc* 33(5):292, 1979.
2. **Ernst JA, Williams JM, Glick MR, et al:** Osmolality of substances used in the intensive care nursery. *Pediatrics* 72(3):347, 1983. Modified with permission from *Pediatrics* © 1983.
3. **Dickerson RN, Melnik G:** Osmolality of oral drug solutions and suspensions. *Am J Hosp Pharm* 45:832, 1988. Originally published © 1988, American Society of Hospital Pharmacists, Inc. All rights reserved. Reprinted with permission (R9958).
4. **Krogh A, Lund CG, Pedersen-Bjergaard K:** The osmotic concentration of human lacrymal fluid. *Acta Physiol Scand* 10:88, 1945.
5. **Hunter FT:** A photoelectric method for the quantitative determination of erythrocyte fragility. *J Clin Invest* 19:691, 1940.
6. **Cadwallader DE, Phillips JR:** Behavior of erythrocytes in various solvent systems: V. Water-liquid amides. *J Pharm Sci* 58(10):1220, 1969.
7. **Phillips JR, Cadwallader DE:** Behavior of erythrocytes in phosphate buffer systems. *J Pharm Sci* 60(7):1033, 1971.
8. **Hadzija BW:** A lecture on colligative properties in an undergraduate curriculum. *Am J Pharm Ed* 59:191, 1995.
9. **Hansen GR:** Diuretic drugs, in *Remington's The Science and Practice of Pharmacy*, 19th ed., AR Gennaro (ed). Easton, PA: Mack, 1995.
10. **White AI, Vincent HC:** Diluting solutions in preparation of adjusted solutions. *J Am Pharm Assoc* 8:406, 1947. Reprinted with permission from the American Pharmaceutical Association.
11. **Sprowls JB:** A further simplification in the use of isotonic diluting solutions. *J Am Pharm Assoc* 10:348, 1949.

Chapter 4

1. **Albert A, Serjeant EP:** *The Determination of Ionization Constants*, 3d ed. New York: Chapman and Hall, 1984.
2. **Foye WO:** *Principles of Medicinal Chemistry*, 3d ed. Philadelphia: Lea & Febiger, 1990.
3. **Newton DW, Kluza RB:** pK_a values of medicinal compounds in pharmacy practice. *Drug Intel Clin Pharm* 12:546, 1978.
4. **Reymond GG, Born JL:** An updated pK_a listing of medicinal compounds. *Drug Intel Clin Pharm* 20:683, 1986.
5. **Berge SR, et al:** Pharmaceutical salts. *J Pharmacol Sci* 66:1, 1977.
6. **Zuckerman CL, Nash RA:** The pH value of common household products using meters, papers and sticks. *IJPC* 2(2):137–8, 1998. Reprinted by permission IJPC © 1998.
7. **Raffanti EF, King JC:** Effect of pH on the stability of sodium ampicillin solutions. *Am J Hosp Pharm* 31:745, 1974.
8. **Yang ST, Wilken LD:** The effects of autoclaving on the stability of physostigmine salicylate in buffer solutions. *J Parent Sci Tech* 42(2):62, 1988.
9. **Tsuda T, Uchiyama M, Sato T, et al:** Mechanism and kinetics of secretin degradation in aqueous solutions. *J Pharm Sci* 79(3):223, 1990.
10. **Pouli N, Antoniadou-Vyzas A, Foscolos GB:** Methocarbamol degradation in aqueous solutions. *J Pharm Sci* 83(4):499, 1994.
11. **Ye JM, Lee GE, Potti GK, et al:** Degradation of antiflammin 2 under acidic conditions. *J Pharm Sci* 85(7):695, 1996.
12. **Wang Y-J, Pan M-H, Cheng A-L, et al:** Stability of curcumin in buffer solutions and characterization of its degradation products. *J Pharm Biomed Anal* 15:1867, 1997.
13. **De Smidt JH, Fokkens JG, Grijssels H, et al:** Dissolution of theophylline monohydrate and anhydrous theophylline in buffer solutions. *J Pharm Sci* 75(5):497, 1986.
14. **Turner RH, Mehta CS, Benet LZ:** Apparent directional permeability coefficient for drug ions: In vitro intestinal perfusion studies. *J Pharm Sci* 59(5):590, 1970.
15. **Benet LZ, Orr JM, Turner RH, et al:** Effect of buffer constituents and time on drug transfer across in vitro rat intestine. *J Pharm Sci* 60(2):234, 1971.
16. **Carlson JA, Mann HJ, Canafax DM:** Effect of pH on disintegration and dissolution of ketoconazole tablets. *Am J Hosp Pharm* 40:1334, 1983.
17. **Cheng S-W, Shanker R, Lindenbaum S:** Thermodynamics and mathematical modeling of the partitioning of chlorpromazine between n-octanol and aqueous buffer. *Pharmaceut Res* 7(8):856, 1990.
18. **Van Slyke DD:** On the measurement of buffer values and on the relationship of buffer value to the dissociation constant of the buffer and the concentration and reaction of the buffer solution. *J Biol Chem* 52:525, 1922.

19. **Hasselbalch KA:** Die Berechnung der Wasserstoffzahl des Blutes aus der freien und gebundenen Kohlensäure Desselben, und die Sauerstoffbindung des Blutes als Funktion der Wasserstoffzahl. *Biochem Z* 48:112, 1917.
20. **Henderson LJ:** A note on the union of the proteins of serum with alkali. *Am J Physiol* 21:169, 1908.
21. **Ellison G, Straumfjord JV, Hummel JP:** Buffer capacities of human blood and plasma. *Clin Chem* 4(6):452, 1958. Reprinted by permission of the American Association for Clinical Chemistry © 1958.
22. **Martin A, Swarbrick J, Cammarata A:** *Physical Pharmacy: Physical Chemical Principles in the Pharmaceutical Sciences.* Philadelphia: Lea & Febiger, 1983. Reprinted by permission of Lippincott, Williams & Wilkins, © 1983.
23. **Riegelman S, Vaughan DG:** Ophthalmic solutions. *J Am Pharm Assoc* 19(8):474, 1958.
24. **Flynn GL:** Isotonicity—colligative properties and dosage form behavior. *J Parent Drug Assoc* 33(5):292, 1979.
25. **Miller MA, et al:** Sulfadiazine urolithiasis during antitoxoplasma therapy. *Drug Invest* 5:334, 1993.
26. **Rahn O, Conn JE:** Effect of increase in acidity on antiseptic efficiency. *Ind Eng Chem* 36:185, 1944.
27. **Stuurman-Bieze AGG, et al:** Biopharmaceutics of rectal administration of drugs in man 2. Effect of particle size on absorption rate and bioavailability. *Int J Pharmacol* 1:337, 1978.
28. **Chin TWF, et al:** Effect of an acidic beverage (Coca-Cola) on absorption of ketoconazole. *Anti-microb Agents Chemother* 39:1671, 1995.
29. U. S. Pharmacopeia 36/National Formulary 31, Rockville, MD: United States Pharmacopeial Convention, Inc., 2014.
30. **Lee YC, Zocharski PD, and Samas B:** An intravenous formulation decision tree for discovery compound formulation development, *Int J Pharmaceutics,* 253:111–119, 2003.
31. **Sek D:** Breaking old habits: Moving away from commonly used buffers in pharmaceuticals. *European Pharmaceutical Review,* 7: 37–41, 2012.
32. **Michaelis L:** Diethylbarbiturate buffer. *J Biol Chem* 87(1):33, 1930. Reprinted by permission of The American Society for Biochemistry & Molecular Biology, © 1921.
33. **Delory GE, King EJ:** A sodium carbonate-bicarbonate buffer for alkaline phosphatases. *Biochem J* 39:245, 1945. Reprinted by permission of the Biochemical Society, © 1945.
34. **McIlvaine TC:** A buffer solution for colorimetric comparison. *J Biol Chem* 49:183, 1921. Reprinted by permission of the American Society for Biochemistry & Molecular Biology, © 1921.
35. **Holmes W:** Silver staining of nerve axons in paraffin sections. *Anat Rec* 86:157, 1943. Reprinted by permission of Wiley-Liss, Inc., a subsidiary of John Wiley & Sons, Inc., © 1943.
36. **Clark WM, Lubs HA:** The colorimetric determination of hydrogen ion concentration and its applications in bacteriology. *J Bacteriol* 2:1, 1917.
37. **Cutie AJ, Sciarrone BJ:** Re-evaluation of pH and tonicity of pharmaceutical buffers at 37°C. *J Pharm Sci* 58(8):990, 1969. Reprinted by permission of Wiley-Liss, Inc., a subsidiary of John Wiley & Sons, Inc.

Chapter 5

1. **Abdou HM:** Dissolution, in *Remington's The Science and Practice of Pharmacy,* 19th ed., AR Gennaro (ed). Easton, PA, Mack, 1995.
2. **Martin A, Bustamante P:** *Physical Pharmacy: Physical Chemical Principles in the Pharmaceutical Sciences.* 4th ed. Philadelphia, Lea & Febiger, 1993.
3. United States Pharmacopeia XXIV/National Formulary XIX. United States Pharmacopeial Convention, Rockville, MD, 2000.
4. **Shargel L, Yu ABC:** *Applied Biopharmaceutics and Pharmacokinetics,* 4th ed. Stamford, CT, Appleton & Lange, 1999.
5. **Hoener B, Benet LZ:** Factors influencing drug absorption and drug availability, in *Modern Pharmaceutics,* 3d ed., GS Banker, CT Rhodes (eds). New York, Marcel Dekker, 1999.

Chapter 6

1. **Lightfoot EN:** *Transport Phenomena and Living Systems: Biomedical Aspects of Momentum and Mass Transport.* New York, Wiley, 1974.
2. **Christensen HN:** *Biological Transport,* 2d ed. Reading, MA, W.A. Benjamin, 1975.
3. **Shargel L, Yu ABC:** *Applied Biopharmaceutics and Pharmacokinetics,* 4th ed. Stamford, CT, Appleton & Lange, 1999.
4. **Martin A, Bustamante P:** *Physical Pharmacy: Physical Chemical Principles in the Pharmaceutical Sciences,* 4th ed. Philadelphia, Lea & Febiger, 1993.
5. **Jantzen GW, Robinson JR:** Sustained- and controlled-release drug delivery systems, in *Modern Pharmaceutics,* 3d ed., GS Banker, CT Rhodes (eds). New York, Marcel Dekker, 1999.
6. **Hoener B, Benet LZ:** Factors influencing drug adbsorption and drug availability, in *Modern Pharmaceutics,* 3d ed., GS Banker, CT Rhodes (eds). New York, Marcel Dekker, 1999.

Chapter 7

1. **Connors K:** Complex formation, in *Remington's The Science and Practice of Pharmacy,* vol. 1, 19th ed., AR Gennaro (ed). Easton, PA: Mack, 1995.
2. **Martin A, Bustamante P:** *Physical Pharmacy: Physical Chemical Principles in the Pharmaceutical Sciences,* 4th ed. Philadelphia: Lea & Febiger, 1993.
3. **Chang R:** *Physical Chemistry for the Chemical and Biological Sciences.* Sausalito, CA, University Science Books, 2000.
4. **Szejtli J:** *Cyclodextrins and Their Inclusion Complexes.* Budapest, Hungary, Akademiai Kiado, 1982.
5. **Bender ML, Komiyama M:** *Cyclodextrin Chemistry.* New York: Springer-Verlag, 1978.
6. **Thompson DO:** Cyclodextrins-enabling excipients: Their present and future use in pharmaceuticals. *Crit Rev Ther Drug Carrier Syst* 14:1, 1997.
7. **Szejtli J:** Industrial applications of cyclodextrins, in *Inclusion Compounds,* vol. 3, JL Atwood, JED Davies, DD McNicol (eds). New York: Academic Press, 1984.

8. **Higuchi T, Connors KA:** Phase-solubility techniques. *Adv Anal Chem Instrum* 4:117, 1965.
9. **Uekama K, Otagiri M:** Cyclodextrins in drug carrier systems. *Crit Rev Ther Drug Carrier Syst* 3:1, 1987.
10. **Bush MT, Alvin JD:** Characterization of drug-protein interactions by classic methods, in *Drug-Protein Binding,* vol. 226, AH Anton, HM Solomon (eds). Annals of the New York Academy of Sciences. New York, New York Academy of Sciences, 1973.
11. **Hummel JP, Dryer WJ:** Measurement of protein-binding phenomena by gel filtration. *Biochem Biophys Acta* 63:530, 1962.
12. **Chignell CF:** Recent advances in methodology: Spectroscopic techniques, in *Drug-Protein Binding,* vol. 226, AH Anton, HM Solomon (eds). Annals of the New York Academy of Sciences. New York, New York Academy of Sciences, 1973.
13. **Peters T:** Serum albumin. *Adv Protein Chem* 37:161, 1985.
14. **Shargel L, Yu ABC:** *Applied Biopharmaceutics and Pharmacokinetics,* 4th ed. Stamford, CT, Appleton & Lange, 1999.

Chapter 9

1. **Tsujii K:** *Surface Activity.* San Diego, Academic Press, 1998.
2. **Hiemenz PC, Rajagopalan R:** *Principles of Colloid and Surface Chemistry.* New York, Marcel Dekker, 1997.
3. **Sinko PJ:** *Martin's Physical Pharmacy and Pharmaceutical Sciences.* Philadelphia, Lippincott, Williams and Wilkins, 2011.
4. **Bummer, PM:** Interfacial phenomena, in *Remington's The Science and Practice of Pharmacy,* 20th ed, LV Allen, AD Adejare, SP Desselle, LA Felton (eds). Easton, PA, Lippincott, Williams and Wilkins, 2000.
5. **Adamson AW:** *Physical Chemistry of Surfaces.* New York, Wiley, 1997.
6. **Florence AT, Attwood D:** *Physicochemical Principles of Pharmacy.* New York, Chapman and Hall, 1988.
7. **Rosen MJ:** *Surfactants and Interfacial Phenomena.* New York, Wiley, 1989.
8. **Gau CH, Zografi G:** Relationships between adsorption and wetting of surfactant solutions. *J Colloid Interface Science* 140:1, 1990.
9. **Schott H:** Colloidal dispersions, in *Remington's The Science and Practice of Pharmacy,* 20th ed, A Gennaro (ed). Easton, PA, Lippincott, Williams and Wilkins, 2000.
10. **Swarbrick J, Rubino JT, Rubino OP:** Coarse dispersions, in *Remington's The Science and Practice of Pharmacy,* 20th ed, A Gennaro (ed). Easton, PA, Lippincott, Williams and Wilkins, 2000.
11. **Kissa E:** Dispersions: Characterization, testing and measurement, in *Surfactant Sciences Series,* AT Hubbard, MJ Schick (eds). New York, Marcel Dekker, 1999.
12. **Cooney DO, Thomason R:** Adsorption of fluoxetine HCl by activated charcoal. *J Pharm Sci* 86:642, 1997.
13. **Burke GM, Wurster DE, Berg MJ, et al:** Surface characterization of activated charcoal by x-ray photoelectron spectroscopy (XPS): Correlation with phenobarbital adsorption data. *Pharm Res* 9:126, 1992.
14. **Wang W:** Instability, stabilization, and formulation of liquid protein pharmaceuticals. *Int J Pharm* 185:129, 1999.
15. **Schwendeman SP, Cardamone M, Klibanov A, Langer R:** *Microparticulate Systems for the Delivery of Proteins and Vaccines,* S Cohen, H Bernstein (eds). New York, Marcel Dekker, 1996.
16. **Johnston TP:** Adsorption of recombinant human granulocyte colony stimulating factor (rhG-CSF) to polyvinyl chloride, polypropylene, and glass: Effect of solvent additives. *PDA J Pharm Sci Technol* 50:238, 1996.
17. **Visor GC, Tsa KP, Duffy J, et al:** Quantitative evaluation of the stability and delivery of interleukin-1B by infusion. *J Parenteral Sci Technol* 44:130, 1990.
18. **Tzannis ST, Hrushesky WJM, Wood PA, et al:** Irreversible inactivation of interleukin 2 in a pump-based delivery environment. *Proc Natl Acad Sci USA* 93:5460, 1996.
19. **Suvajittanont W, McGuire J, Bothwell MK:** Adsorption of thermonospora fusca E(5) cellulase on silanized silica. *Biotechnol Bioeng* 67:12, 2000.
20. **Brang J, Langkjaer L:** Insulin formulation and delivery, in *Protein Delivery, Physical Systems,* LM Sanders, RW Hadren (eds). New York, Plenum, 1997.
21. **Sefton MV, Antonacci GM:** Adsorption isotherms of insulin onto various materials. *Diabetes* 33:674, 1984.
22. **Bam NB, Randolph TW, Cleland JL:** Stability of protein formulations: Investigation of surfactant effects by a novel EPR spectroscopic technique. *Pharm Res* 12:2, 1995.
23. **Sluzky V, Tamada JA, Klibanov A, Langer R:** Kinetics of insulin aggregation in aqueous solutions upon agitation in the presence of hydrophobic surfaces. *Proc Natl Acad Sci USA* 88:9277, 1991.
24. **Fatouros A, Sjostrom B:** Recombinant factor VIII SQ influence of formulation parameters on structure and surface adsorption. *Int J Pharm* 194:69, 2000.
25. **Bam NB, Cleland JL, Yang J, et al:** Tween protects recombinant human growth hormone against agitation-induced damage via hydrophobic interactions. *J Pharm Sci* 87:1554, 1998.
26. **Hawe A, Friess W:** Development of HSA-free formulations for a hydrophobic cytokine with improved stability. *Eur J Pharm Biophar* 68: 169–82, 2008.
27. **Modi NB, Veng-Petersen P, Wuster DE, et al:** Phenobarbital removal characteristics of three bands of activated charcoal: A system analysis approach. *Pharm Res* 11:318, 1994.
28. **Wuster DE, Burke GM, Berg MJ, et al:** Phenobarbital adsorption from simulated intestinal fluid, USP, and simulated gastric fluid, USP, by two activated charcoals. *Pharm Res* 5:183, 1988.
29. **Davies JT, Rideal EK:** *Interfacial Phenomena,* 2d ed. London, Academic Press, 1963.
30. **Wallwork SC, Grant DJW:** *Physical Pharmacy.* London, Longman, 1983.
31. **Muster TH, Prestidge CA:** Face specific surface properties of pharmaceutical crystals. *J Pharm Sci* 91:1432, 2002.
32. **Bandyopadhyay R, Grant DJ:** Influence of crystal habit on the surface free energy and interparticulate bonding of L-lysine monohydrochloride dihydrate. *Pharm Dev Techn* 5:27, 2000.

33. **Prestidge CA, Tsatouhas G:** Wettability of morphine sulfate powders. *Inter J Pharm* 198:201, 2000.
34. **Duncan-Hewitt W, Nisman R:** Investigations of the surface energy of pharmaceutical materials from contact angle sedimentations and adhesion measurements. *J Adhes Sci Technol* 7:263, 1993.
35. **Yang YW, Zografi G, Miller E:** Capillary flow phenomena and wettability in porous media. II: Dynamic flow studies. *J Colloid Interface Science* 122:35, 1988.

Chapter 10

1. **Schott H:** Rheology, in *Remington's The Science and Practice of Pharmacy,* vol 1, 19th ed, AR Gennaro (ed). Easton, PA, Mack, 1995.
2. **Martin A, Bustamante P:** *Physical Pharmacy: Physical Chemical Principles in the Pharmaceutical Sciences,* 4th ed. Philadelphia, Lea & Febiger, 1993.
3. **Tabibi SE, Rhodes CT:** Disperse systems, in *Modern Pharmaceutics,* 3d ed, GS Banker, CT Rhodes (eds). New York, Marcel Dekker, 1999.
4. **Flory PJ:** *Principles of Polymer Chemistry.* Ithaca, NY, Cornell University Press, 1953.
5. **Merrill EW:** Rheology of blood. *Phys Rev* 49:863, 1969.
6. **Schmid-Schonbein H:** Blood rheology and physiology of microcirculation. *Ric Clin Lab* 11:13, 1981.
7. **Lowe GD:** Blood rheology in general medicine and surgery. *Baillieres Clin Haematol* 1:827, 1987.
8. **Marriott C, Gregory NP:** Mucus physiology and pathology, in *Bioadhesive Drug Delivery Systems,* V Lenaerts, R Gurny (eds). Boca Raton, FL, CRC Press, 1990.
9. **Slomiany BL, Murty VLN, Piotrowski J, Slomiany A:** Effect of antiulcer agents on the physicochemical properties of gastric mucus, in *Mucus and Related Topics,* E Chantler, NA Ratcliffe (eds). Cambridge, UK, The Company of Biologist, 1989.
10. **Chantler E, Sharma R, Sharman D:** Changes in cervical mucus that prevent penetration by spermatozoa, in *Mucus and Related Topics,* E Chantler, NA Ratcliffe (eds). Cambridge, UK, The Company of Biologist, 1989.
11. **Schurz J, Ribitsch V:** Rheology of synovial fluid. *Biorheology* 24:385, 1987.
12. **Balazs EA, Denlinger JL:** Viscosupplementation: A new concept in the treatment of osteoarthritis. *J Rhematoid Suppl* 39:3, 1993.
13. **Rippie EG:** Mixing, in *The Theory and Practice of Industrial Pharmacy,* 2d ed, L Lachman, HA Lieberman, JL Koenig (eds). Philadelphia, Lea & Febiger, 1976.
14. **Briceno MI:** Rheology of suspensions and emulsions, in *Pharmaceutical Emulsions and Suspensions,* F Nielloud, G Marti-Mestres (eds). New York, Marcel Dekker, 2000.
15. **Thompson JE:** *A Practical Guide to Contemporary Pharmacy Practice.* Baltimore, Williams & Wilkins, 1998.

Index

A

ABC transporters, 7
Absolute viscosity, 200
Absorption, 8–10, 103, 104f, 127–131
Acacia gum, 168, 212
Accelerated stability testing, 232–233
Acetaminophen, 29t, 66t
Acetaminophen drops, 51t
Acetaminophen elixir, 51t
Acetaminophen liquid, 72t
Acetanilide, 68t
Acetate, 30t
Acetic acid, 66t, 183
Acetone, 90t, 95, 201t
Acetylsalicylic acid, 66t, 207, 208t, 222f
Acid
   Arrhenius classification, 62
   Bronsted-Lowry concept, 62–63
   Lewis electronic theory, 63
   strong, 65
Acid-base reaction, 34, 35f
Acid-base titration, 76
Acid phthalate, 80
Acidemia, 22
Acidic ionization constant, 66. *See also* Ionization constants
Activated charcoal, 182
Activated complex, 230
Active transport, 7, 119–120

Activity, 75
Activity coefficient, 75
Acyl isethionates, 185
N-acyl sarcosinates, 185
N-acyl taurines, 185
Adhesive forces, 15
ADME, 8. *See also* LADME
Administration routes, 3, 4t
Adsorption, 180–189
   CMC, 187–189
   defined, 180
   Gibbs equation, 187
   HLB, 186
   isotherms, 180–181, 194–196
   liquid interfaces, 184–189
   solid interfaces, 180–181
   solid-liquid interfaces, 181–184
   surface active agents, 184–186
Adsorption isotherms, 180–181, 194–196
Aerosol, 2t
Affinity of binding, 147
Agar solution, 210
Albumin, 7, 146–147, 146t, 151, 207
Alginate, 212
Alkaline borate, 80
Alkalol, 72t
N-alkyl betaines, 186
Alkyl polygulcosides, 186
Alkylbenzyldimethyl ammonium salts, 185

Alkyltrimethyl ammonium salts, 185
α-cyclodextrin, 140t, 153t
α-D-Glucose, 66t
α-helix, 20
α1-acid glycoprotein, 7, 146t, 147
Aluminum hydroxide, 92t
Alzet, 132
Amantadine HCl solution, 51t
Amino acid transporter, 120t
Amino acids, 11
Ammonia, 68t, 136
Amorphization, 36t
Amorphous dispersion, 31
Amorphous forms, 6
Amorphous solids, 30–31
Amoxicillin suspension, 51t
Amphipathic helix, 20
Amphiphiles, 184
Ampholytes, 74
Amphoteric electrolyte, 74
Ampicillin, 66t
Ampicillin trihydrate, 105
Anhydrous salts, 90t
Anion, 61
Anion exchanger, 141
Anionic surface active agents, 184–185
Anticoagulants, 207
Apomorphine, 68t
Apothecary measure (U.S. fluid system), 249
Apothecary weight, 249

257

Apparent partition coefficient, 107, 108
Apparent viscosity, 200
Apparent volume of distribution (Vd), 11–12
Apple juice, 72t
Aqueous gelatin, 210
Aripiprazole polymorphs and solvates, 27f
Arrhenius classification of acids/bases, 62
Arrhenius equation, 229, 231
Arrhenius plot, 231f, 233f
Arterial blood gases, 22t
Arthritis, 209
Ascorbic acid, 66t
Aspirin, 89t
Association colloids, 160–161
Association constant (K_a), 143, 147, 151, 152
Atropine, 61, 68t, 89t
Atropine sulfate, 89t
Avoirdupois weight, 249
Axetil, 31

B
BAL-lead complex, 139
Barbiturate poisoning, 108
Barbituric acid, 66t
Barium carbonate, 92t
Barium oxalate, 92t
Base
 Arrhenius classification, 62
 Bronsted-Lowry concept, 62–63
 Lewis electronic theory, 63
 strong, 65
Bentonite, 215
Benzene, 175t
Benzocaine, 68t
Benzoic acid, 66t, 109
Benzyl alcohol, 90t
Benzyl penicillin, 66t
BET isotherm, 181, 196
β-cyclodextrin, 140t, 153t
Beta lactam antibiotics, 243
Bidentate, 136
Bile transporter, 119

Biliary section, 12
Binary solution, 41
Binding affinity, 147
Binding equilibria, 152–154
Bingham, Eugene, 197
Bioavailability, 3, 103
Biological fluids
 colligative properties, 47t
 osmolality, 50t
Biological membrane, 122
Biopharmaceutics, 1
Blood, 74, 206–207
Blood-brain barrier, 121
Blood gases, 22
Body osmolality, 48
Boiling point (BP), 23, 31
Boiling point elevation, 46
Borate buffers, 82
Boric acid, 66t, 85t
Boric acid-sodium hydroxide buffers, 83
Boric acid solution, 51
Boyle's law, 19
BP, 23, 31
Breaking, 163
Bromide, 30t
Bronchial mucus, 208
Bronsted-Lowry classification of acids/bases, 62–63
Brownian motion, 162–163
Buccal route of administration, 4t
Buffer capacity, 73
Buffer effect, 73
Buffer equation, 72
Buffer solutions, 72–74, 80–85
 borate buffers, 82
 boric acid-sodium hydroxide buffers, 83
 buffer capacity, 73
 buffer equation, 72
 disodium phosphate and citric acid buffers, 82

other buffer systems, 85t
 phosphate-sodium hydroxide buffers, 83
 phthalate-hydrochloric acid buffers, 83
 phthalate-sodium hydroxide buffers, 83
 physiologic buffer systems, 74
 potassium chloride-hydrochloric acid buffers, 82–83
 preparation of buffers, 73
 sodium carbonate-bicarbonate buffers, 82
 Sorensen's buffers, 83–84, 84t
 USP buffer systems, 80
 veronal buffers, 82
Buffer value, 73
Burows solution, 72t
Burst effect, 125
Butyl p-aminobenzoate, 122t

C
Caffeine, 29t, 68t, 87
Caking, 163
Calcium, 30t
Calcium carbonate, 91, 92t
Calcium channel blockers, 207
Calcium fluoride, 92t
Calcium glubionate syrup, 51t
Calcium oxalate, 92t
Calcium phosphate, 92t
Calcium sulfate, 92t
Camphor, 89t, 96t
Capillary penetration method, 194
Capillary rise (depression) method, 192–193
Capillary viscometer, 217–218
Capsule, 2t
Carbomer, 32t
Carbonic acid, 66t
Carbonic anhydrase, 139
Carboplatin, 138, 138f

Carbopol resins, 213, 214*t*
Carboxypeptidase, 139
Carpobol, 206
Carrageenan, 212
Casein micelle, 160
Castor oil, 201*t*
Cation, 61
Cation exchanger, 141
Cationic surface active agents, 185
Cefazolin, 66*t*
Cefuroxime, 31
Cefuroxime axetil, 39
Cellular transport, 120
Cellulose, 210, 211*f*
Cellulose derivatives, 210–212
Centistoke (cs), 201
Central nervous system (CNS), 11
Cephalothin, 66*t*
Cephradine, 66*t*
Cervical mucus, 208
Cetyl alcohol, 96*t*
Changes in state, 31–33
Charles' law, 21
Chelate, 137
Chemical kinetics, 221–243
    enzyme catalysis reactions, 234–236
    first-order absorption, 237
    first-order reaction, 225–226, 228–229
    hydrolysis, 222–223
    order of reaction, 223–229
    oxidation, 223
    pharmacokinetics, 236–238
    pseudo reaction order, 226–227
    pseudo-zero-order reactions (suspensions), 227
    radioactivity, 238
    stability and shelf life of drugs, 229–234
    zero-order absorption, 236–237
    zero-order reaction, 224
Chemical potential, 118
Chemical stability, 34–35
Chemically cross-linked gels, 210
Chemisorption, 180
Chitosan, 210
Chloramphenicol, 104
Chloramphenicol palmitate, 27, 29*f*, 29*t*
Chlorcyclizine, 68*t*
Chlordiazepoxide, 68*t*
Chloroform, 175*t*
Chlorpheniramine, 68*t*
Chlorpromazine, 68*t*
Cholesterol-menthol, 33*t*
Cholestyramine, 143
Cimetidine, 29*t*
Cimetidine solution, 51*t*
Circular dichroism (CD), 145
Cisplatin, 138, 138*f*
Citrate, 30*t*
Citric acid, 66*t*
Clausius Clapeyron equation, 23
Clays, 215
Clearance, 13
Clindamycin, 68*t*
Club soda, 72*t*
CMC, 187–189
Coarse dispersions, 161–162
Coarse emulsions, 161, 168
Coarse suspensions, 161–162
Coaxial-cylinder viscometer, 219, 219*f*
Cobalt, 139
Cocaine, 68*t*
Cocrystal, 26*f*, 29–30
Cocrystal former, 30
Codeine, 68*t*, 89*t*
Coefficient of viscosity, 199
Coffee, instant, 72*t*
Cohesive forces, 15
Colestipol, 143
Colligative properties, 46, 47–50
Collision diameter, 16
Collision theory of reaction rates, 229–231
Colloid solution, 159
Colloidal dispersions, 159–161
Colloidal oncotic pressure (COP), 48
Commit, 151
Complexation, 7, 135–146
    coordination complexes, 136–137
    copper and cobalt complexes, 139
    cyclodextrin complexes, 140–141, 151–152
    defined, 135
    ion-exchange resins, 141–143
    molecular complexes, 137–138
    platinum complexes, 138–139
    toxic heavy metal complexes, 139
    zinc complexes, 139
Concentration expressions, 42–44
Concentration gradient, 118
Concerta, 132*t*
Cone-and-plate rotational viscometer, 219–220, 220*f*
Conjugate acid-base pairs, 63, 63*t*
Conjugation reactions, 12
Contact angle, 177–179, 193–194
Convection, 118
Coordination complexes, 136–137
Coordination number, 136
Copolymer resins, 213, 214*t*
Copper and cobalt complexes, 139
Cottonseed oil, 90*t*
Cough syrup, 72*t*
Cp_{ss}, 13
Cream, 2*t*
Critical micelle concentration (CMC), 160, 187–189
Crystal bonding, 26*t*
Crystal field theory, 137
Crystalline forms, 6
Crystalline solids, 25–30
    cocrystals, 29–30
    lattice units, 25, 25*f*
    polymorphs, 26–27
    salt crystals, 26*f*, 29
    solvates and hydrates, 27–29

Crystallization, 36, 36t
Crystalloid solution, 159
Crystals, 3
cs, 201
Cube-root dissolution rate constant, 99
Cubic system, 25, 25f
Cupric sulfide, 92t
Cyanocobalamin, 139, 139f
Cyclodextrin complexes, 140–141, 151–152
Cyclosporine, 161
Cylinder method (dissolution testing), 101–102
CYP, 12
Cystic fibrosis, 208
Cytochrome, 137
Cytochrome P450 (CYP), 12
Cytoplasmic proteins, 20

D
Dalton's law, 22
Debrox ear drops, 72t
Debye and Huckel equation, 76
Debye forces, 17
Deferoxamine, 138, 138f
Definite integral, 246
Deflocculated particles, 167
Degradation, 34–35
Degradation reactions, 221, 222–223
Degree of hysteresis, 205
Delavirdine mesylate salt, 37
Delsym, 142
Dependent variable, 245
Descriptive solubility terms, 87, 88t
Di-divalent electrolyte, 47t
Dialysis membrane, 144
Diazepam, 68t, 94, 153t
Dielectric properties, 90, 90t
Diethylamine, 30t
Differentiation and integration functions, 246–247
Diffuse double-layer, 164
Diffuse layer, 164, 165f, 179
Diffusion, 118
Diffusion coefficient, 122, 122t, 125
Diffusion layer model, 97–98
Diffusion through a membrane, 121–125
Diffusional matrix delivery system, 126–127, 128t
Diffusional release, 125–127
Diffusional reservoir delivery system, 126, 127t
Digitoxin, 153t
Digoxin, 96t, 153t
Digoxin elixir, 51t
Di-univalent electrolyte, 47t
Dilatant flow, 204–205
2,3-dimercaptopropanol (BAL)-lead complex, 139
N,N-Dimethylacetamide, 90t
Diphenhydramine, 68t
Diphenhydramine syrup, 72t
Diphenylhexatriene, 201
Dipolar forces, 20
Dipole-dipole forces, 17
Dipole-dipole interactions, 144
Dipole-induced dipole forces, 17
Dipole moment, 90
Disodium citrate, 85t
Disodium phosphate and citric acid buffers, 82
Dispersed systems, 157–171
  aggregation, 163–164
  Brownian motion, 162–163
  classification by particle size, 159–162
  classification by phases, 157–158
  coarse dispersions, 161–162
  colloidal dispersions, 159–161
  dispersion particle growth, 163–168
  dispersion uniformity, 162–163
  DLVO forces, 164–166
  flocculation, 167, 168f
  gravitational forces, 163
  non-DLVO forces, 166
  Ostwald ripening, 166–167
  product formulation, 167–168
  zeta potential, 166
Dispersion forces, 17
Dispersion uniformity, 162–163
Dissolution, 6, 97–106
  absorption, 103, 104f
  bioavailability, 103
  crystalline state, 104–105
  defined, 97
  diffusion layer model, 97–98
  gastric emptying and intestinal transit times, 106
  Hixson-Crowell cube-root equation, 99
  ionized versus unionized forms, 103
  Noyes-Whitney equation, 98
  particle size, 103–104
  pH, 106
  quality control test, 102
  semisolid dosage forms, 105–106
  solid dosage forms, 105
  suspensions and emulsions, 105
  testing. See USP dissolution testing
Dissolution calibrators, 102
Distek, Inc., 99
Distek 2100B United States Pharmacopeia automated dissolution system, 100f
Distilled vinegar, 72t
Distribution, 10–12, 127–131
Distribution sites, 11
Ditropan XL, 132t
DLVO forces, 164–166
DLVO theory, 164
dM/dt, 98, 103, 123
Docusate sodium syrup, 51t
Dopamine, 130
Dosage forms, 2–3, 2t
Dose dumping, 126
Double layer, 179, 179f
Double reciprocal Lineweaver-Burk plot, 234f
Double-reciprocal method, 154
Drug absorption, 8–10, 103, 104f, 127–131

Drug complexation. *See* Complexation
Drug-cyclodextrin complexes, 142*f,* 153*t*
Drug degradation, 34–35
Drug degradation reactions, 221, 222–223
Drug diffusion, 126
Drug disposition, 3–7
Drug dissolution. *See* Dissolution
Drug distribution, 10–12, 127–131
Drug dosage forms, 2–3, 2*t*
Drug entry and distribution, 2
Drug excretion, 12–13
Drug metabolism, 12
Drug metabolism activity, 23
Drug permeation, 108
Drug-protein binding analysis, 152–156
Drug release by diffusion, 125–127
Drug release by osmosis, 131–132
Drug solubility. *See* Solubility
Drug targets, 1–2
Drug transporters, 7
Drug:polymer ratio, 31
Drying, 36*t*
Du Nouy ring method, 192
Dynamic dialysis, 145
Dynamic viscosity, 219
Dynes, 175

E
E, 52
EDTA, 137*f*
Efflux transporter, 9
Elasticity, 198
Elastomeric infusion pump, 217
Electrical double layer, 179–180
Electrolyte
 amphoteric, 74
 approximate distribution, 48*t*
 defined, 61
 ionization, 65–70
 L_{iso} value, 47*t*

strong. *See* Strong electrolytes
weak. *See* Weak electrolytes
Electrostatic interaction, 141, 144
Electrostatic repulsive forces, 164, 165
Elimination half-life, 13
Emptying time, 106
Emulsion breaking, 163
Emulsions
 coarse dispersion, as, 161
 dissolution, 105
 preservation of, 109
 rheology, 209
Enantiotropic system, 36, 37*f*
Endocytosis, 120
Endosome, 120
Enema, 72*t*
Energy of activation, 230, 231
English systems (mathematical tables), 249
Enterohepatic circulation, 9–10
Enterohepatic cycling, 10
Enzyme catalysis reactions, 234–236
Ephedrine, 68*t*
Ephedrine hydrochloride, 71
Epidural injection, 4*t*
Epithelium, 12
Equations, 245
Equilibrium dialysis, 144–145
Erythrocytes, 206, 206*t*
Erythromycin, 68*t*
Erythromycin ethylsuccinate suspension, 51*t*
Ethanol, 201*t*
Ether, 90*t*
Ethyl acetate, 90*t*
Ethylcellulose, 126
Ethylenediaminetetraacetic acid (EDTA), 137*f*
Ethynodiol diacetate, 122*t*
Etravirine, 31
Eudragit, 31
Eutectic mixture, 33
Eutectic point, 33
Excretion of drugs, 108–109
Excretion sites, 12–13

Exocytosis, 120
Exponential function, 246
Extravasation, 59
Eye drops, 209
Eye fluid, 74

F
f_u, 152
Facilitated diffusion, 7
Facilitated transport, 118–119
Falling sphere viscometer, 217
Fatty acid alkanolamides, 186
Fenbufen, 153*t*
Ferric hydroxide, 92*t*
Ferrous hydroxide, 92*t*
Ferrous sulfate liquid, 51*t*
Ferrous sulfate solution, 72*t*
Fibrinogen, 207
Fick's law of diffusion, 122
First-order absorption, 237
First-order reaction, 225–226, 228–229
First-order transport, 124
First-pass effect, 128
First-pass metabolism, 10
Flocculation, 167, 168*f*
Floccules, 167
Flow. *See* Rheology
Flow-through cell dissolution apparatus, 101
Flurbiprofen, 153*t*
Fluocinolone acetonide, 122*t*
Fluorescence spectroscopy, 146
5-fluorouracil, 120
Flux, 122
Folic acid, 66*t*
Formic acid, 66*t*
Formulators, 167
Fraction unbound (f_u), 152
Franz cell, 106
Free energy, 163
Freely soluble, 88*t*
Freezing point depression, 46, 47
Freezing point depression method, 52
Freundlich isotherm, 181, 196
Frost-free freezer, 21

Fructose, 61
Functionalization reactions, 12
Furosemide, 66t
Furosemide solution, 51t

G

γ-cyclodextrin, 140t, 153t
Gas cylinder, 22
Gas masks, 180
Gaseous state, 19–22
Gastric emptying, 10
Gastric emptying time, 106
Gastrointestinal pH, 9
Gastrointestinal therapeutic system (GITS), 131, 132
Gatorade, 72t
Gel, 2t
Gel filtration, 145
Gellan gum, 210
Gels, 210
General gas equation, 21
General partition law, 106, 107
Gentamicin complex, 68t
Gibbs adsorption equation, 187
GITS, 131, 132
Glass capillary viscometer, 217, 218f
Glass transition (Tg) temperature, 30
Glasses, 30
Globular conformation, 203
Glucose, 11
Glucose transporter 1 (GLUT1), 11
Glucotrol XL, 132t
Glucuronide metabolites, 10
GLUT1, 11
Glycerin, 90t, 201t
Glycerol, 61
Glycine, 66t, 68t, 74
Goniometric method, 194
Gouy-Chapman layer, 179, 179f
Gouy-Chapman-Stern model, 164
Graphical method of determining order of reaction, 227–229
Gravitational forces, 163
Gum arabic, 212

H

Habitrol, 132
Half-life, 13, 224
Haloperidol concentrate, 51t
Hanson Research, 99
Heat of fusion, 33
Heat of solution, 89, 90t
Heat of vaporization, 31
Heme proteins, 137
Hemocyanin, 139
Hemoglobin, 137
Hemolysis, 51
Henderson-Hasselbalch equation, 72, 108, 129
Henry's law, 22
Hepatic portal system, 9
Heptane, 175t
Heteromeric crystal, 26
Hexadecane, 175t, 177t
Hexadentate, 136
Hexagonal system, 25f
Hexane, 175t
Higuchi equation, 126
Higuchi square-root-of-time relationship, 106
Hixson-Crowell cube-root equation, 99
HLB, 186
Homogeneous, 41
Homomeric crystal, 26, 26f
Homopolymer resins, 213, 214t
HPMC, 31
Human milk to plasma concentration (m/p) ratio, 109
Hyaluronic acid, 208–209
Hydrate, 26f, 27–29
Hydrated cocrystal, 26f
Hydration, 36t
Hydration forces, 166
Hydrochloric acid, 61, 80
Hydrochloride, 30t
Hydrochloromethazide HCl, 236
Hydrocolloids, 159
Hydrocortisone, 153t
Hydroflumethazide HCl, 236
Hydrogen bonding, 18–19, 20, 90
Hydrogen peroxide, 72t

Hydrolysis, 70, 222–223
Hydrophile-lipophile balance (HLB), 186
Hydrophilic colloids, 159
Hydrophilic metabolites, 10
Hydrophobic attractive forces, 19
Hydrophobic bonding, 182
Hydrophobic colloids, 160
Hydrophobic effect, 19
Hydrophobic interactions, 19, 144
Hydroquinone, 68t
Hydroxyethyl cellulose, 211, 211f
Hydroxypropyl cellulose, 211, 211f
Hydroxypropyl methylcellulose, 211–212, 211f
Hypertonic aqueous solution, 51
Hypertonic solution, 49, 121
Hypotonic aqueous solution, 51
Hypotonic solution, 49, 121

I

Ibuprofen, 153t
Ideal solution, 45
Imperial measure (British), 249
In vivo erythromycin breath test, 23
Indefinite integral, 246
Independent variable, 245
Indomethacin, 66t
Induced dipole-induced dipole forces, 17
Influx transporter, 9
Insoluble, 88t
Instant coffee, 72t
Insulin, 139
Insulin glargine, 115
Integration, 246–247
Interface, 173, 173t
Interfacial phenomena, 173–196
  adsorption. *See* Adsorption
  contact angle, 177–179, 193–194
  defined, 191

electrical double layer, 179–180
spreading coefficient, 176–177
surface tension, 173–176, 192–193
Interfacial tension, 174, 175t
Intermolecular attractive forces, 16–19
Intermolecular binding forces, 15–16
Intermolecular forces, 15–19
Intestinal transit time, 106
Intramuscular injection, 4t
Intranasal route of administration, 4t
Intraosseous injection, 4t
Intraperitoneal infusion, 4t
Intrasynovial injection, 4t
Intrathecal injection, 4t
Intravenous injection, 4t
Intravenous mannitol solutions, 59
Intravitreal injection, 4t
Iodine tincture, 72t
Ion-dipole forces, 17–18
Ion exchange, 141, 182
Ion-exchange chromatography, 142
Ion-exchange resins, 141–143
Ion-induced dipole, 18
Ion-pair formation, 119
Ion pairing, 182
Ion trapping, 131
Ionic equilibria and buffers, 61–85
  acid-base titration and titration curve, 76, 77f
  acids and bases, 62–63
  activity and activity coefficient, 75–76
  buffers. See Buffer solutions
  ionization, 61–62
  ionization of amphoteric electrolytes, 74
  ionization of electrolytes, 65–70
  ionization of polyprotic acids, 75
  ionization of salts, 70–71
  ionization of water, 63–65

Ionic strength, 76
Ionization, 6, 61–62
Ionization constants
  defined, 66
  use, 63
  weak acids, 66–67t
  weak bases, 68–69t
Ionization state, 130
Iota carrageenan, 212
Ipecac syrup, 72t
Isoamyl alcohol, 175t
Isoelectric point, 74
Isoosmotic solution, 50, 51
Isoproterenol, 68t
Isosorbide dinitrate, 153t
Isotonic buffer solutions, 52t
Isotonic solution, 49, 121
Isotonic solutions, 50–54
Isotonicity, 132
Itraconazole, 30, 31

J
Jello, 210

K
K_a, 65, 143, 147, 151, 152
K_b, 65
K_d, 106
K_{sp}, 91
Kaletra, 28
Kanamycin, 68t
Kappa carrageenan, 212
Kayexalate, 143
Keesom forces, 17
Ketchup, 203
Ketoprofen, 153t
Kidney, 12
Kinematic viscosity, 201
Kinetic energy (KE), 19
Kinetics, 117–118
Kolex nasal spray, 72t
Kollidon, 213

L
L_{iso} value, 47, 47t
L-type amino acid transporter (LAT1), 11
Lactic acid, 66t
Lactulose syrup, 51t

LADME
  absorption, 8–10
  distribution, 10–12
  excretion, 12–13
  liberation, 8
  metabolism, 12
Lag-time effect, 124–125
Lambda carrageenan, 212
Langmuir isotherm, 180–181, 194–195
LAT1, 11
Lead iodate, 92t
Lead sulfate, 92t
Lemon juice, 72t
Lennard-Jones potential, 15, 16
Leukocytes, 206, 206t
Levodopa, 130
Lewis acid-base pair, 63
Lewis electronic theory, 63
Liberation, 8
Lidocaine, 68t
Ligand, 135
Ligand-binding sites, 143
Like dissolves like, 91
Lincomycin, 68t
Linear measure, 249
Lineweaver-Burk equation, 234
Lipid/water partition coefficient, 108
Lipophilic drugs, 6
Lipoproteins, 146t, 147
Liquid homogeneous aqueous systems, 42
Liquid state, 23–24
Liquid to gas conversion, 31–33
Listerine, 72t
Lithium carbonate, 92t
Lithium citrate syrup, 51t
Liver, 12
Logarithmic function, 245
London forces, 17, 20
Lopinavir/ritanovir, 31
Lubricating agents, 105
Lungs, 174
Lyophilic colloids, 159
Lyophobic colloids, 159
Lysine, 30t
Lysolecithin, 208t

M

m/p ratio, 109
Macromolecular dispersions, 159
Macromolecule, 159, 168
Magnesium, 30t
Magnesium carbonate, 92t
Magnesium citrate solution, 72t
Magnesium hydroxide, 91
Magnesium stearate, 71
Magnesium sulfate, 43, 44
Maleate, 30t
Mallard reaction, 34, 35f
Mandelic acid, 66t
Mannitol, 90t, 96t
Mass transport, 7, 117–134
    active transport, 119–120
    cellular transport, 120
    defined, 117
    diffusion through a membrane, 121–125
    drug absorption and distribution, 127–131
    drug release by diffusion, 125–127
    drug release by osmosis, 131–132
    facilitated transport, 118–119
    Fick's law of diffusion, 122
    isotonicity, 132
    lag-time and burst effects, 124–125
    passive transport, 118
    sink condition approximation, 123–124
    solute transport, 118–120
    solvent transport, 120–121
Mathematical concepts
    differentiation and integration functions, 246–247
    equations, 245
    exponential function, 246
    logarithmic function, 245
Mathematical constants, 248
Mathematical tables
    English systems, 249
    mathematical constants, 248
    metric conversion table, 248
    metric system, 249
Matrix system, 126–127, 128t
Medroxyprogesterone acetate, 122t
Megace, 160
Megestrol acetate, 160
Meglumine, 30t
Membrane, 121
Membrane and metabolic barriers, 3
Membrane-bound transporters, 9
Membrane passage, 7
Menthol, camphor, phenol thymol, and/or chloral hydrate, 33t
Meperidine, 68t
Mercuric sulfide, 92t
Mercurochrome, 72t
Mercurous chloride, 92t
Mesylate, 30t
Metabolic sites, 12
Metabolism, 12
Metformin, 134
Methanol, 90t
Methapyrilene, 68t
Methicillin, 66t
Methocarbamol, 96t
Methylcellulose, 32t, 90t, 211, 211f
Methyldopa suspension, 51t
Methylene chloride intermolecular bonds, 17f
Metric conversion table, 248
Metric system, 249
Micelles, 184, 185f, 188
Michaelis-Menten constant, 234
Michaelis-Menten equation, 234
Michaelis-Menten kinetic profile, 119
Microemulsion, 160f, 161
Microfluidity, 201
Micronization, 90–91, 104
Microviscosity, 201
Milligrams per deciliter, 43t
Miscible, 87, 88, 89t
Mixing process, 209
Modified sessile drop method, 194
Moexipril, 34f
Molality, 43t, 44
Molarity, 42, 43, 43t
Mole fraction, 43t
Molecular complexes, 137–138
Molecular orbital theory, 137
Molecular weight (MW), 42
Molecules, 3
Monoalkyl phosphates, 185
Monoclinic system, 25f
Monomer, 31
Monotropic system, 36, 37f
Morphine, 68t, 89t
Mucin, 207
Mucus, 207–208
Mucus glycoprotein, 207
Multidentate, 136
Multiple emulsions, 161
MW, 42
Myoglobin, 137

N

Nafcillin, 66t
Natural clays, 215
Natural gums, 212
Natural membrane, 121
Natural polymers, 31
Nelfinavir mesylate, 31
Neoral, 161
Neutralized phthalate, 80
Newtonian flow, 198–201
Nicotine gum, 134
Nicotine transdermal patch, 132
Nicotinic acid, 66t
Nifedipine, 29t
Nitrofurantoin, 66t
Nitroglycerin, 24
Non-DLVO forces, 166
Non-Newtonian flow, 202–205
Nonelectrolyte
    defined, 61
    L_{iso} value, 47t
    partitioning, 107
    solubility, 96
    solutions, 45t
Nonelectrolyte solubility comparisons, 96t
Nonionic surface active agents, 185–186

N

Normal saline solution, 72t
Normality, 43, 43t
Norvir, 28
Noyes-Whitney equation, 98
Nuclear magnetic resonance (NMR), 145
Nystatin suspension, 51t

O

OAT, 12
OATP, 7, 9
OCT, 12
OCT2, 134
n-Octanol, 175t, 177t
Ointment, 2t
Ointments, 209
Oleic acid, 175t
Oligopeptide transporter, 120t
Olive oil, 201t
Omeprazole, 3, 119
Ophthalmic drops, 209–210
Ophthalmic route of administration, 4t
Ophthalmic solution, 42
Optical rotatory dispersion (ORD), 145
Oral emulsions, 161
Oral osmotic delivery systems, 132t
Oral route of administration, 4t
Oral solution, 42
Order of reaction, 223–229
Organic anion transporter (OAT), 12
Organic anion transporting peptide (OATP), 7, 9
Organic cation transporter (OCT), 12
Organic cation transporter-s (OCT2), 134
Organic resins, 141
Organic weak acids, 65–68
Organic weak bases, 68–70
Organogel, 215
OROS, 131
Orthorhombic system, 25f
Osmolality, 43t, 48, 49t, 50t, 51t
Osmolarity, 43t, 49
Osmosis, 46, 120, 121, 121f
Osmotic effect, 166n
Osmotic pressure, 46, 120
Ostwald ripening, 166–167
Otic solution, 42
Oxacillin, 66t
Oxalic acid, 66t
Oxidation, 223
Oxytetracycline, 68t

P

P-glycoprotein (Pgp), 7, 9, 120
P-glycoprotein efflux system, 120t
Paddle and shaft assembly (dissolution testing), 100–101
Paddle over disk (dissolution testing), 101
Papaverine, 68t
Paracellular route, 128
Parenteral solution, 42
Parkinson's disease, 130
Partial pressure, 22
Particle dissolution. *See* Dissolution
Particle flocculation, 167, 168f
Particle size
  dispersed systems, 162
  dissolution, 103–104
  solubility, 90–91
Partition coefficient, 6, 115, 123, 128
Partition law, 106, 107
Partitioning, 106–109
  drug permeation, 108
  excretion of drugs, 108–109
  general concepts, 106
  nonelectrolytes, 107
  pH-partition hypothesis, 108
  preservation of emulsions, 109
  strong electrolytes, 106
  weak electrolytes, 107–108
Passive diffusion, 7, 9
Passive transport, 118
PCO_2, 22, 22t
PEG, 31
Pennkinetic system, 142
Pentobarbital, 66t, 93
Pentoxifylline, 207
Pepsi, 72t
Peptide intermolecular hydrogen bonding, 19f
Percent by volume, 43t
Percent by weight, 43t
Percent weight in volume, 43t
Permeability coefficient, 124
Permutite, 142
Pgp, 7, 9, 120
pH
  dissolution, 106
  liquids, 72t
  reaction rate, 233–234
pH-partition hypothesis, 6, 62, 108, 129
pH-rate profiles, 233
Phagocytosis, 120
Pharmaceutical buffers. *See* Buffer solutions
Pharmaceutical dispersions, 157. *See also* Dispersed systems
Pharmaceutical polymers, 31, 32t
Pharmaceutical solids. *See* Solids
Pharmaceutical solutions, 41, 42. *See also* Physical properties of solutions
Pharmaceutical suspensions. *See* Suspensions
Pharmaceutical transport systems. *See* Mass transport
Pharmacokinetics, 8, 236–238
Phase 1 reactions, 12
Phase 2 reactions, 12
Phase-solubility diagram, 151
Phase-solubility profile, 151
Phenethicillin, 66t
Phenobarbital, 61, 66t, 89t
Phenobarbital poisoning, 130
Phenobarbital sodium, 29t, 89t
Phenobarbital solubility, 96t
Phenol
  dielectric constant, 90t
  heat of solution, 90t
  ionization constant, 66t
  miscibility, 95, 95f
  solubility, 89t

Phenolphthalein, 66t
Phenylephrine, 68t
Phenylpropanolamine, 69t
Phenytoin, 29t, 67t
Phenytoin sodium suspension, 51t
Phosphate, 30t, 80
Phosphate-sodium hydroxide buffers, 83
Phosphate transporter, 120t
Phosphoric acid, 67t, 75
Phthalate-hydrochloric acid buffers, 83
Phthalate-sodium hydroxide buffers, 83
Physical adsorption, 180. See also Adsorption
Physical properties of solutions, 41–59
  advantages/disadvantages of solutions, 42
  classification of aqueous solution systems, 45
  colligative properties, 46
  concentration expressions, 42–44
  E value, 52
  general considerations, 41–42
  isotonic solutions, 50–54
  physiological applications of colligative properties, 47–50
Physical stability, 36–37
Physical stability transformations, 36t
Physicochemical complementarity, 144
Physicochemical factors
  complexation and protein binding, 7
  ionization, 6
  mass transport and membrane passage, 7
  partition coefficient, 6
  solid state properties, 6
  solubility and dissolution, 6
Physiologic buffer systems, 74
Physostigmine, 69t

Pilocarpine, 69t
Pinocytosis, 120
pK_a, 63, 65
pKa, 6, 9, 106
pK_b, 65
pKw, 21
Plasma, 206t
Plasma concentration *versus* time, 5f
Plasma gas concentrations, 22
Plasma protein binding, 7, 10, 130, 146–147, 152–156
Plasma proteins, 146–147
Plasma volume expansion, 159
Plastic flow, 202
Platelets, 206
Platinum complexes, 138–139
Pluronic copolymers, 214–215
Pluronic F-127, 210, 214, 215
Pluronics, 214
Po_2, 22, 22t
Point of equilibrium, 117
Poiseuille's law, 217
Polarity, 90
Poloxamers, 214
Poly (d,l-lactide), 32t
Poly(acrylic acid), 213f
Poly(acrylic acid) resins, 213–214
Polydentate, 136
Polyethylene (PE), 32t
Polyethylene glycol, 32t
Polyethylene glycol, 90t, 400
Polymer gels, 210
Polymeric binders, 105
Polymeric macromolecules, 166, 168
Polymeric solids, 31, 32t
Polymethacylate, 32t
Polymethylmethacrylate (PMMA), 32t
Polymorph, 6, 26–27
Polymorphic drugs, 29t
Polymorphism, 26, 104
Polymyxin B, 69t
Polyoxyethylene alkyl ethers, 185–186
Polyoxyethylene glycol, 32t
Polyoxyethylene sorbitan fatty acid esters, 186

Polypropylene (PP), 32t
Polyprotic acid, 75
Polysaccharides, 159
Polyvinyl alcohol, 32t
Poly(*N*-vinylpyrrolidone) (PVP), 213–214, 214f
Potassium, 30t
Potassium chloride, 85t
Potassium chloride-hydrochloric acid buffers, 82–83
Potassium phosphate, 52t, 85t
Potential energy diagram
  attractive forces, 16f
  gaseous molecules, 16f
  repulsive forces, 16f
Povidone-iodine, 137
Povidone-iodine solution, 72t
Practically insoluble, 88t
Prednisolone, 153t
Preservation of emulsions, 109
Prilosec, 119
Primary minimum, 166, 168f
Procaine, 69t
Procardia XL, 132t
Processes affecting fate of drug, 5f
Progesterone, 29t, 153t
Promazine, 69t
Promethazine, 69t
N-Propanol, 90t
Propionic acid, 67t
Propylene glycol, 90t
Prostaglandin E_1, 153t
Prostaglandin E_1 analog, 208t
Protein binding, 7, 147f
Protein folding, 20
Protein-ligand interaction, 143–146
Protein storage and handling, 21
Protein targets, 1
Protein unfolding, 144f
Proton-pump inhibitors, 119
Pseudo reaction order, 226–227
Pseudo-zero-order reactions (suspensions), 227
Pseudoplastic, thixotropic system, 167
Pseudoplastic flow, 203–204

Pulmonary route of
    administration, 4t
PVP, 213–214, 214f
Pyridine, 69t

Q
Q, 102, 106
Quality control test, 102
Quaternary ammonium
    compounds, 185
Quinapril hydrochloride, 31
Quinidine, 69t
Quinine, 69t

R
Radioactivity, 238
Radioactivity decay, 238
Ranitidine, 39
Raoult's law, 45, 46
Rate-limiting step, 10, 103
Rate of shear, 199
Rational correlation time, 201
Reaction, 36t
Ready-to-use enema, 72t
Real solution, 45
Reciprocating cylinder
    (dissolution testing), 101
Reciprocating disk method
    (dissolution testing), 102
Rectal route of administration, 4t
Reiner, Markus, 197
Renagel, 143
Renal excretion, 12
Renal secretion, 120
Repulsive forces, 15
Reserpine, 69t
Reservoir system, 126, 127t
Respiratory alkalosis, 22
Respiratory distress
    syndrome, 174
Rheogram, 200
Rheology, 197–220
    blood, 206–207
    defined, 197
    dilatant flow, 204–205
    importance, 197
    main components, 197–198
    mixing process, 209
    mucus, 207–208

Newtonian flow, 198–201
non-Newtonian flow, 202–205
ointments, 209
ophthalmic drops, 209–210
plastic flow, 202
polymer gels, 210
pseudoplastic flow, 203–204
suspensions and
    emulsions, 209
synovial fluid, 208–209
thixotropy, 205
viscoelasticity, 206
viscosity. See Viscosity
Rheometer, 219
Rheumatoid arthritis, 209
Rhombohedral system, 25f
Ritonavir, 28
Robitussin DM, 72t
Rotating basket system, 100
Rotational viscometer, 219
Rouleaux, 207
Routes of drug administration,
    3, 4t

S
s/p ratio, 109
Saccharin, 67t
Salicylates, 122t
Salicylic acid, 67t
Salicylic acid intermolecular
    bonds, 19f
Salt, 70
Salt cocrystal, 26f, 30
Salt crystals, 26f, 29
Salt hydrate, 26f
Salt hydrate cocrystal, 26f
Salts of strong acids and weak
    bases, 71
Salts of strong acids/bases, 70
Salts of weak acids and strong
    bases, 70–71
Sanger model, 121
Saturated solution
    concentration, 87
Scatchard method, 155–156
Sebum, 176
Secondary minimum, 167,
    168f
Semipermeable membrane, 120

Semisolid dosage forms,
    105–106
Semisynthetic polymers, 31
Sephadex beads, 145
Shear plane, 165f, 166
Shear stress, 198
Shear thickening, 204, 209
Shear thinning, 203, 205,
    209, 217
Shelf life, 229–234
Sickle cell anemia, 39
Silicone rubber, 122
Silly putty, 206
Silver bromide, 92t
Silver chloride, 89t, 92t
Silver hydroxide, 92t
Sinemet, 130
Sink conditions, 123
Skin, 121–122
SLC transporters, 7
Slightly soluble, 88t
Sodium, 30t
Sodium acetate, 70
Sodium alginate, 210, 212–213,
    213f
Sodium carbenicillin, 90t
Sodium carbonate, 89t
Sodium carbonate-bicarbonate
    buffers, 82
Sodium carboxymethylcellulose,
    211, 211f
Sodium chloride equivalent
    method, 53
Sodium chloride equivalent
    values (E), 52
Sodium hydroxide, 85t
Sodium phosphate, 52t, 85t
Sodium salt resins, 213, 214t
Solid dosage forms, 105
Solid state chemical
    degradation, 35
Solid state properties, 6
Solid to liquid conversion, 33
Solids, 24–31
    amorphous, 30–31
    crystalline, 25–30
    overview, 25f
    polymeric, 31, 32t
    stability, 34–37

Solubility, 6, 87–97
    adsorption, 184
    defined, 87
    descriptive terms, 87, 88t
    drug permeation, 108
    general rules, 91
    hydrogen bonding, 90
    liquids in liquids, 95
    nonelectrolytes, 96
    particle size, 90–91
    polarity, 90
    solubility product, 91–92, 92t
    strong electrolytes, 95
    temperature, 89
    weak acids, 92–93
    weak bases, 93–94
    weak electrolytes, 95–96
    weak electrolytes in buffer solutions, 96–97
Solubility curves, 88f
Solubility product, 91–92, 92t
Soluble, 88t
Solute, 41
Solute transport, 118–120
Solution, 2t, 41, 42t. *See also* Physical properties of solutions
Solvate, 26f, 27–29
Solvent, 41
Solvent correction curve, 76
Solvent transport, 120–121
Solvolysis, 221
Sorbic acid, 109
Sorbitan fatty acid esters, 186
Sorensen's buffers, 83–84, 84t
Sotax, Inc., 101
Sparingly soluble, 88t
Specific area, 181
Spectroscopic methods, 145
Spironolactone, 153t
Spreading coefficient, 176–177
Sprowl's method, 54
Square-root-of-time relationship, 106
Stability and shelf life of drugs, 229–234
Stability of solids, 34–37

States of matter
    changes in state, 31–33
    gaseous state, 19–22
    liquid state, 23–24
    solid state, 24–31
Steady state concentration, 13
Stearic acid, 67t
Steric complementarity, 144
Steric forces, 166
Stern layer, 164, 165f, 179, 179f
Stoke (s), 201
Stoke's law, 163, 218
Stomach mucus viscosity, 208t
Stratum corneum, 128
Strong acids, 65
Strong bases, 65
Strong electrolytes
    defined, 61
    partitioning, 106
    solubility, 95
    solutions, 45t
Strychnine, 69t
Subcutaneous injection, 4t
Sublingual route of administration, 4t
Substrate, 135
Substrate concentration, 234
Succinic acid, 67t
Sucralfate, 208t
Sucrose, 61
Sulfadiazine, 61, 67t, 89t
Sulfapyridine, 67t
Sulfate, 30t, 185
Sulfathiazole, 67t
Sulfonates, 185
Supercooled liquids, 30
Suppository, 2t
Surfactants, 184, 191
Surface, 173, 173t
Surface active agents, 174, 184–186
Surface active molecule, 184, 184f
Surface tension, 24, 24t, 173–176, 192–193
Surfactant micelle, 160, 160f
Suspensions, 2t
    coarse dispersion, as, 161–162
    dilatant rheology, 204–205

    dissolution, 105
    pseudo-zero-order reactions, 227
    rheology, 209
Sweat to plasma concentration (s/p) ratio, 109
Synovial fluid, 208–209
Synthetic membrane, 122
Synthetic polymers, 31

T
$t_{1/2}$, 224
Tablet, 2t
Tablet disintegration, 105
Tartaric, 67t
Tartrate, 30t
Tea, 72t
Tears, 74
Temperature
    centigrade, Fahrenheit, Kelvin scales, 249
    reaction rate, 231–232
    shelf life, 229–232
    solubility, 89
    viscosity, 201
Tensiometric methods, 194
Testosterone, 153t
Testosterone-menthol, 33t
Tetracycline, 69t
Tetragonal system, 25f
Theobromine, 69t
Theophylline, 29t, 69t
Theophylline solution, 51t
Therapeutic proteins, 183
Thermodynamic relationships between polymorphs, 37t
Thermodynamics, 117
Thiopental, 153t
Thiourea, 69t
Thixotropy, 205
Thrombocytes, 206, 206t
Thrombolytic agents, 207
Timoptic XE, 210
Tissue binding, 10
Titration curve, 76, 77f
Tolbutamide, 153t
Tomato sauce, 203
Topical emulsions, 161
Topical route of administration, 4t

Topical solution, 42
Toxic heavy metal complexes, 139
Tragacanth, 212
Transacylation reaction, 34, 35f
Transcellular route, 128
Transdermal patches, 128
Transdermal route of administration, 4t
Transdermal systems, 126
Transformation, 36t
Transporters, 7
Triamcinolone, 153t
Trichloroacetic, 67t
Triclinic system, 25, 25f
Tromethamine, 30t
True solid state chemical degradation, 35
Tussionex, 142
Tyndall cone, 161
Tyndall effect, 161

U
Ultracentrifuge, 145
Ultrafiltration, 145
Ultramicroscopy, 161
Ultraviolet (UV) spectroscopy, 145
Uni-divalent electrolyte, 47t
Uni-univalent electrolyte, 47t
Unidentate, 136
Unit cell, 25
Uptake transporter, 9
Urea, 33, 61, 69t
Uric acid, 67t
Urine pH, 108–109
USP buffer systems, 80
USP dissolution testing, 97, 99–102
  acceptance criteria, 102
  calibrators, 102
  cylinder method, 101–102
  flow-through cell, 101
  overview, 101f
  paddle, 100–101
  paddle over disk, 101
  reciprocating cylinder, 101
  reciprocating disk method, 102
  rotating basket, 100

V
Vaginal route of administration, 4t
Valence bond theory, 137
van der Waals attraction, 168f
van der Waals forces, 17–18, 144, 164, 165
VanKel, 99
van't Hoff equation, 46
Vapor pressure (VP), 23
Vapor pressure composition, 45f
Vapor pressure lowering, 46
Vd, 11–12
Veegum, 215
Veronal buffers, 82
Very slightly soluble, 88t
Very soluble, 88t
Vinegar, distilled, 72t
Viscoelasticity, 198, 206
Viscometer, 219
Viscosity
  absolute, 200
  apparent, 200
  cellulose derivatives, 210–212
  clays, 215
  defined, 197
  measurement of, 217–220
  natural gums, 212
  Newtonian fluids, 200–201
  pluronic copolymers, 214–215
  poly(acrylic acid) resins, 213–214
  sodium alginate, 212–213, 213f
  temperature, 201
Viscosity modifiers, 210–215
Visible absorption spectroscopy, 145
Visine eye drops, 72t
Vitamin B_{12}, 139, 139f
Vitamin D_2 solution, 51t
Vitamin E drops, 51t
Vitamin K, 153t
Volmax, 132t
VP, 23

W
W_{adh}, 177, 177f
W_{coh}, 177, 177f
Wada, Akiyoshi, 20
Warfarin, 67t, 147
Washburn equation, 194
Water
  dielectric constant, 90t
  diffusion coefficient, 122t
  surface tension, 175t
  viscosity, 201t
Water accessible backbone, 20
Water-soluble polymers, 210
WBV, 206
Weak acid drugs, 30t
Weak acids, 65–68
  ionization constants, 66–67t
  solubility, 92–93
Weak base drugs, 30t
Weak bases, 68–70
  ionization constants, 68–69t
  solubility, 93–94
Weak electrolytes
  defined, 61
  L_{iso} value, 47t
  partitioning, 107–108
  solubility, 95–96
  solutions, 45t
Wetting agents, 178
Whipping cream, 204
White-Vincent equation, 53, 54
White-Vincent method, 53
White wine, 72t
Whole blood, 206t
Whole blood viscosity (WBV), 206
Wilhemy plate method, 193
Work of adhesion (W_{adh}), 177, 177f
Work of cohesion (W_{coh}), 177, 177f

X
Xanthan, 212
Xenobiotics, 9

Y
Yield value, 202
Young's equation, 178, 179

Z
Zafirlukast, 31
Zeolites, 141
Zero-order absorption, 236–237
Zero-order process, 124
Zero-order reaction, 224
Zeta potential, 166, 179–180
Zinc, 30t, 139
Zinc complexes, 139
Zinc finger, 139
Zinc hydroxide, 92t
Zinc oxalate, 92t
Zinc sulfide, 92t
Zwitterion, 74
Zwitterionic surface active agents, 186

CPSIA information can be obtained
at www.ICGtesting.com
Printed in the USA
LVOW02*1041310817
546856LV00001B/1/P